MODERN POWER SYSTEMS

John R. Neuenswander

PRINCIPAL ENGINEER, DETROIT EDISON COMPANY

FORMERLY ASSISTANT PROFESSOR OF ELECTRICAL ENGINEERING
KANSAS STATE UNIVERSITY

INTERNATIONAL TEXTBOOK COMPANY

THE INTERNATIONAL SERIES IN ELECTRICAL ENGINEERING

Robert F. Lambert —— consulting editor

PROFESSOR OF ELECTRICAL ENGINEERING
UNIVERSITY OF MINNESOTA

I-C-RK

ISBN 0-7002-2289-8

Library of Congress Catalog Card Number 71-151640

INTEXT EDUCATIONAL PUBLISHERS
666 FIFTH AVENUE
NEW YORK, N.Y. 10019

TO my wife **PATRICIA**

my sons **DAN, JOE,** and **ROB**

my daughters **JOAN** and **ANNETTE**

PREFACE

"Modern Power Systems" treats four areas of major concern to the electrical power systems engineer in the solving of the power network problems. These areas are (1) Short Circuit Studies, (2) Load Flow Studies, (3) Economic Dispatch and Unit Commitment, and (4) System Stability.

With the advent of the digital computer and new methods of network solution (matrices, numerical methods, etc.) technical articles abound in the field of power system analysis. However, their message is often conveyed in a spotty manner, at times being couched in difficult terms. This text should help to bridge the gap which often exists between the power systems engineer and the rapidly moving field which he has chosen. The book was designed for the electrical engineering student at the senior and/or graduate level for at least a two semester sequence in power system analysis. The book should also be most useful to the electrical engineer presently involved with power network calculations. It is also likely that chapters of this book involving basic network theory and numerical methods could be useful as a supplementary text in course work related to applied numerical methods or general network analysis.

In presenting the newer methods, a real effort has been made to *stress engineering concepts* as opposed to the purely mathematical generalizations. In order to achieve the four basic objectives, the groundwork is laid in Chapters 1–6 and Appendixes A–C. The first four chapters cover topics such as network reductions and representation of lines, generators, and transformers. Also, basic material on per unit notation, one line diagrams, and concepts of analyzer boards are given. Chapters 5 and 6 lay the groundwork for solving network equations, also dealing with network topology and transformation methods. The appendix covers line parameters, circle diagrams, matrices, and determinants. The remainder of the text (Chapters 7–14) is devoted to meeting the four aforementioned objectives of the text. Note a new emphasis is placed upon the bus impedance matrix (and its corresponding "rake" equivalent circuit) for use in the power network problems. The author regards this tool as one of the most powerful concepts in the field of system analysis today. For this reason, numerous applications are scattered throughout the text which utilize this concept.

There is an intentional *blending* of old and new principles. A de-emphasis today of power applications in the circuit theory and energy conversion coursework has required a rather thorough review of certain of the older principles. While this may seem trivial to the experienced power engineer, he may at the same time appreciate the appendix coverage of matrices which is "old hat" to the younger set. The intention is to permit both the new and old schools to understand and apply the more sophisticated techniques as presented in the later chapters.

The author gratefully acknowledges suggestions given by those persons who reviewed the manuscript. Special thanks are due to Messrs. PhilipT. Holz and Robert B. Cline, Detroit Edison Company, Dr. Mark Enns, University of Michigan, Dr. Vernon Albertson, University of Minnesota, and Dr. J. A. Bennett, Queens University, Kingston, Ontario. I also wish to express my deep appreciation to the Detroit Edison Company for the encouragement and assitance given to me, especially in the early stages of the book's development.

John R. Neuenswander

Detroit, Michigan
February, 1971

CONTENTS

chapter 4

POWER NETWORK REPRESENTATIONS 44

chapter 5

BASIC NETWORK SOLUTIONS 75

chapter 6

NETWORK SOLUTIONS WITH MATRIX TRANSFORMATIONS 101

chapter 7

SYMMETRICAL FAULT STUDIES 121

chapter **13**

ECONOMIC DISPATCH AND UNIT COMMITMENT 297

chapter **14**

SYSTEM STABILITY 332

appendix **A**

LINE PARAMETERS 369

chapter 1

INTRODUCTION

1-1. The Old and the New

If one key word were chosen to best describe recent events in the electrical power industry, that word would be "change" Of course, the same might be said for all of the engineering sciences, since scientific progress has not been identified solely with any one particular area. We find that progress in one field seems to reinforce and bolster that in another. It is much as if one were to represent the achievements of the engineering sciences as a wide platform, serving humanity on a given level. This platform is supported by numerous vertical poles, upon which the weight is distributed. The poles, or supports, represent the various facets of science, both pure and applied. Assume one pole is removed and replaced with another of slightly greater length. While no visible change may be noted at first, this incremental change may be sufficient as to cause adjacent poles to be quite inadequate. Any sizable increase in the length of one support would call for extra weight and attention upon this support until the load is redistributed. While it may have originally been difficult to remove adjacent supports, they can now easily be removed and lengthened (indeed they must be) due to the increased contribution of the lengthened member. This process of adjustment and readjustment must take place continually to accommodate the smoothest upward progress.

In recent years this upward movement has occurred at an almost unbelievable pace. For example, improvement of engineering materials aids the development of computer science. Advancements in computer science bolster the applied and theoretical sciences. Nuclear breakthroughs have been so rapid in coming as to leave visible gaps with respect to the possibilities of growth in other related areas. Standing in the midst of this mushroom of knowledge, the engineer may be often reminded by onlookers of such prophetic statements as "In the last days—knowledge shall increase" and "Ever learning, and never able to come to the knowledge of the truth." It is traditionally understood, that when the engineer, in his work, uses such terms as progress, advancement, or betterment his primary reference frame is confined to the field of engineering endeavor

—social uses and abuses notwithstanding. In the light of this reference frame, progress has been rapid. Some have said that our store of knowledge is doubling about every fifteen years.

The power industry finds itself in the very midst of this growing situation today. Statistics continue to show the industry more than doubling its capacity and production every ten years. Possibilities of improvement are endless, and these improvements are rapidly taking place. The digital computer, for example, has changed not only our method of solving problems but some of our very concepts of mathematics. A few short years ago a trial-and-error approach to problem solving was often frowned upon as being tiring, tedious, and inexact. The theoretical mathematician (and engineer as well) often preferred the direct approach. Approximations were a matter of necessity. Now witness the age of the computer, where trial-and-error solutions are its very foundation. This change in philosophy concerning the acceptance of indirect methods is easily understood when one realizes the time and effort involved in a hand-calculated, trial-and-error design procedure. After three passes, the engineer may have given up or compromised with the best set of parameters. Now compare this laborious procedure with the relatively short time it takes for the digital computer to traverse through one cycle of a trial-and-error loop, or even a few thousand passes.

Next, consider the specific case of the power systems engineer. A somewhat indirect approach (for the solution of otherwise too difficult problems) has been available since the late 1920's. Reference is made here to the a-c and d-c calculating boards. These boards are scaled-down replicas of a large power system. The boards consist of adjustable voltage sources and adjustable impedances which can be used to represent a large network on a per unit, per phase basis. The parameters provided are connected in accordance with the actual network and measurements are readily available. While these analyzer boards serve well for making load-flow studies, short-circuit studies, and system-stability studies, they are rapidly being replaced by the digital computer today. Such factors as problem time, accuracy and economy have paved the way for the latter. This text will not attempt to gear computer methods to one particular computer or scheme of programming. Rather, the basic concepts will be stressed, all of which can be readily adapted to the particular circumstance.

1-2. General Objectives

Technical articles covering subjects of modern network calculations abound today for the engineer. However, it is needless to say that their message is often conveyed in anything but storybook fashion. It is hoped that this text may serve to aid in closing the gap that exists between the new power engineer and the rapidly moving field which he has chosen.

It is the aim of this text to present methods applicable to four major areas of study. They are:

1. Short-circuit studies
2. Load-flow studies
3. Economic load dispatching
4. System stability studies

With the advent of the digital computer, the a-c and d-c analyzer boards are rapidly giving way to computer techniques. The boards are, however, treated in some detail in Chapter 4 in order to give a better physical concept of the system.

The primary concern of the text is to familiarize the reader with some of the more modern methods of solution. Rather than to isolate general mathematical methods from the four basic problems at hand, these methods have been incorporated with application wherever possible. This is done both for the sake of concept and continuity of thought. Broad generalizations are sometimes discarded in favor of specific applications as it is recognized that, to the engineer, there is nothing sacred about a general treatment, merely for the sake of being general.

The author is not so concerned with such time-proven methods as determining line parameters or the construction of circle diagrams. For this reason, these subjects are moved to the Appendix. The network problem is uppermost here—both how to represent the given parameters in the network and how to solve the power-network problem.

All of the older power principles are by no means cast off, however. In fact, there are numerous instances throughout the text in which a special effort has been made to lay such basic background material. This is deemed necessary as there is some deemphasis today of the power applications in both the circuit theory and energy-conversion courses. While such review may offend the intelligence of the experienced power engineer, he might at the same time appreciate the Appendix coverage of matrices and determinants, which is old stuff to the younger set. There is then an intentional blending of old and new principles, intended to bridge the gap which presently exists.

1-3. Utilizing the Appendix

The Appendix covers four basic subjects, any one of which could be incorporated into the course material, depending upon the particular needs of the class or individual. Appendix coverage includes:

A. Line parameters

B. Circle diagrams

C. Fundamentals of determinants and matrices

Use of the Appendix will depend not only upon reader background but upon the course objective. If time permits, a two-or-three semester coverage in power systems analysis could include all text material with appropriate use of the Appendix. However, it may be that the class objective is to get on with the "meat" of the network problems as rapidly as possible. In such a case, more time can be spent with the later chapters (10–14), and it will be to the class advantage to treat the Appendix material but briefly. Since matrix methods are emphasized throughout the text, material of Appendix C is to be considered prerequisite. Appendix A regarding line parameters is useful but, for the most part, the system problems in this book (stability, load flow, etc.) will assume line parameters given.

1-4. A New Emphasis

The development of the bus impedance matrix (and its corresponding equivalent circuit) represents one of the most helpful concepts to the field of power system analysis of recent years. For this reason a train of application utilizing this concept is found scattered throughout the textbook. It is first mentioned in Chapter 2 along with the subject of equivalent circuits. Some mention is made of the equivalent in Chapter 4, where the analyzer board is used to obtain the matrix. Next, the matrix appears in its glory in Chapter 10 to take the limelight as a most natural tool for the calculation of currents and bus voltages under short-circuit conditions. Time is also spent in Chapter 10 in developing the means for arriving at the matrix by computer. In the short-circuit application, the bus impedance matrix is also sometimes termed the "short-circuit matrix."

Chapter 11 makes use of this bus impedance matrix and its equivalent for the determination of bus voltage changes when switching shunt capacitors or injecting vars at generators and synchronous condensers. Chapter 12 employs the bus impedance matrix in one of several methods treated for solving system load-flow problems. Chapter 13 makes use of the matrix in a method for minimizing system line losses and fuel costs. Application can be given to the matrix in solving the stability problem. However, space is naturally limited in a single text of this nature, so a more conventional approach was chosen at this point.

One reason stands out in explaining the reluctance on the part of many engineers to comprehend and utilize the bus impedance matrix for use in systems problems. Much of the technical literature in an attempt for brevity has understandably approached the subject from the mathematical viewpoint altogether. Comments are often made that such methods are couched in the most difficult terms. While such was never the intent, the short and precise mathematical approach sometimes failed to give sufficient concept to the engineer. Therefore, wherever the bus impedance matrix is utilized in this text, a special effort has been made to stress the physical concept of the matrix, which in this case entails the use of the so called "rake" equivalent circuit, to be explained in more detail in Secs. 2-4, 2-5, 4-12, and 10-1. Once the equivalent concept is realized, the applications are much more easily comprehended.

1-5. Paving the Way—Chapters 2–6

In order to achieve the four basic objectives set forth in Sec. 1-2, certain groundwork must be laid. This is done in Chapters 2–6. A very brief survey of these chapters is given here. Chapters 2 and 3 take up such topics as basic network reductions, transmission-line equivalents, and four-terminal networks. Chapter 4 is most basic to all of system calculations in that it reviews the equivalent representation of lines, transformers, and generators, then combines these equivalents for use in the larger network problem. From here, Chapter 4 treats such basic material as per unit notation and one-line diagrams. The a-c and d-c analyzer boards are given some attention for the benefit of the system concept which they lend to the power engineer. It is understood, as explained in Sec. 1-1, that while the analyzer boards have proven to be a most useful tool, they are rapidly giving way to the digital methods.

Chapter 5 deals with some of the conventional methods for setting up simultaneous equations in the solution of power system problems. The common mesh-impedance method is a special case of the loop impedance method. The tree-link approach is emphasized for the purpose of assuring the proper number of independent voltage or current equations. An effort is made to simplify to the student the cut-set approach.

Since transformation methods are often employed in the calculation of power networks, Chapter 6 is devoted to this technique. If time is at a premium in the course work, however, it is possible to omit Chapter 6 without undue handicap to the student for the chapters that follow. Several numerical example problems are presented to aid in the understanding of the transformation concepts.

1-6. Short-Circuit Calculations

In order to protect the lines and equipment from undue currents and/or abnormal voltages in case of short circuits, protective devices such as relays and circuit breakers are provided. To obtain proper settings of these relays, the values of such currents and voltages must be known. Chapters 7 and 8 treat the conventional means of fault calculations utilizing symmetrical components. The four basic types of faults are: the symmetrical or three-phase fault, the line-to-ground fault, line-to-line fault, and double line-to-ground fault. Items related to the fault problem which in themselves tend to complicate the general solutions of chapters seven and eight are forestalled to Chapter 9. They include such topics as the balancing-ampere-turns technique for transformer loading, sequence impedance of lines, generators, and transformers, as well as the added complication of transformer phase shift. Chapter 10 is devoted to the bus impedance matrix method of fault solution. As previously pointed out, it goes beyond the mere use of the matrix and offers a method for obtaining the matrix as well. The power of this matrix is readily seen when it is realized that once the impedance matrix (or its equivalent circuit) is found, fault currents and bus voltages are immediately attainable for a fault on any bus in the system. Equations for attaining this information from the matrix are derived in Chapter 10.

1-7. Concepts of Real and Reactive Power Flow

Chapter 11 attempts first to give a conceptual view of the relationships between voltage and the flow of real and reactive power. It is emphasized that reactive volt-amperes flow (in a network composed primarily of reactance elements) between two points is a direct function of the difference in voltage magnitudes of these points. The watt flow between two such points is, at the same time, basically a function of the relative phase angles of the voltages. Methods for adjusting such flows are discussed. The practical concept of circulating real or reactive power between paralleled transformers and generators is presented. Use of the bus impedance matrix in determining incremental voltage changes due to injection or redistribution of reactive volt-amperes is treated. Methods for paralleling of automatic load-tap-changing transformers are considered. Finally, a practical look is given to the subject of overexcitation protection in the large generator transformer. The possibility of transformer overexcitation can become quite an additional constraint upon reactive flow in a load flow or dispatch problem.

1-8. Load-Flow Studies

In the planning of power systems, care is taken to predict in advance the system voltage levels, and the loading of lines and equipment. Therefore, such information is necessary for all points in the system both for normal operation as well as various open-line conditions. Longhand calculations would be most difficult as this involves a complex network of many generators, lines, transformers, or loads. Chapter 12 can be considered as composed of two separate but related parts. Sections 1–8 include a number of mathematical methods for the solution of simultaneous equations. Later, portions of the chapter (Secs. 9–14) have singled out certain of these mathematical methods and applied them specifically to the load-flow problem. Again, if time is at a premium, it is suggested that Secs. 12-6, 12-7 and 12-9 through 12-14 can be taken with the omission of the others. Material of Sec. 3-3 should be familiar to the student before moving into Sec. 12-9.

1-9. Load Forecast, Unit Commitment, and Economic Dispatch

The day of dispatching load by the "seat of the pants" is rapidly coming to a close for the larger power utilities. Too much saving is to be gained in minimizing costs through lower line losses, start-up and shutdown costs, fuel costs, and maintenance costs. This cost reduction is best realized through the determination of three basic factors.

1. A reasonable *short-term forecast* or load prediction. This prediction must consider weather effects such as temperature, humidity, light, wind velocity. Some knowledge of industrial and residential load demand is also essential. A brief coverage of the subject of forecasting is included in Chapter 13 sufficient to indicate the importance of forecasting to the total problem of economics.
2. Economic scheduling of generators (*unit commitment*). The object of this phase of the problem is to guarantee at all times the optimum combination of generators on the line to provide the hourly load demand plus system reserve. "Spinning reserve" requirements must be maintained to account for forecast errors and possible forced outages. Some attention must be given to the operating constraints of generators, lines, and transformers with respect to loading capability and system stability. Careful geographic scheduling of reserve is necessary in case of partial isolation of the system during a critical line outage. To commit units economically involves the minimizing of costs which include the effect of start-up and shutdown costs, fuel costs, line losses, and maintenance costs. Obviously, it is essential to a meaningful scheduling program that a good short-term forecast program is also available.
3. *Economic dispatch*. This refers to the most economic loading of

the generators, once they are committed or running and on the line. In the past, this third factor has received more attention than has the commitment problem. Economic dispatch accomplishes the minimizing of generator fuel costs and system line losses. As a new load increment is added to the system, the online computer is able to determine which generator can most economically take on the added load.

chapter **2**

SELECTED NETWORK CONCEPTS

2-1. Purpose

The purpose of this chapter is to provide review and reference in order to insure a more complete understanding of what is to follow. It is understood that the backgrounds of various readers will differ to some extent, and yet certain essential tools are necessary in the study of modern power systems. For example, a summary of four-terminal networks is included to provide a background for the coverage of power transmission lines. It will be seen that emphasis is placed upon a relatively new type of equivalent circuit in this chapter and others to follow. It is referred to as a "rake" equivalent; when applied to the simplified four-terminal network it will be called the V equivalent. The reason for this new emphasis is explained in the practicality of this equivalent for specific power systems calculations.

2-2. Network Reductions

The simple series equivalent of a passive, linear, bilateral network as seen between two terminals can in general be found by one or more of the following operations: (1) addition of series impedances, (2) combination of parallel impedances, and (3) Y-Δ conversions.

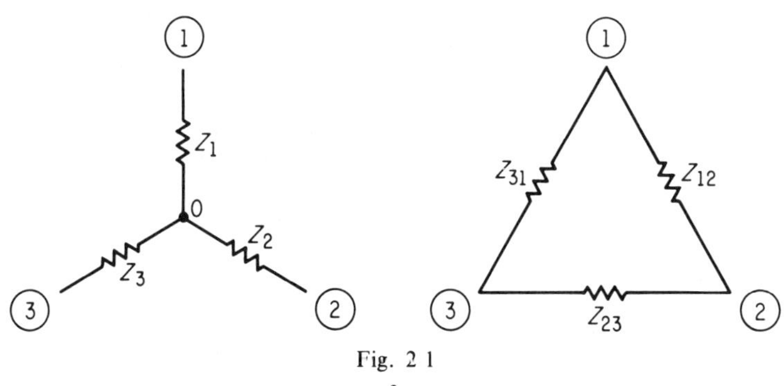

Fig. 2 1

In recalling the Y-Δ transformations, the following relations hold (refer to Fig. 2-1).

$$Z_{12} = \frac{Z_1Z_2 + Z_3Z_1 + Z_2Z_3}{Z_3} \tag{2-1}$$

$$= Z_1Z_2\left(\frac{1}{Z_1} + \frac{1}{Z_2} + \frac{1}{Z_3}\right)$$

$$= Z_1Z_2\Sigma \frac{1}{Z_y}$$

where $\Sigma \frac{1}{Z_y}$ is the sum of the reciprocals of the impedances in the wye.

$$Z_{23} = \frac{Z_1Z_2 + Z_3Z_1 + Z_2Z_3}{Z_1} \tag{2-2}$$

$$= Z_2Z_3\Sigma \frac{1}{Z_y}$$

$$Z_{31} = \frac{Z_1Z_2 + Z_3Z_1 + Z_2Z_3}{Z_2} \tag{2-3}$$

$$= Z_3Z_1\Sigma \frac{1}{Z_y}$$

Note that node 0 was eliminated in the process.

For converting Δ to Y, the following equations are useful.

$$Z_1 = \frac{Z_{12}Z_{31}}{Z_{12} + Z_{31} + Z_{23}} = \frac{Z_{12}Z_{31}}{\Sigma Z_\Delta} \tag{2-4}$$

where ΣZ_Δ is the sum of the delta impedances.

$$Z_2 = \frac{Z_{12}Z_{23}}{\Sigma Z_\Delta} \tag{2-5}$$

$$Z_3 = \frac{Z_{31}Z_{23}}{\Sigma Z_\Delta} \tag{2-6}$$

The Y-Δ transformation may be extended to a star-mesh conversion where more than three impedances terminate on node 0 of the star.

In general, this conversion results in a mesh with an impedance between every possible pair of nodes (except 0). Thus, node 0 is eliminated. The conversion is as follows:

$$Z_{mn} = Z_mZ_nY_{oo} \tag{2-7}$$

where Y_{oo} is the total admittance of all branches surrounding node 0 or

$$Y_{oo} = \frac{1}{Z_1} + \frac{1}{Z_2} + \frac{1}{Z_3} + \frac{1}{Z_m} + \frac{1}{Z_n}$$

for the example shown in Fig. 2-2.

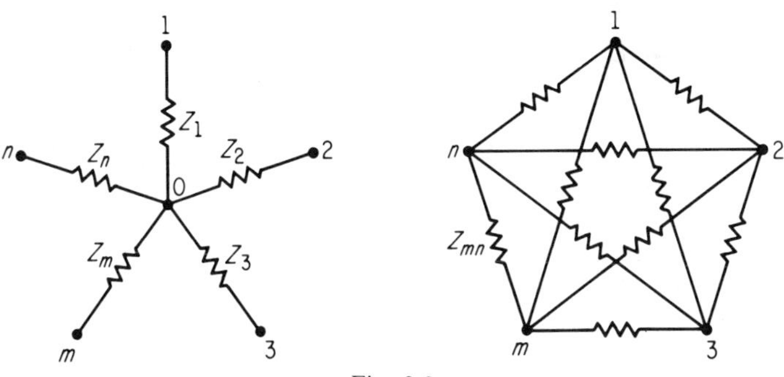

Fig. 2-2

It is said in three terminal networks that every Δ has its equivalent Y. Also, we have noted that the star connection of $n + 1$ nodes has its equivalent mesh of n nodes. However, it does *not* follow that every mesh circuit of n nodes has an equivalent star of the simple $n + 1$ variety. Take the case of a passive network with four nodes including the reference node 0. This passive network may be represented by the mesh equivalent of Fig. 2-3a. It can easily be shown that, in general, Fig. 2-3b is *not* the equivalent of Fig. 2-3a. Assume, for example that one lone current enters the network through node 0 and leaves through node 1, with nodes 2 and 3 open-circuited. It is reasoned that if Fig. 2-3b were the equivalent of Fig. 2-3a, then nodes 2 and 3 would not, in general, be at the same potential. Yet the star connection places them at the same potential under the conditions described. Therefore, the two circuits are not equivalent.

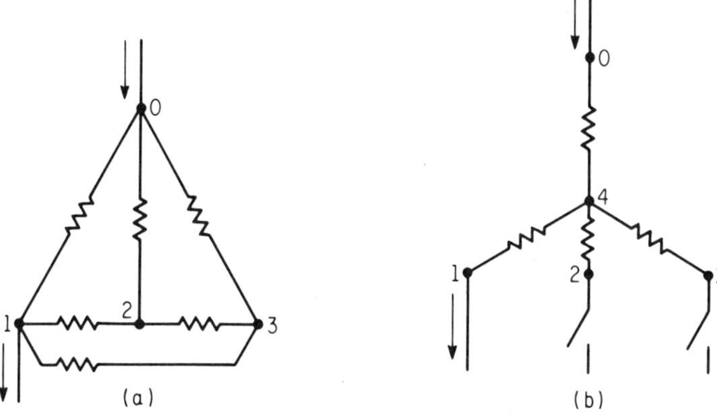

Fig. 2-3

2-3. Four-Terminal Networks

Any linear, passive, bilateral, two-port network may be represented by a π or T circuit. A third equivalent might be mentioned here which we will call the V equivalent; z_{12} of this V equivalent may be treated as a mutual or transfer impedance between two lines of self-impedance z_{11} and z_{22}. Refer to Figs. 2-4b and 2-4c relating impedances of the V equivalent

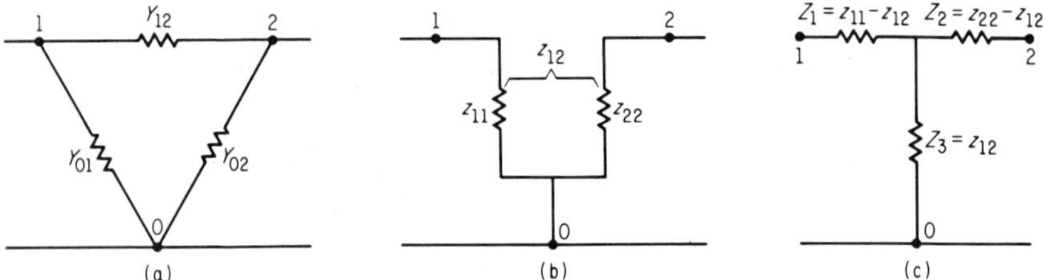

Fig. 2-4. (a) π equivalent. (b) V equivalent. (c) T equivalent.

to those of the T equivalent. It is later shown in Sec. 2-4 that an equivalent of the general nature of Fig. 2-4b may be expanded to any number of nodes with one reference node. The resulting "rake" equivalent will be especially helpful in the power applications of short-circuit and load-flow problems.

A four-terminal (two port) network will be generalized as shown in Fig. 2-5. It is possible to express the input voltage and current in terms of

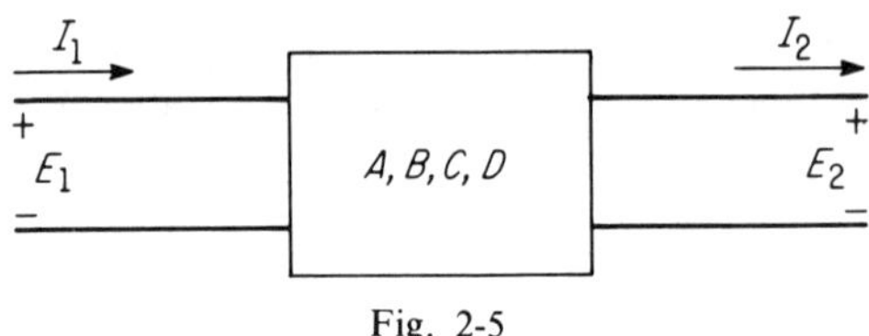

Fig. 2-5

the output values as follows:

$$E_1 = AE_2 + BI_2 \tag{2-8}$$

$$I_1 = CE_2 + DI_2 \tag{2-9}$$

While matrix theory will be summarized in the Appendix, it will be assumed here that the student is familiar with basic matrix manipulation:

$$\begin{bmatrix} E_1 \\ I_1 \end{bmatrix} = \begin{bmatrix} A & B \\ C & D \end{bmatrix} \begin{bmatrix} E_2 \\ I_2 \end{bmatrix} \tag{2-10}$$

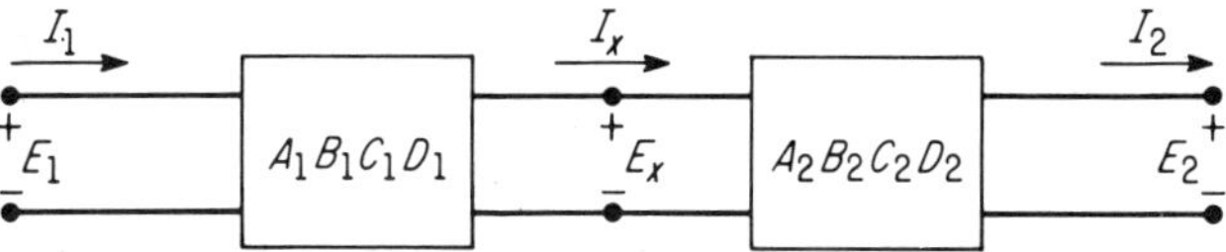

Fig. 2-6. Networks in cascade.

For two circuits in cascade as shown in Fig. 2-6 the matrix becomes

$$\begin{bmatrix} E_1 \\ I_1 \end{bmatrix} = \begin{bmatrix} A_1 & B_1 \\ C_1 & D_1 \end{bmatrix} \begin{bmatrix} E_x \\ I_x \end{bmatrix} \quad \text{and} \quad \begin{bmatrix} E_x \\ I_x \end{bmatrix} = \begin{bmatrix} A_2 & B_2 \\ C_2 & D_2 \end{bmatrix} \begin{bmatrix} E_2 \\ I_2 \end{bmatrix}$$

By substitution,

$$\begin{bmatrix} E_1 \\ I_1 \end{bmatrix} = \begin{bmatrix} A_1 & B_1 \\ C_1 & D_1 \end{bmatrix} \begin{bmatrix} A_2 & B_2 \\ C_2 & D_2 \end{bmatrix} \begin{bmatrix} E_2 \\ I_2 \end{bmatrix}$$

A convenient method for handling two parallel networks is found in making use of the π equivalents associated with the y parameters. (See Fig. 2-7.)

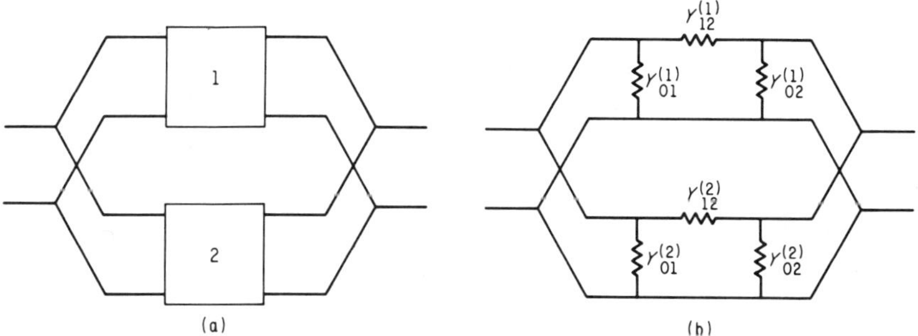

Fig. 2-7. (a) Networks in parallel. (b) Parallel network with y parameters.

It is seen that corresponding admittances of Fig. 2-7b are in parallel and may be combined to form one equivalent network. This is accomplished merely by adding corresponding admittances together. In other words, $Y_{12} = Y_{12}^{(1)} + Y_{12}^{(2)}$, $Y_{01} = Y_{01}^{(1)} + Y_{01}^{(2)}$ and $Y_{02} = Y_{02}^{(1)} + Y_{02}^{(2)}$. Values of the above admittances in terms of $ABCD$ constants are given in a later section.

Once again consider Eq. 2-10. The $ABCD$ constants are, in general, complex and may be calculated from design data or determined by test. This matrix of $ABCD$ constants will be termed the a matrix. We would obtain expressions for E_2 and I_2 by premultiplying both sides by $[a]^{-1}$.

$$\begin{bmatrix} E_1 \\ I_1 \end{bmatrix} = \begin{bmatrix} A & B \\ C & D \end{bmatrix} \begin{bmatrix} E_2 \\ I_2 \end{bmatrix} = [a] \begin{bmatrix} E_2 \\ I_2 \end{bmatrix}; \qquad \begin{bmatrix} E_2 \\ I_2 \end{bmatrix} = [a]^{-1} \begin{bmatrix} E_1 \\ I_1 \end{bmatrix}$$

Here,

$$[a]^{-1} = \frac{1}{\Delta} \begin{bmatrix} D & -B \\ -C & A \end{bmatrix}$$

where $\Delta = AD - BC = 1$.

The proof of the fact that $\Delta = 1$ is taken up in detail in numerous texts involving general network theory.

Other forms for expressing input and output relationships will be briefly mentioned.

$$\begin{bmatrix} E_1 \\ E_2 \end{bmatrix} = \begin{bmatrix} z_{11} & z_{12} \\ z_{21} & z_{22} \end{bmatrix} \begin{bmatrix} I_1 \\ I_2 \end{bmatrix} = [z] \begin{bmatrix} I_1 \\ I_2 \end{bmatrix} \tag{2-11}$$

The z parameters above may be easily related to Figs. 2-4b and 2-4c. Taking the T equivalent, for example, the value of $Z_1 = z_{11} - z_{12}$, $Z_2 = z_{22} - z_{12}$, $Z_3 = z_{12}$. This will satisfy Kirchoff's equations where the current directions are assumed as in Fig. 2-8.

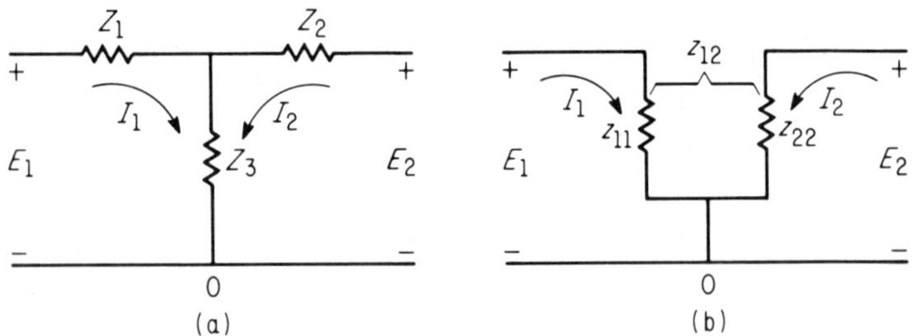

Fig. 2-8. (a) T circuit. (b) V circuit.

The physical interpretation of the z parameters is:

$$z_{11} = \frac{E_1}{I_1} \quad \text{with terminal 2 open, } I_2 = 0$$

$$z_{22} = \frac{E_2}{I_2} \quad \text{with terminal 1 open, } I_1 = 0$$

$$z_{12} = \frac{E_1}{I_2} \quad \text{with terminal 1 open, } I_1 = 0$$

$$z_{21} = \frac{E_2}{I_1} \quad \text{with terminal 2 open, } I_2 = 0$$

For passive bilateral networks $z_{21} = z_{12}$.

Another form of expressing input and output quantities utilizes the so-called y parameters, where

$$\begin{bmatrix} I_1 \\ I_2 \end{bmatrix} = \begin{bmatrix} y_{11}\, y_{12} \\ y_{21}\, y_{22} \end{bmatrix} \begin{bmatrix} E_1 \\ E_2 \end{bmatrix} = [y] \begin{bmatrix} E_1 \\ E_2 \end{bmatrix} \qquad (2\text{-}12)$$

Of course, in this case the y matrix could be obtained by inverting the previous z matrix. It might be worthwhile to relate the parameters of the y matrix with those of the π equivalent of Fig. 2-4a. Refer to Fig. 2-9.

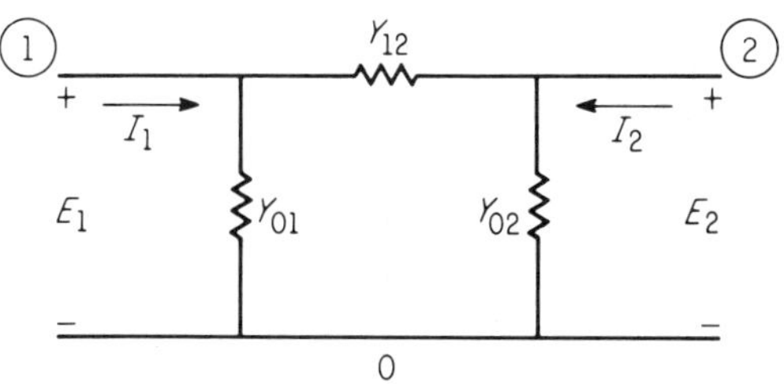

Fig. 2-9. π circuit.

For current directions as shown, $y_{11} = Y_{01} + Y_{12}$, $y_{22} = Y_{12} + Y_{02}$, and $y_{12} = -Y_{12} = y_{21}$. This follows since in any nodal solution the diagonal elements of the admittance matrix represent the total admittance surrounding a node, while the off-diagonal element of position 1-2 for example represents the negative admittance between nodes 1 and 2.

Still another form utilizes the h parameters, where

$$\begin{bmatrix} E_1 \\ I_2 \end{bmatrix} = \begin{bmatrix} h_{11}\, h_{12} \\ h_{21}\, h_{22} \end{bmatrix} \begin{bmatrix} E_2 \\ I_1 \end{bmatrix} \quad \text{or} \quad \begin{bmatrix} E_1 \\ I_2 \end{bmatrix} = [h] \begin{bmatrix} E_2 \\ I_1 \end{bmatrix} \qquad (2\text{-}13)$$

Still other forms are possible. However, for power networks, the a parameters ($ABCD$ constants) are most widely used. Specific $ABCD$ constants for π, T, and series networks are taken up in Chapter 3.

2-4. The Rake Equivalent for the Bus Impedance Matrix

There are two per phase equivalents in wide use for the n-node (or n-bus) system with reference node 0. The two widely used equivalents will be termed the *mesh equivalent* which is related to the *bus admittance matrix* and the *rake equivalent* which is related to the *bus impedance matrix*. The rake equivalent may also be termed the bus impedance matrix equivalent or, for the short circuit application, the short-circuit matrix equivalent. As later pointed out (Sec. 10-1), the term "rake equivalent" was first suggested to the author by Robert Cline of Detroit Edison Company. This convenient term has since been used by the author in preference to the more cumbersome titles. The term seems most appropriate in view of the fact that equivalents such as the star or delta are named in accordance with their shape and appearance.

It is recalled from Sec. 2-2 that the mesh equivalent has an im-

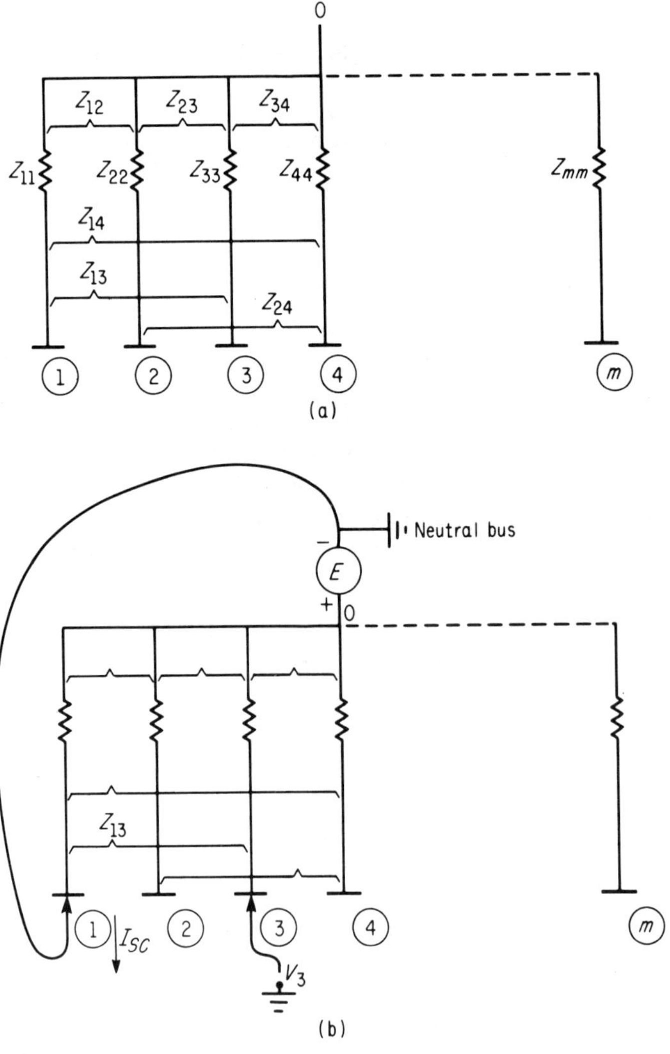

Fig. 2-10. (a) Rake equivalent for passive network with reference node 0. (A per phase diagram.) (b) Example of the short circuit equivalent with generation added and a symmetrical three-phase fault on bus 1.

pedance joining each bus to every other bus (node). However, the rake equivalent has one impedance each from bus 0 to every other bus. These impedances (Z_{11}, Z_{22}, Z_{33}, Z_{44}) are termed *self-* or *driving-point impedances.* In addition to the self-impedances, there is mutual impedance between each self-impedance branch and every other branch. These are represented in Fig. 2-10a by the bracketed quantities (Z_{12}, Z_{23}, Z_{31}, Z_{14}, Z_{24}, Z_{34}). These mutual impedances are also called *transfer impedances.*

A number of advantages are to be gained in the use of rake equivalent for short-circuit studies. Three of these advantages will be given in this section.

1. The equivalent itself is derived from an orderly calculation procedure and its resulting short-circuit matrix. Chapter 10 deals in greater detail with the short-circuit matrix method, its justification and its use.
2. If generation is added at bus 0 of Fig. 2-10b, and a symmetrical three-phase short is placed on bus 1, the short-circuit current (I_{sc}) per phase is found immediately as E/Z_{11}. Also, the voltage to neutral for every other bus is easily found. For example, it is obvious that the voltage on bus 3 for a fault on 1 is $E - I_{sc}Z_{13}$.
3. Network reduction is no problem whatsoever once the equivalent is in this form. For example, if buses 3 and 4 are of no interest, and will not be faulted or loaded in the study, branches 3 and 4 may be simply eliminated, leaving a two-bus equivalent. Justification for this is seen in that if no current flows from these branches, the branches cannot possibly affect the remaining equivalent. To further stress the value of this network reduction, a short-circuit matrix or equivalent could contain several hundred nodes which could by inspection be immediately reduced to an equivalent containing only a relatively few buses of interest.

2-5. The Rake-Mesh Equivalent Conversions

The problem here is to determine the mesh equivalent from a given rake equivalent and vice versa. Refer to Fig. 2-11 which represents a three-bus network plus the reference bus. First, the rake-to-mesh conversion will be given, following the justification for the procedure. Note the parameters of the mesh equivalent are shown in admittance form and are not the reciprocals of the Z parameters of Fig. 2-11a.

The rake equivalent to mesh equivalent is as follows:

1. Write out the impedance array for Fig. 2-11a where the diagonal elements are the self-impedances and the off-diagonal elements are the mutual impedances. The result is termed the bus im-

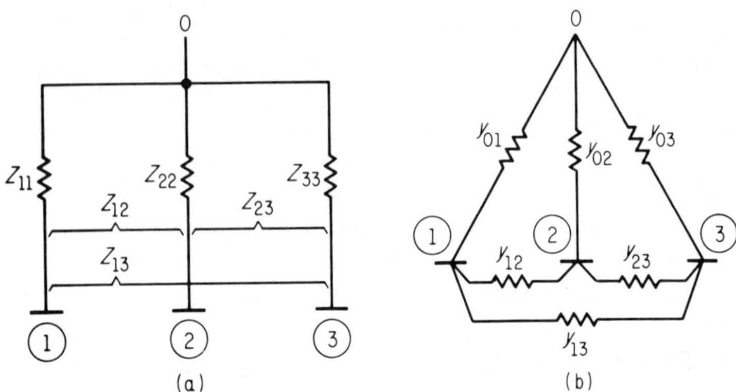

Fig. 2-11. (a) Rake equivalent for one phase, corresponding to Eq. 2-14. (b) Mesh equivalent of Fig. 2-11a.

pedance matrix Z_{bus}, where

$$Z_{\text{bus}} = \begin{bmatrix} Z_{11} & Z_{12} & Z_{13} \\ Z_{21} & Z_{22} & Z_{23} \\ Z_{31} & Z_{32} & Z_{33} \end{bmatrix} \tag{2-14}$$

2. Obtain the inverse of Z_{bus}, yielding

$$Z_{\text{bus}}^{-1} = \begin{bmatrix} Y_{11} & Y_{12} & Y_{13} \\ Y_{21} & Y_{22} & Y_{23} \\ Y_{31} & Y_{32} & Y_{33} \end{bmatrix} = Y_{\text{bus}} \tag{2-15}$$

For students unfamiliar with the obtaining of a matrix inverse, reference is made to the Appendix for a conventional or adjoint method of inversion. The conventional method is often most cumbersome. For alternate methods of obtaining the inverse, refer to the early sections of Chapter 12.

The resulting inverted matrix of Y elements in Y_{bus} are related to the y's of the mesh equivalent as follows:

$$\left.\begin{aligned} Y_{11} &= y_{01} + y_{12} + y_{13} \\ Y_{22} &= y_{02} + y_{23} + y_{21} \\ Y_{33} &= y_{03} + y_{31} + y_{32} \\ Y_{12} &= -y_{12} \\ Y_{13} &= -y_{13} \\ Y_{23} &= -y_{23} \end{aligned}\right\} \tag{2-16}$$

Solving for the y values,

$$y_{01} = Y_{11} - y_{12} - y_{13}$$

or

$$\left.\begin{aligned} y_{01} &= Y_{11} + Y_{12} + Y_{13} \\ y_{02} &= Y_{22} + Y_{21} + Y_{23} \\ y_{03} &= Y_{33} + Y_{31} + Y_{32} \end{aligned}\right\} \qquad (2\text{-}17)$$

A more general form of n buses (plus the reference bus) would require essentially the same procedure, where

$$\left.\begin{aligned} y_{ok} &= \sum_{m=1}^{m=n} Y_{km} \\ y_{mn} &= -Y_{mn} \end{aligned}\right\} \qquad (2\text{-}18)$$

The justification for this conversion process is made possible by reviewing the forms of Kirchhoff's equations for the two equivalents. The equivalents are redrawn in Fig. 2-12 and it is assumed that voltages E_1,

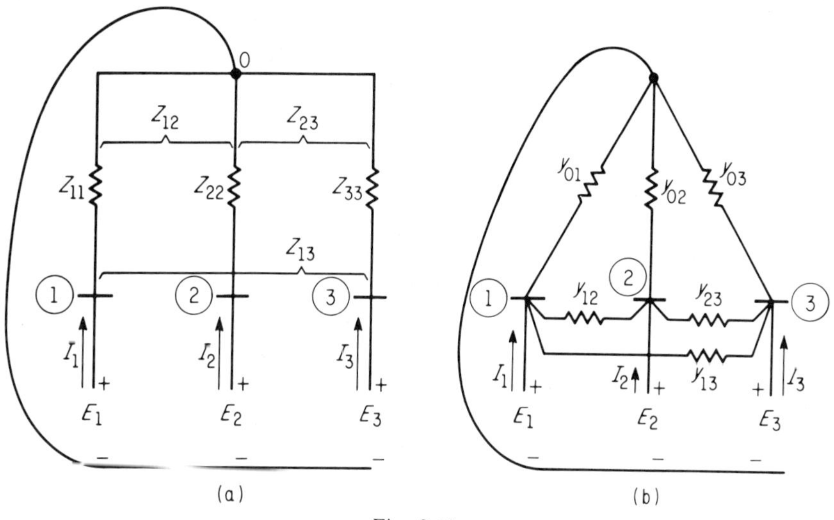

Fig. 2-12

E_2, and E_3 are present on the nodes with currents I_1, I_2, and I_3 flowing into the nodes. For Fig. 2-12a Kirchhoff's three voltage equations expressed in matrix form are

$$\begin{bmatrix} E_1 \\ E_2 \\ E_3 \end{bmatrix} = \underbrace{\begin{bmatrix} Z_{11} & Z_{12} & Z_{13} \\ Z_{21} & Z_{22} & Z_{23} \\ Z_{31} & Z_{32} & Z_{33} \end{bmatrix}}_{Z_{\text{bus}}} \begin{bmatrix} I_1 \\ I_2 \\ I_3 \end{bmatrix} \qquad (2\text{-}19)$$

The three current equations for 2-12b expressed in matrix form are

$$\begin{bmatrix} I_1 \\ I_2 \\ I_3 \end{bmatrix} = \overbrace{\begin{bmatrix} Y_{11} & Y_{12} & Y_{13} \\ Y_{21} & Y_{22} & Y_{23} \\ Y_{31} & Y_{32} & Y_{33} \end{bmatrix}}^{Y_{\text{bus}}} \begin{bmatrix} E_1 \\ E_2 \\ E_3 \end{bmatrix} \tag{2-20}$$

Now the E's and the I's of Eq. 2-19 of the rake equivalent are identical with the E 's and the I's of Eq. 2-20 for the mesh equivalent. Premultiplying Eq. 2-19 by Z_{bus}^{-1} will yield

$$[Z_{\text{bus}}]^{-1}[E] = [I] \tag{2-21}$$

which is the identical form of Eq. 2-20, where $Z_{\text{bus}}^{-1} = Y_{\text{bus}}$. Regarding the relationships of Eq. 2-16, it is known that in the writing of the nodal equations of 2-20, the value of the diagonal element Y_{kk} includes all of the admittance surrounding node k and the off diagonal element Y_{kp} is the negative of the admittance value linking node k to node p.

The *mesh-equivalent-to-rake-equivalent conversion* is obtained with equal ease by looking at the equations of 2-16. Here the y's are known and the Y's can be found. The inversion of Y_{bus} will then yield the bus impedance matrix and the corresponding elements of the rake equivalent circuit.

It is sometimes convenient to obtain the rake equivalent from the mesh equivalent directly from the analyzer board as will be explained in Secs. 4-13 and 4-14. The reverse process is not so easily accomplished on the board however, because of the difficulty of representing the many mutual impedance terms of the rake equivalent.

Methods for obtaining the rake equivalent of a network (by computer) are discussed in Chapter 10.

Problems

2-1. (a) Find one equivalent delta impedance of the circuit to the right of points a, b, and c for the circuit of Prob. 2-1.

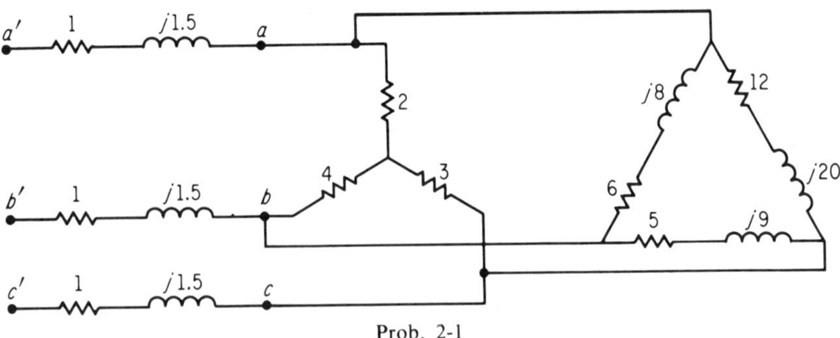

Prob. 2-1

(b) Find the equivalent Y impedance of the circuit to the right of a', b', and c'.

2-2. Convert the star of Prob. 2-2 to its mesh equivalent, thus eliminating node 0.

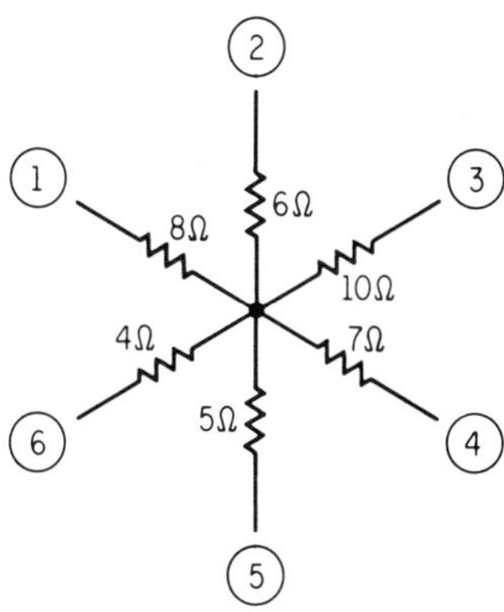

Prob. 2-2

2-3. Two lines have only one end in common with mutual impedance between them as shown in Prob. 2-3. Find the π and T equivalents between these two points.

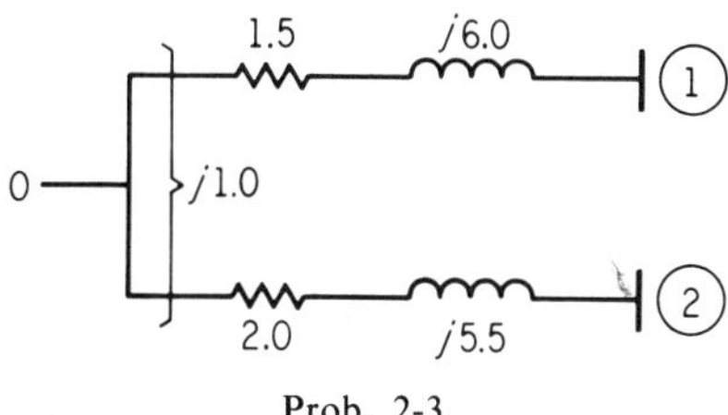

Prob. 2-3

2-4. Two four-terminal networks are connected in cascade. Their individual $ABCD$ constants are as follows:

$$A_1 = 1.0,\ B_1 = 50 \text{ ohms},\ C_1 = 0,\ D_1 = 1.0$$
$$A_2 = 0.9\ \underline{/2.0^\circ},\ B_2 = 150\ \underline{/79^\circ} \text{ ohms}$$
$$C_2 = 9 \times 10^{-4}\ \underline{/91^\circ} \text{ mho},\ D_2 = 0.9\ \underline{/2.0^\circ}$$

(a) Find the equivalent $ABCD$ constants for the combination.

(b) If $E_2 = 115{,}000\ \underline{/0^\circ}$ volts and $I_2 = 350\ \underline{/-30^\circ}$ amp, find E_1 and I_1. Subscripts 1 and 2 are as indicated in Fig. 2-6.

(c) Invert the a matrix for the combination. Now find E_2 and I_2 where $E_1 = 130{,}000\ \underline{/0^\circ}$ and $I_1 = 400\ \underline{/20^\circ}$.

2-5. Two 60-Hz T networks are in parallel. Their impedance values are given where subscripting corresponds to Fig. 2.8a. For network I:

$$Z_1 = 11 + j70 \text{ ohm} \qquad Z_2 = 9 + j75 \qquad Z_3 = 4{,}000 \text{ ohms}$$

For network II:

$$Z_1 = 17 + j110 \qquad Z_2 = 15 + j120 \qquad Z_3 = 4{,}000 \text{ ohms}$$

Find the π and T equivalents for the combination.

2-6. The d-c generator of Prob. 2-6 is supplying a network containing four separate load buses. Resistances values for the rake equivalent are given in Prob. 2-6.

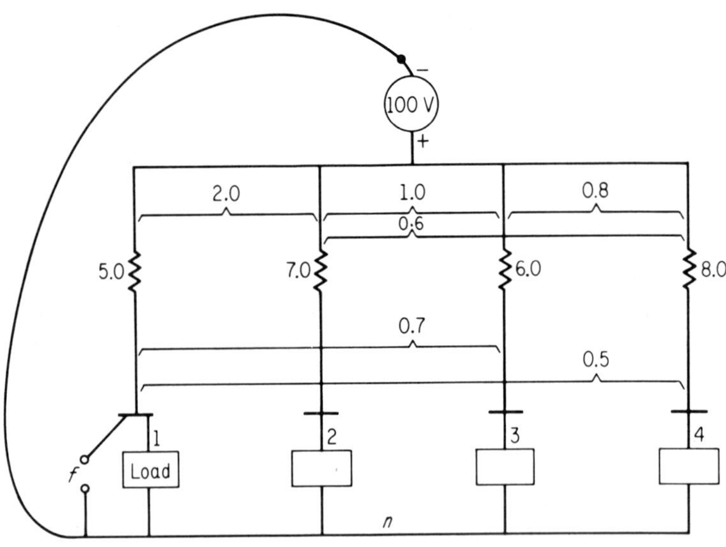

Prob. 2-6

(a) What fault current would be expected if bus 1 were shorted to n at f? Neglect all load currents, considering them small with respect to the fault current.

(b) Find the voltages expected from point n to buses 2, 3, and 4 with bus 1 faulted. Again, neglect the effects of load currents.

(c) What are the fault currents with the short placed alternately on buses 2, 3, and 4?

2-7. (a) Write out the 4×4 Z_{bus} matrix for the rake equivalent of Prob. 2-6.

(b) Convert Z_{bus} to Y_{bus}. For the inversion procedure refer to the Appendix or Sec. 12-5.

(c) Knowing Y_{bus}, find the mesh equivalent to replace the rake equivalent of Prob. 2-6.

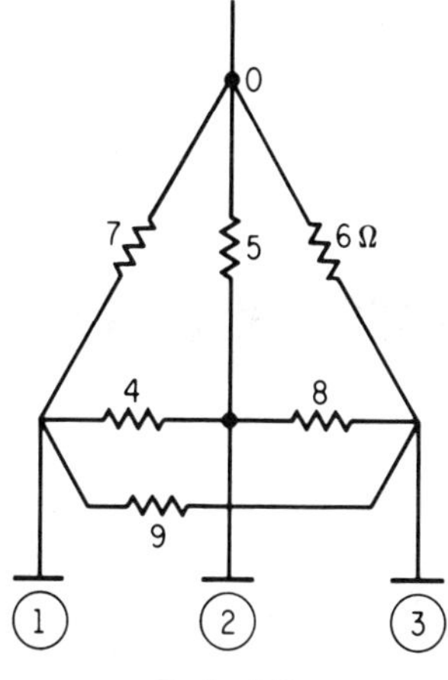

Prob. 2-8

2-8. (a) From the resistive network of Prob. 2-8, write the Y_{bus} matrix where node 0 is the reference node.

(b) Convert Y_{bus} to Z_{bus}.

(c) Draw the rake equivalent for Z_{bus}.

chapter **3**

POWER TRANSMISSION LINES

3-1. Introduction

Some deviation is made at this point from the standard presentation of power systems analysis. This is seen in the omission of material involved with calculation of line parameters. The author refers to a coverage of the time-proven principles of calculation of line inductance, capacitance, and resistance. This information is available in the Appendix and could, of course, be covered at this stage. However, often this material may already have been a part of the student's background in transmission lines or network theory. If not, it is still possible that a study of power systems might progress with more continuity (toward the actual power network problems) by isolating this specialized material. Our network problems of later chapters will for the most part assume that line parameters are known. Sequence of coverage is a matter of individual or group objectives. It has been the experience of some professors that the student "may not see the forest for the trees" and may easily lose interest in the challenging power system problems before he emerges from such detailed procedure. Therefore we shall proceed toward a better understanding of the balanced three-phase transmission line with known parameters.

3-2. Three-Phase Transmission—Review

A three-phase generator could of course be used to provide power for three separate single-phase loads, requiring a total of six lines from source to load. By way of review, it is recalled that the three voltages generated in the coils of the generator are equal in magnitude but are 120 degrees apart in phase angle. The separate distribution of this power as demonstrated in Fig. 3-1a would be impractical, since for balanced loading it is seen from the phasor diagram that $I_1 + I_2 + I_3 = 0$. Therefore, n and 0 points could be connected and return paths could be joined together in one common return, or the return line might be eliminated

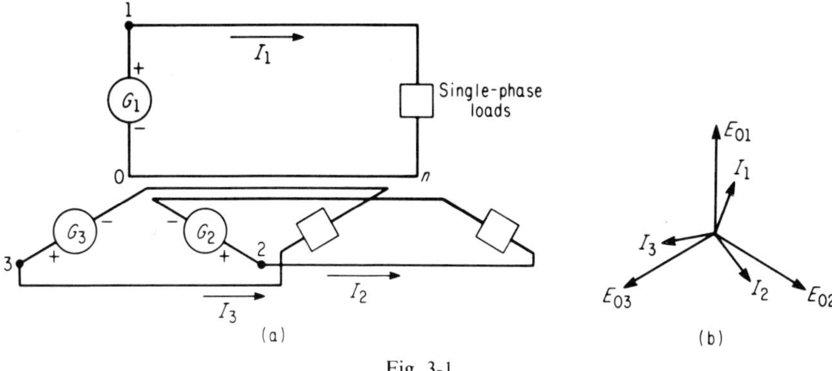

Fig. 3-1

altogether. Some obvious advantages gained in transmitting power by this resulting Y connection are (1) fewer conductors necessary, (2) less I^2R loss, and (3) less IZ line drop from source to load. Other advantages of three-phase power are gained insofar as machinery size, cost, and efficiency are concerned. Of course, the delta connection will yield these same general advantages.

In many of the problems of three-phase systems, a balanced condition may be assumed. For this reason it is convenient to simplify the circuit by placing it on a per phase or single-phase basis. For this purpose the Y-connection is most easily visualized.

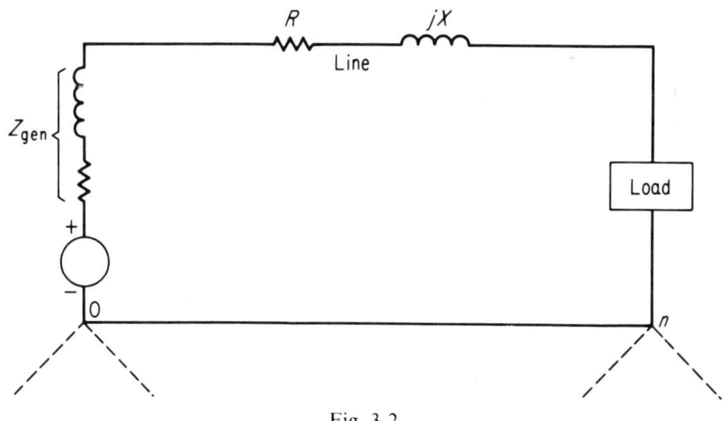

Fig. 3-2

The line of Fig. 3-2 is represented by a lumped R and X in series, only one phase of the balanced Y being considered. It is understood that the other two phases carry currents of the same magnitude and contain impedances equal to that of the phase considered. The return line may not actually exist, but regardless, we may include the line as a zero-impedance connection between n and 0 (which are at the same potential, since $I_n = I_1 + I_2 + I_3 = 0$ for a balanced system).

To further simplify the network, all values of voltages, currents, impedances, and power may be expressed in per unit values. Per unit values will not be used in this chapter, however, but will be dealt with in Chapter 4 with the treatment of power-network representations.

3-3. Reviewing the Concept of the Power Triangle

The power triangle method for solving simple three-phase problems is at times most useful. Sometimes it is found to be preferable to more conventional approaches where Kirchhoff's current and voltage laws are used. A review of this method is given here. Refer to Figs. 3-3a and 3-3b.

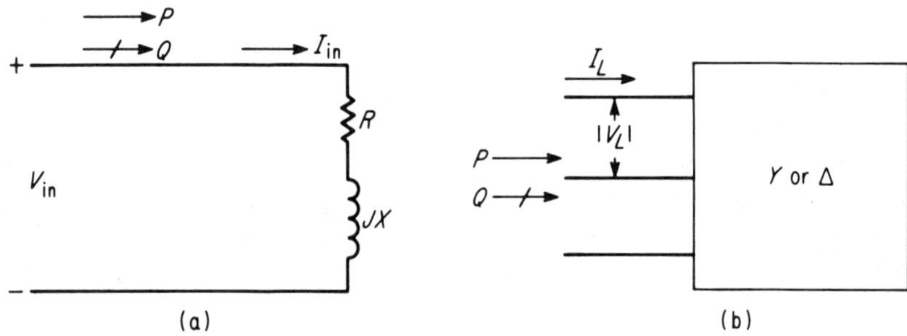

Fig. 3-3. (a) Single-phase circuit. (b) Balanced three-phase circuit.

First, for the single-phase circuit,

$$P = |I_R|^2 R = |V_R|^2/R = |V_R||I_R| = |V_{in}||I_{in}| \cos\theta \text{ (watts)} \tag{3-1}$$

$$Q = |I_x|^2 X = |V_x|^2/X = |V_x||I_x| = |V_{in}||I_{in}| \sin\theta \text{ (vars)} \tag{3-2}$$

$$VA = |I_z|^2 Z = |V_z|^2/Z = |V_{in}||I_{in}| \text{ (volt-amperes)} \tag{3-3}$$

where $\theta = \arctan X/R$.

For the balanced *three-phase* circuit of Fig. 3-3b,

$$P_{in} = 3P_P = 3|V_P||I_P| \cos\theta_P$$

where the P subscript indicates a phase value.

$$Q_{in} = 3Q_P = 3|V_P||I_P| \sin\theta_P$$

$$VA_{in} = 3(VA)_P = 3|V_{in}||I_{in}|$$

In terms of line values:

$$P_{in} = \sqrt{3}\,|V_L||I_L|\cos\theta \tag{3-4}$$

$$Q_{in} = \sqrt{3}\,|V_L||I_L|\sin\theta \tag{3-5}$$

$$VA_{in} = \sqrt{3}\,|V_L||I_L| \tag{3-6}$$

From these relationships the power triangle is apparent, similar to the impedance triangle. *Inductive vars will be considered positive* and capacitive vars negative. One way to express this relationship in terms of current and voltage phasors and on a per phase basis is

$$P + jQ = V_P I_P^* \tag{3-7}$$

or

$$P - jQ = V_P^* I_P \tag{3-8}$$

where the conjugate is indicated by *.

For those familiar with the power-triangle method, it will be recalled from circuit theory that this method offered a strong alternate approach to the steady-state solution of certain circuit problems, as opposed to the use of node-current equations or loop-voltage equations. Underlying this method is the fact that, regardless of whether circuit elements are in series or parallel (or whatever the circuit configuration), the total real (or reactive) power supplied to a system is the algebraic sum of the individual real (or reactive) powers utilized in the system. To illustrate this fact,

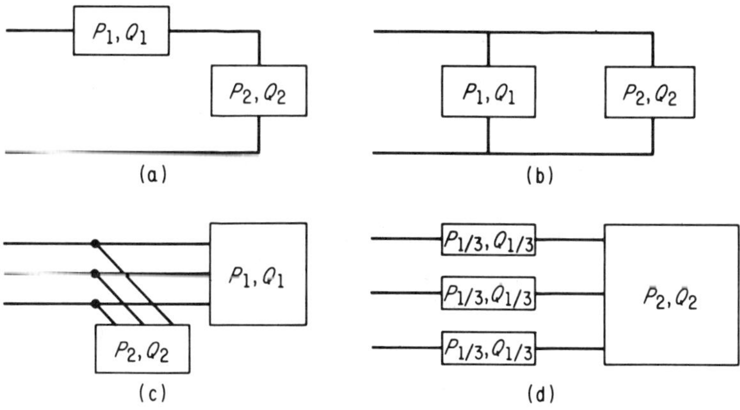

Fig. 3-4. Circuits for demonstrating Eqs. 3-9 through 3-11.

Figs. 3-4a–d are given. In each case, the input to the system is expressed by

$$P_{in} = P_1 + P_2 \tag{3-9}$$

$$Q_{in} = Q_1 + Q_2 \tag{3-10}$$

$$VA_{in} = \sqrt{P_{in}^2 + Q_{in}^2} \tag{3-11}$$

An example follows to serve as a review of the principles and use of the power-triangle method.

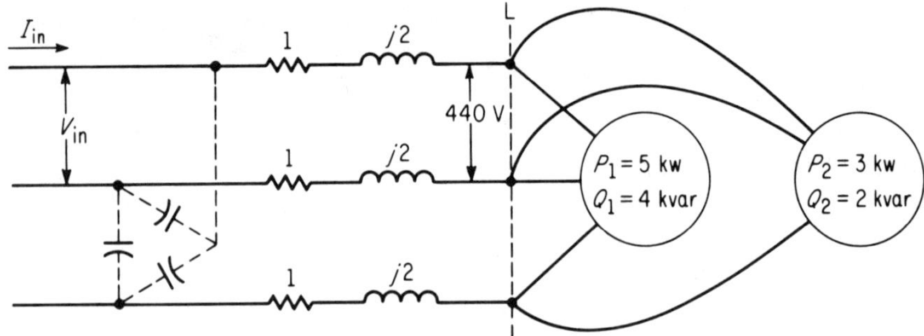

Fig. 3-5. Schematic for Example 3-1. The capacitor bank applies only to *Solution* (*b*) of the problem.

Example 3-1. Given induction motor loads 1 and 2 in parallel as shown in Fig. 3-5. These balanced loads are served from a balanced three-phase line of $1 + j2$ ohms. The junction (L) from which the loads are paralleled reads 440 volts line-to-line.

(a) Find $|V_{in}|$, $|I_{in}|$, and input power factor using the power-triangle method. Also, sketch the power triangle for the system.

(b) It is proposed that a Δ-connected capacitor bank be connected to the input terminals for power-factor correction. Find the capacitance values per phase of the Δ which would be necessary to bring the input power factor to 0.9 lagging. Also, find the new reduced value of input current. Assume V_{in} is held at the value found in part (a).

Solution (a)

$$P_L = P_1 + P_2 = 5 + 3 = 8 \text{ kw}$$

$$Q_L = Q_1 + Q_2 = 4 + 2 = 6 \text{ kvar}$$

$$VA_L = \sqrt{P_L^2 + Q_L^2} = \sqrt{8^2 + 6^2} = 10 \text{ kva}$$

or

$$VA_L = \sqrt{3}|V_L||I_L| = 10{,}000 \text{ volt-amperes}$$

$$I_L = I_{in} = \frac{10{,}000}{\sqrt{3} \times 440} = \underline{\underline{13.13}} \text{ amp}$$

$$P_{line} = 3|I_{in}|^2 R = 3(13.13)^2(1) = 518 \text{ watts}$$

$$Q_{line} = 3|I_{in}|^2 X = 3(13.13)^2(2) = 1{,}036 \text{ vars}$$

$$P_{in} = P_{line} + P_L = 518 + 8{,}000 = 8{,}518 \text{ watts}$$

$$Q_{in} = Q_{line} + Q_L = 1036 + 6{,}000 = 7{,}036 \text{ vars}$$

$$VA_{in} = \sqrt{P_{in}^2 + Q_{in}^2} = \sqrt{(8.52)^2 + (7.04)^2} = 11.05 \text{ kva}$$

$$\sqrt{3}|V_{in}||I_{in}| = 11{,}050; \quad V_{in} = \frac{11{,}050}{\sqrt{3} \times 13.13} = \underline{\underline{486}} \text{ volts}$$

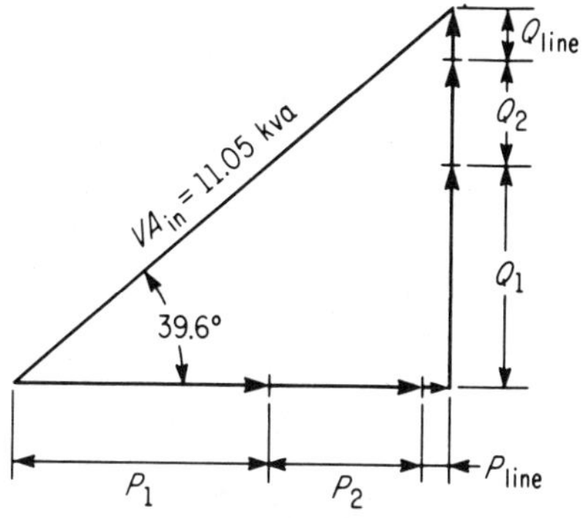

Fig. 3-6. Power triangle for *Solution* (*a*).

The power triangle of Fig. 3-6 shows

$$\theta_{in} = \arctan Q_{in}/P_{in}$$
$$= 39.6^\circ$$

Power factor = cos 39.6° = $\underline{\underline{0.771}}$ (lagging).

Solution (b): To correct the power factor to 0.9, the new θ_{in} must be reduced from $\theta = 39.6$ to $\theta = \cos^{-1}(.9) = 25.8^\circ$. Negative (capacitive)

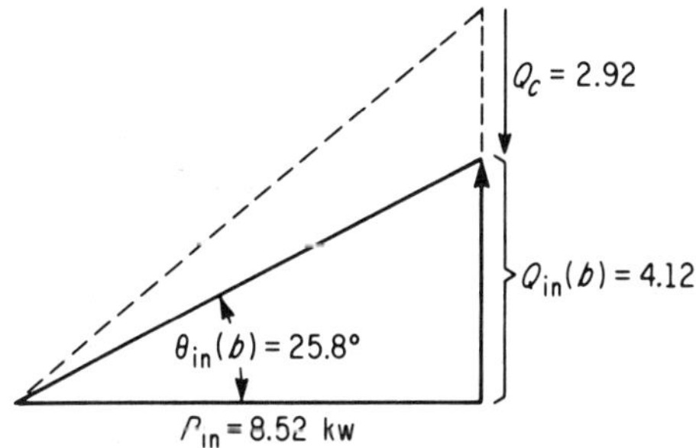

Fig. 3-7. Power triangle for *Solution* (*b*).

vars are added as shown in Fig. 3-7.

$$Q_{in}(b) = P_{in} \tan \theta_{in}(b)$$
$$= 8.52 \tan 25.8^\circ = 4.12 \quad \text{kvar}$$
$$Q_{added} = Q_{in}(b) - Q_{in}(a)$$
$$= 4.12 - 7.04 = -2.92 \quad \text{kvar}$$
$$Q_c = Q_{added}/3 = \frac{2.92}{3} = 0.973 \quad \text{kvar}$$
$$= 973 \quad \text{var}$$
$$Q_c = |V_c|^2/X_c$$

or

$$X_c = |V_c|^2/Q_c = \frac{(486)^2}{973} = 243 \quad \text{ohms}$$

$$C = \frac{1}{2\pi f x_c} = \frac{1}{2\pi(60)(243)} = \underline{\underline{10.9}} \quad \mu\text{f}$$

The new and reduced value of input current is

$$I'_{\text{in}} = \frac{P_{\text{in}}}{\sqrt{3}\,|V_{\text{in}}| \times \text{p.f.}} = \frac{8518}{3 \times 486 \times 0.9}$$

$$= \underline{\underline{11.25}} \quad \text{amperes}$$

3-4. Short Transmission Lines

Short-transmission-line analysis can safely apply to 60 Hz lines up to 30 miles in length. Actually no definite length can be associated with the long, medium, or short lines, as this factor depends upon both the type of problem and the characteristics of the particular line in question. The equivalent circuit is shown in Fig. 3-8 as a simple, lumped series im-

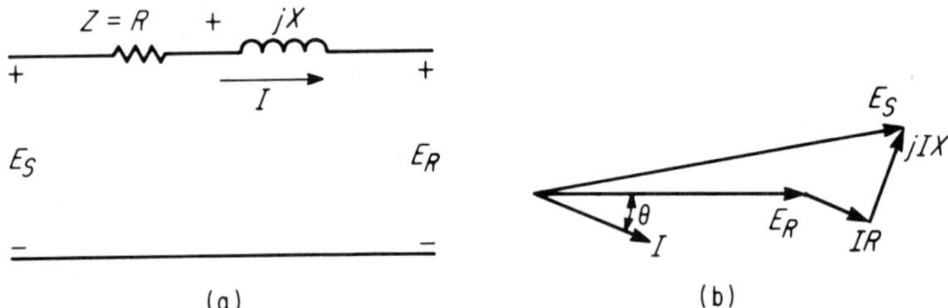

Fig. 3-8. (a) The short line. (b) Phasor diagram for the short line.

pedance. Voltage regulation of the line may be defined as the change in voltage at the receiving end of a line (expressed as per unit or percent of full-load voltage) in going from no load to full load. Sending voltage (E_s) is assumed to remain constant and load power factor is given. This may be expressed by the equation

$$\text{Percent voltage regulation} = \frac{|E_s| - |E_R|}{|E_R|} \times 100 \qquad (3\text{-}12)$$

In general $X > R$, and regulation will be poorer for inductive loads than for resistive loads. In the case of capacitive loads, regulation may become negative. This is best demonstrated by the vector diagram of Fig. 3-9 where it is seen that $E_R > E_S$. In some applications the short line will be represented by a simple reactance with R neglected entirely. There are times when even the long line is represented in this manner, in order to

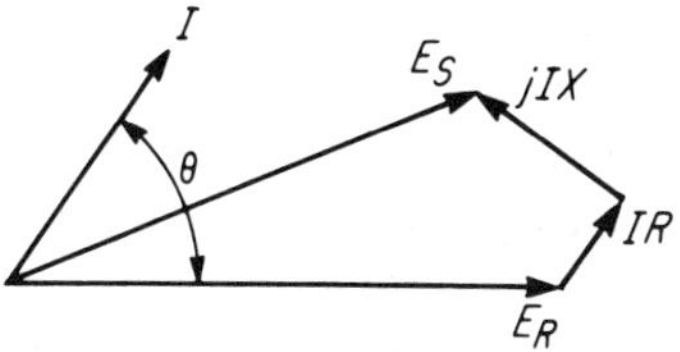

Fig. 3-9

simplify the problem. Short-circuit studies often make this approximation where the inaccuracy introduced is not of great magnitude.

The generalized constants (*a* parameters) for Fig. 3-8 can be determined by inspection. Recalling the equation

$$\begin{bmatrix} E_S \\ I_S \end{bmatrix} = \begin{bmatrix} A & B \\ C & D \end{bmatrix} \begin{bmatrix} E_R \\ I_R \end{bmatrix}$$

From this relationship we know that

$$A = 1, \quad B = Z, \quad C = 0, \quad D = 1 \tag{3-13}$$

In other words,

$$E_S = (1)E_R + (Z)I_R$$

$$I_S = (0)E_R + (1)I_R$$

There is the possibility that a short three-phase line may contain enough mutual inductance with an adjacent three-phase line to alter the simple series impedance representation. One-line diagrams are shown in Fig. 3-10. The lines may terminate on four separate buses as indicated in Fig. 3-10a, or they may have one end common to both lines (Fig. 3-10b).

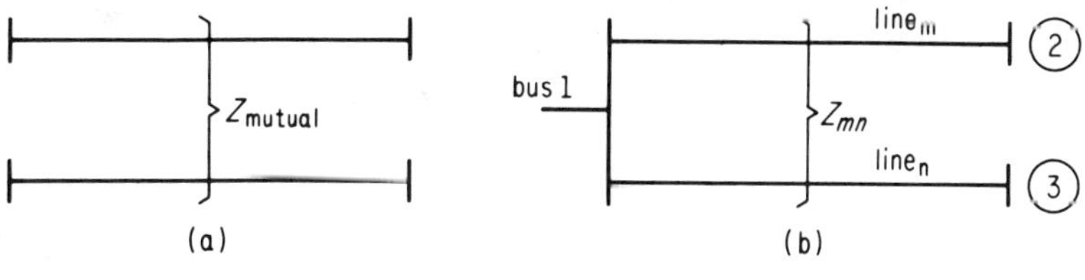

Fig. 3-10

The equivalent circuit for Fig. 3-10b can be shown either of two ways (Figs. 3-11a or 3-11b).

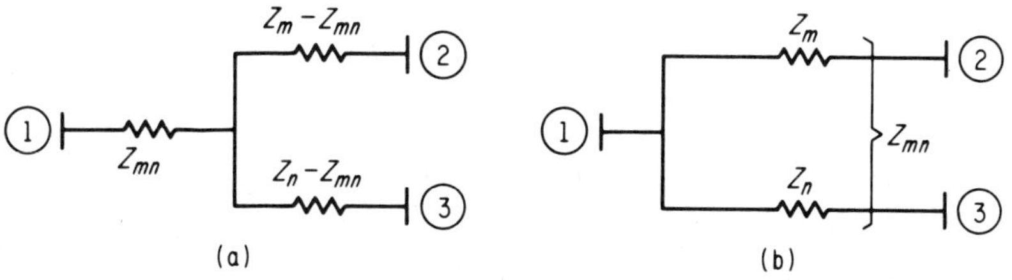

Fig. 3-11

3-5. π and T Equivalents for the "Medium" Line

As the length of line increases, the simple series impedance representation yields more inaccuracy, since shunt admittance has been neglected entirely. In order to improve the line representation for a line of medium length, a better approximation is arrived at through either the π or the T connection. These types were mentioned in Chapter 2, but will be considered here strictly in terms of the a parameters ($ABCD$ constants). The dividing line between the "medium" and the "long" line will generally fall in the vicinity of 150 to 200 miles.

The π circuit takes the lumped series Z as the product of the z per unit length times the length of the line. The total shunt admittance per

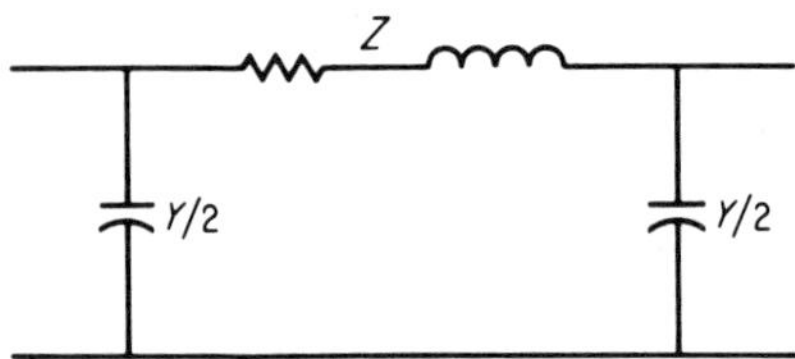

Fig. 3-12. Nominal π circuit.

phase to neutral for the total length of line is divided in half, as shown in Fig. 3-12. The $ABCD$ constants are given as

$$A = 1 + \frac{ZY}{2}, \quad B = Z, \quad C = Y + \frac{ZY^2}{4}, \quad D = 1 + \frac{ZY}{2} \tag{3-14}$$

The above expressions are easily verified by writing out Kirchhoff's equations for the circuit:

$$E_s = (E_R Y/2 + I_R) Z + E_R$$

or

$$E_s = \underbrace{(YZ/2 + 1)}_{A} E_R + \underbrace{Z}_{B} I_R \tag{3-15}$$

$$I_S = E_S(Y/2) + E_R Y/2 + I_R$$

Substituting Eq. 3-15 for E_S,

$$I_S = [(YZ/2 + 1) E_R + ZI_R] Y/2 + \frac{E_R Y}{2} + I_R$$

or

$$I_S = \underbrace{\left(\frac{Y^2 Z}{4} + Y\right)}_{C} \cdot E_R + \underbrace{\left(\frac{ZY}{2} + 1\right)}_{D} \cdot I_R \tag{3-16}$$

The nominal T may be used to approximate the effect of shunt capacitance and series impedance by splitting Z in half on either side of the total shunt admittance (see Fig. 3-13).

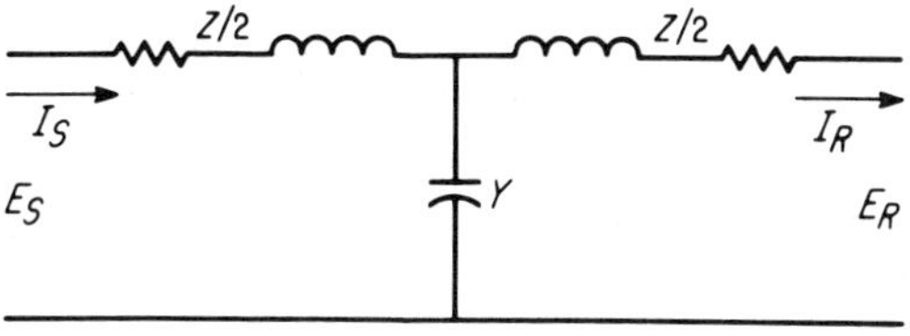

Fig. 3-13. Nominal T circuit.

The $ABCD$ constants are:

$$A = 1 + ZY/2, \quad B = Z + \frac{YZ^2}{4}, \quad C = Y, \quad D = 1 + \frac{YZ}{2} \tag{3-17}$$

In order to prove the relationships above, again write Kirchhoff's equation as

$$\begin{aligned} E_S &= I_S Z/2 + E_R + I_R Z/2 \\ &= [I_R + (E_R + I_R Z/2)Y]Z/2 + E_R + I_R Z/2 \end{aligned}$$

$$E_S = \underbrace{\left(1 + \frac{ZY}{2}\right)}_{A} E_R + \underbrace{\left(Z + \frac{Z^2 Y}{4}\right)}_{B} I_R \tag{3-18}$$

$$I_S = I_R + (E_R + I_R Z/2)\,Y$$

$$I_S = \underbrace{Y}_{C} E_R + \underbrace{\left(1 + \frac{ZY}{2}\right)}_{D} I_R \tag{3-19}$$

It may be easily shown through the use of Y-Δ conversions that the nominal π and T circuits are not equivalent. In either case they represent merely an approximation of the line. The line representation will more closely approach the actual line if one breaks the line up into segments, cascading the segments as was shown earlier in Fig. 2-6. The $ABCD$ matrix for the cascaded case is again the product of all the segment a matrices,

$$\begin{bmatrix} A & B \\ C & D \end{bmatrix}_{\text{total}} = \begin{bmatrix} a_1 \end{bmatrix} \begin{bmatrix} a_2 \end{bmatrix} \cdots \begin{bmatrix} a_n \end{bmatrix} \tag{3-20}$$

Now the result of cascading of T equivalents will more nearly equal the result of cascading the π equivalents than was the case with the simple nominal π or T.

3-6. The Long Transmission Line

When a better representation is needed for the long line than the nominal π or T, a more sophisticated solution is necessary. One must look at the incremental line length and consider the exact effect of distributed capacitance and its relationship to the incremental line impedance. To be exact, one must take an infinite number of these line segments, which requires the solution of appropriate differential equations. Figure 3-14 shows the line with one of the incremental (dx) portions at a distance

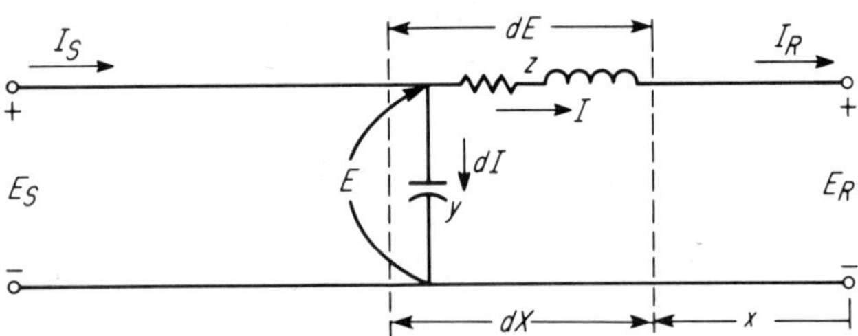

Fig. 3-14. Three-phase long line (per phase basis).

(x) from the receiving end. The sinusoidal steady-state consideration is presented. Here

z = impedance per unit length and y = admittance per unit length
Voltage across the incremental of line = $dE = Iz\,dx$
Incremental charging current $dI = Ey\,dx$

or

$$\frac{dE}{dx} = Iz \tag{3-21}$$

and

$$\frac{dI}{dx} = Ey \tag{3-22}$$

Differentiating Eqs. 3-21 and 3-22 with respect to x,

$$\frac{d^2E}{dx^2} = z\,\frac{dI}{dx} \tag{3-23}$$

$$\frac{d^2I}{dx^2} = y\,\frac{dE}{dx} \tag{3-24}$$

Substitute the value of dI/dx in Eq. 3-22 into Eq. 3-23, and substitute the value of $\frac{dE}{dx}$ in Eq. 3-21 into Eq. 3-24:

$$\frac{d^2E}{dx^2} = zEy \tag{3-25}$$

$$\frac{d^2I}{dx^2} = yIz \tag{3-26}$$

Now Eqs. 3-25 and 3-26 must be solved for E and I respectively in terms of the variable x. Laplace transforms will be used at this point. The equations might be said to exhibit the duality principle, where E, z, and y correspond to I, y, and z respectively. This holds even for the initial conditions, where $x = 0$, $I = I_R$, and $E = E_R$. Since these equations are similar, only Eq. 3-25 will be solved and results for Eq. 3-26 can be written out by inspection (dual of Eq. 3-25). Again,

$$\frac{d^2E(x)}{dx^2} - zEy = 0$$

Taking the Laplace transform of the above equation,

$$s^2E(s) - sE(x_0) - E'(x_0) - zyE(s) = 0 \tag{3-27}$$

when $x = 0$, $E = E_R$ and $I = I_R$.

$$E'(x_0) = \left.\frac{dE}{dx}\right]_{x=0} = Iz\Big]_{x=0} = I_Rz \tag{3-28}$$

Substituting the results of Eq. 3-28 into Eq. 3-27 yields

$$s^2E(s) - sE_R - I_Rz - zyE(s) = 0$$

Solving for $E(s)$:

$$(s^2 - zy)E(s) = sE_R + zI_R$$

$$E(s) = \frac{sE_R + zI_R}{s^2 - zy}$$

or

$$E(s) = E_R\frac{s}{s^2 - zy} + zI_R\frac{1}{s^2 - zy} \tag{3-29}$$

Now obtaining the inverse transform from the transform tables yields the following solution in terms of the variable X:

$$E(x) = \underbrace{\left[\cosh\sqrt{zy}\,x\right]}_{A}E_R + \underbrace{\left[\sqrt{\frac{z}{y}}\sinh\sqrt{zy}\,x\right]}_{B}I_R \tag{3-30}$$

As previously stated, the expression for $I(x)$ may now be found by inspecting the form of Eq. 3-30:

$$I(x) = (\cosh\sqrt{zy}\,x)I_R = \left(\sqrt{\frac{y}{z}}\sinh\sqrt{zy}\,x\right)E_R$$

or

$$I(x) = \underbrace{\left(\sqrt{\frac{y}{z}} \sinh \sqrt{zy}\,x\right)}_{C} E_R + \underbrace{\left(\cosh \sqrt{zy}\,x\right)}_{D} I_R \tag{3-31}$$

We will define $\sqrt{zy} = m$ as the propagation constant and the value of $\sqrt{\dfrac{z}{y}}$ as the characteristic impedance (Z_0) of the line. Also $Z_0 = 1/Y_0$. Rewriting Eqs. 3-30 and 3-31,

$$E(x) = (\cosh mx)\,E_R + (Z_0 \sinh mx) I_R \tag{3-32}$$

$$I(x) = (Y_0 \sinh mx)\,E_R + (\cosh mx)\,I_R \tag{3-33}$$

Placing Eqs. 3-32 and 3-33 in matrix form,

$$\begin{bmatrix} E(x) \\ I(x) \end{bmatrix} = \underbrace{\begin{bmatrix} \cosh mx & Z_0 \sinh mx \\ Y_0 \sinh mx & \cosh mx \end{bmatrix}}_{a \text{ parameters } (ABCD \text{ constants})} \begin{bmatrix} E_R \\ I_R \end{bmatrix} \tag{3-34}$$

While Eq. 3-34 is concise and correct, gaining a concept for interpretation from this form is somewhat difficult. Hyperbolic functions have a way of escaping the imagination. First it will be realized that the propagation constant (m) is complex. Let $m = u + jv$. Also we know that

$$\sinh mx = \frac{e^{mx} - e^{-mx}}{2}$$

and

$$\cosh mx = \frac{e^{mx} + e^{-mx}}{2}$$

Making the above substitutions into Eqs. 3-32 and 3-33 yields

$$E(x) = \frac{e^{mx} + e^{-mx}}{2} E_R + Z_0 \frac{e^{mx} - e^{-mx}}{2} I_R$$

$$= \frac{e^{mx}}{2}(E_R + Z_0 I_R) + \frac{e^{-mx}}{2}(E_R - Z_0 I_R$$

$$E(x) = \frac{e^{ux} e^{jvx}}{2}(E_R + Z_0 I_R) + \frac{e^{-ux} e^{-jvx}}{2}(E_R - Z_0 I_R) \tag{3-35}$$

$$I(x) = \frac{e^{mx} - e^{-mx}}{2} Y_0 E_R + \frac{e^{mx} + e^{-mx}}{2} I_R$$

$$= \frac{e^{mx}}{2}(Y_0 E_R + I_R) + \frac{e^{-mx}}{2}(I_R - Y_0 E_R)$$

$$I(x) = \frac{e^{ux} e^{jvx}}{2}(I_R + Y_0 E_R) + \frac{e^{-ux} e^{-jvx}}{2}(I_R - Y_0 E_R) \tag{3-36}$$

While Eqs. 3-35 and 3-36 are not as compact as the matrix form of Eq. 3-34 they are easier to interpret. Both Eqs. 3-35 and 3-36 are of the same form, therefore only the voltage equation will be investigated (Eq. 3-35). The terms $E_R + I_R Z_0$ and $E_R - I_R Z_0$ are *phasors*, and operating upon these phasors are exponential factors. The factors of e^{ux} and e^{-ux} may be considered as *magnitude operators*, since they effect only the magnitudes of the two terms of the equation. As x increases (moving away from the receiving end) the first term becomes larger because of e^{ux} and second term smaller because of e^{-ux}.

The factors of e^{jvx} and e^{-jvx} are rotational operators which shift the phase angles on the phasors as x varies. The terms $e^{ux}e^{jvx}(E_R + I_R Z_0)/2$ and $e^{-ux}e^{-jvx}(E_R - I_R Z_0)/2$ are called the *incident wave* and *reflected wave* respectively. Note that when $x = 0$, the only difference between the two terms is in the sign of the $I_R Z_0$ phasor. Of course $E(x_0) = E_R$ in this case. The two terms of the expression for $E(x)$ behave like traveling waves as we move along the line. There is a continual phase shift upon the phasors (as we increase the distance x) as well as magnitude change. It is analogous to a disturbance in the water at some sending point. When the ripple (ever decaying in magnitude) reaches the bank or receiving point it moves back toward the sending point as a reflected wave. The magnitude of the incident wave (incident upon the load) is greater at the sending end, while that of the reflected wave is greater at the receiving end. The total voltage at a given point on the line is the sum of the two superimposed voltages. When the two terms of the expression of Eq. 3-35 are 180° out of phase, a cancellation takes place at that particular position. The cancellation will be at least partial, depending upon the value of load impedance and the value of the magnitude operators e^{ux} and e^{-ux}. The cancellation would be complete in the special case where $u = o$ and $I_R = 0$. In this case the cancellation would occur when $vx = \pi/2$ radians or at one-quarter wavelength, where the reflected and incident waves are 180° out of phase. In general, this phenomenon of voltage cancellation is not such an important factor at the power frequency as will be shown in the next paragraph.

One can easily determine the wavelength (λ) or distance x on the line necessary to accomplish a phase shift of 2π radians for the incident and reflected waves. This occurs when $e^{jvx} = e^{j2\pi}$ or since the phase shift $\theta = vx$, then

$$2\pi = v\lambda$$

or

$$\lambda = 2\pi/v \tag{3-37}$$

Now we know that the value of λ can be related to the velocity of propagation (V) in mile per second by the equation

$$\lambda = V/f \tag{3-38}$$

An assumption could be made to determine a rough approximation for the wavelength of the 60 Hz line. If it were assumed that the velocity of propagation V approached the speed of light (186,000 miles per second) then it would follow from Eq. 3-38 that λ is approximately 3,100 miles. From this approximation it could be further deduced that on a 100-mile power line, the reflected and incident waves are shifted only about 12°. For example,

$$\begin{aligned} \theta &= vx \qquad (3\text{-}39) \\ &= \frac{2\pi}{\lambda}\,x \\ &\doteq 2\pi\,\frac{100}{3100} = .202 \text{ radian} \\ &\doteq 11.6^\circ \end{aligned}$$

This demonstrates that the 100-mile power line still falls far short of reaching the quarter-wave conditions mentioned in the preceding paragraph.

Again we will refer to the *open-circuited line*. Here $I_R = 0$ and Eqs. 3-35 and 3-36 reduce to

$$E(x) = \frac{E_R}{2}\,e^{ux}e^{jvx} + \frac{E_R}{2}\,e^{-ux}e^{-jvx}$$

$$I(x) = \frac{E_R Y_0}{2}\,e^{ux}e^{jvx} - \frac{E_R Y_0}{2}\,e^{-ux}e^{-jvx}$$

At the receiving end both the incident and reflected waves are equal to $E_R/2$ and add to E_R. The forward and reflected current waves are equal in magnitude but opposite in sign and therefore add to zero at the receiving end, as we would expect in the open-circuited case.

Equations for the *short-circuited line* reduce to the following:

$$E(x) = \frac{I_R Z_0}{2}\,e^{ux}e^{jvx} - \frac{I_R Z_0}{2}\,e^{-ux}e^{-jvx}$$

$$I(x) = \frac{I_R}{2}\,e^{ux}e^{jvx} + \frac{I_R}{2}\,e^{-ux}e^{-jvx}$$

At the receiving end it is seen that the reflected and incident voltages are equal in magnitude but out of phase by 180°, therefore adding to zero. The reflected and incident current waves are equal in magnitude and in phase at the receiving end and therefore add to I_R.

An *infinite line* is one which is terminated in its characteristic impedance (Z_0), in which case the reflected wave is eliminated for both $E(x)$ and $I(x)$. For this condition Eqs. 3-35 and 3-36 reduce to

$$E(x) = I_R Z_0 e^{ux} e^{jvx} + (0)\, e^{-ux} e^{-jvx}$$
$$I(x) = I_R e^{ux} e^{jvx} + (0)\, e^{-ux} e^{-jvx}$$

This condition of infinite line (also called the *flat line*) is not usually found in power applications.

3-7. Long-Line Equivalents

Having included the separate effects of all of the incremental line elements, it is now possible to use the resulting $ABCD$ constants (in hyperbolic form) in order to obtain an accurate long-line equivalent for the steady-state circuit. Two circuits will be developed, the π and the T.

In Sec. 3-5 certain relationships were derived, relating the $ABCD$ constants of a line to the parameters of the π and T circuits. Knowing the $ABCD$ (hyperbolic) constants, the more accurate Z and Y values are possible.

The π *circuit* relationships for A and B of Eq. 3.14 are $A = 1 + Z_\pi Y_\pi/2$; $B = Z_\pi$. It was also found (see Eq. 3-34) that $A = \cosh mx$ and $B = Z_0 \sinh mx$. By simple substitution,

$$Z_\pi = Z_0 \sinh mx \tag{3-40}$$

and

$$1 + \frac{Z_\pi Y_\pi}{2} = \cosh mx \tag{3-41}$$

Substituting Z_π of Eq. 3-40 into Eq. 3-41 yields

$$\frac{Y\pi}{2} = \frac{\cosh mx - 1}{Z_0 \sinh mx} \tag{3-42}$$

or

$$\frac{Y\pi}{2} = \frac{1}{Z_0} \tanh \frac{mx}{2} \tag{3-43}$$

Obtaining Eq. 3-43 from Eq. 3-42 requires the identity

$$\tanh \frac{mx}{2} = \frac{\cosh mx - 1}{\sinh mx}$$

This identity is verified as

$$\frac{\cosh mx - 1}{\sinh mx} = \frac{\cosh mx - (\cosh^2 mx - \sinh^2 mx)}{\sinh mx}$$

$$= \frac{(e^{mx} - e^{-mx})\frac{1}{2} - [(e^{mx} - e^{-mx})\frac{1}{2}]^2 + [(e^{mx} - e^{-mx})\frac{1}{2}]^2}{(e^{mx} - e^{-mx})\frac{1}{2}}$$

$$= \frac{e^{mx} + e^{-mx} - 2e^{-mx}e^{mx}}{e^{mx} - e^{-mx}}$$

$$= \frac{(e^{mx/2} - e^{-mx/2})^2}{(e^{mx/2} - e^{-mx/2})(e^{mx/2} + e^{-mx/2})} = \frac{e^{mx/2} - e^{-mx/2}}{e^{mx/2} + e^{-mx/2}}$$

The π equivalent is seen in Fig. 3-15.

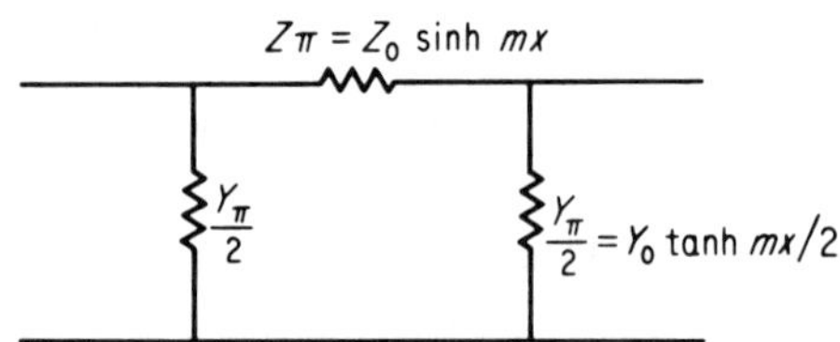

Fig. 3-15. Long-line π equivalent.

The *T equivalent* obtainable from the long-line *ABCD* constants is represented in Fig. 3-16. This circuit is readily found by comparing re-

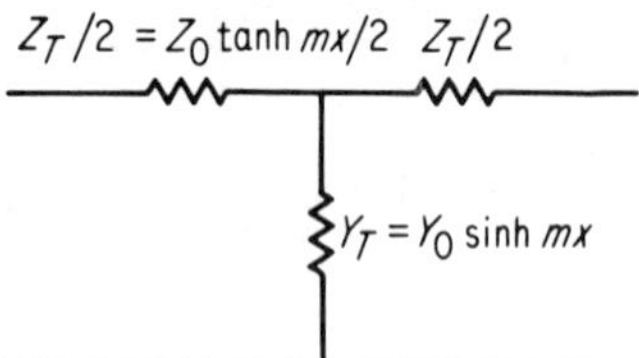

Fig. 3-16. Long-line T equivalent.

lationships of Eq. 3-17 with the *ABCD* constants of Eq. 3-34. Results of this comparison yield

$$Y_T = Y_0 \sinh mx \tag{3-44}$$

and

$$A = 1 + \frac{Z_T Y_T}{2} = \cosh mx$$

or

$$\frac{Z_T}{2} = \frac{\cosh mx - 1}{Y_0 \sinh mx}$$

$$\frac{Z_T}{2} = Z_0 \tanh \frac{mx}{2} \tag{3-45}$$

Equation 3-45 is obtained from Eq. 3-44 by using the same identity used for obtaining Eq. 3-43 from Eq. 3-42.

3-8. Comparison of Long-Line Equivalents with the Approximate π and T Equivalents of Sec. 3-4

The value of the characteristic impedance (Z_0) can be expressed in terms of the propagation constant (m) and the series line impedance (Z) of the approximate π and T of Figs. 3-12 and 3-13.

$$Z_0 = \sqrt{z/y} = \sqrt{z/y \cdot \frac{x^2 z}{x^2 z}} = \sqrt{\frac{x^2 z^2}{x^2 yz}} = \sqrt{\frac{Z^2}{x^2 m^2}}$$

$$Z_0 = \frac{Z}{mx} \tag{3-46}$$

Z_0 can also be expressed in terms of m and the total shunt admittance Y of Figs. 3-12 and 3-13:

$$Z_0 = \sqrt{z/y} = \sqrt{\frac{z}{y} \cdot \frac{yx^2}{yx^2}} = \sqrt{\frac{m^2 x^2}{Y^2}}$$

$$Z_0 = \frac{mx}{Y} \tag{3-47}$$

If the values for Z_0 expressed in Eqs. 3-46 and 3-47 are substituted into Eqs. 3-40, 3-43, 3-44 and 3-45 the following results are obtained:

$$Z_\pi = Z\left(\frac{\sinh mx}{mx}\right) \tag{3-48}$$

$$\frac{Y_\pi}{2} = Y/2\left(\frac{\tanh mx/2}{mx/2}\right) \tag{3-49}$$

$$\frac{Z_T}{2} = Z/2\left(\frac{\tanh mx/2}{mx/2}\right) \tag{3-50}$$

$$Y_T = Y\left(\frac{\sinh mx}{mx}\right) \tag{3-51}$$

In each of Eqs. 3-48 through 3-51 the quantities in parentheses are cor-

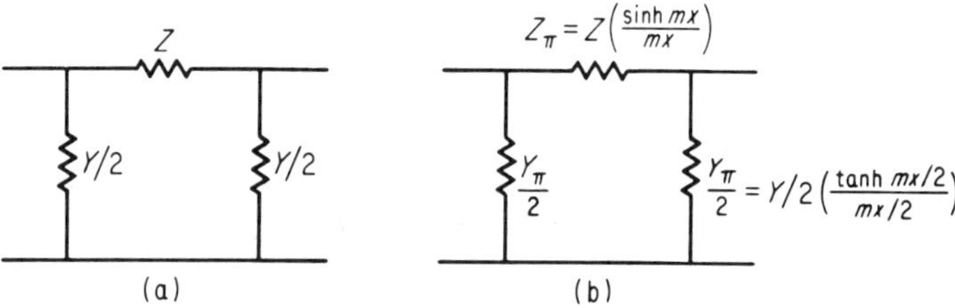

Fig 3-17. (a) Approximate equivalent using lumped parameters. (b) Hyperbolic equivalent in terms of lumped parameters of Fig. 3-17a.

rection factors acting upon the less accurate lumped parameters of the approximate and T equivalents to obtain the hyperbolic parameters. Figures 3-17 and 3-18 demonstrate these comparisons.

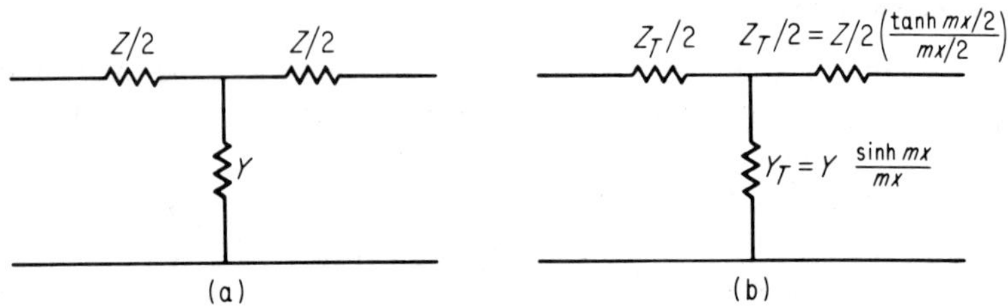

Fig. 3-18. (a) Approximate T equivalent using lumped parameters. (b) Hyperbolic equivalent in terms of lumped parameters of Fig. 3-18a.

Problems

3-1. Two balanced three-phase loads are paralleled as shown in Prob. 3-1. All known information is on the drawing.

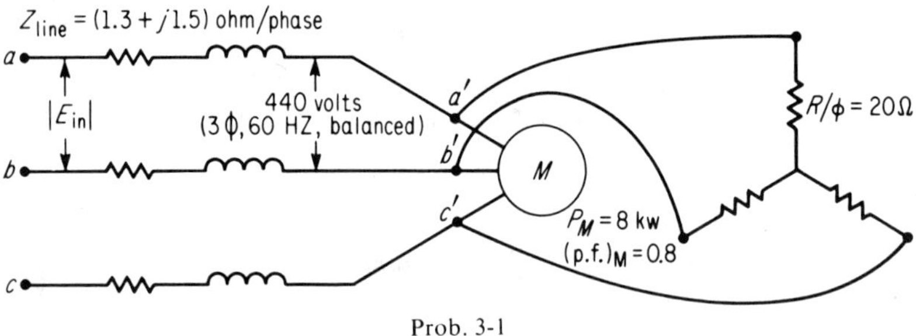

Prob. 3-1

(a) Find the values of $|E_{in}|$, P_{in}, $|I_{in}|$ and input power factor. Use the power-triangle method.

(b) Show the three-phase phasor diagrams for the input terminals. Include six voltage phasors (3 line and 3 phase) and three current phasors, taking E_{na} as the reference voltage, with an *a-b-c* phase sequence.

(c) Find the value of delta capacitors to place across the input terminals in order to correct input power factor to unity. Assume the input voltage stays constant.

3-2. A short three-phase transmission line has a series line impedance per phase of $0.5 + j0.7$ ohms. The receiving-end line-to-neutral voltage of phase a is $2770 \angle 0°$ volts and the full-load receiving end current is $200 \angle -45°$ amperes.

(a) What is the voltage regulation of the line and what are the $ABCD$ constants of the line? Show the phasor diagram of E_s, E_r, and I.

(b) Change the current I_r to $200 \angle +30°$ and repeat part (a).

3-3. Three balanced three-phase motor loads are connected in parallel across a 240-volt line. Their loads are as follows:

M_1: 8 KW input at 0.92 p.f. lagging
M_2: 5 KVA input at 0.8 p.f. lagging
M_3: 10-hp output at 90 percent efficiency and 0.85 input p.f., lagging

(a) Find the total input power, input line current (magnitude), and input power factor. Use the power-triangle method.
(b) Find the individual motor currents (magnitudes).
(c) Find the total KVAR capacity of the delta capacitor bank that would be required to correct input power factor to 0.95.

3-4. The total series self-impedance per phase of a four-wire, three-phase transmission line (m) is $2 + j3$ ohms. This line has one end in common with a second four-wire, three-phase line n. Line n has a self impedance of $3 + j4$ ohms per phase. A mutual impedance exists between like phases of these lines which equals $j1$ ohms per phase. A voltage of $E_{0a} = 2400 \underline{/0^\circ}$ volts is present (line to neutral) on phase a at the common terminal (bus 1).

(a) Sketch both the V and the T equivalents for one phase of this three-bus arrangement, with impedance values included.
(b) Balanced currents flow in the m lines between the common bus 1 and bus 2. The a-phase current in this line is $100 \underline{/-30^\circ}$ amperes. Likewise, the a-phase current of the balanced line n is passing $70 \underline{/0^\circ}$ amperes from bus 1 to bus 3. Find the magnitude of phase voltages present at terminal (or buses) 2 and 3.

(*Note.* This problem is largely academic in nature with respect to the relatively large magnitude of mutual impedance given as will be shown in Chapter 9. However, mutual flux between sets of three-phase lines which carry balanced loads or balanced faults will often nearly cancel one another. Such is *not* the case for mutual flux of zero sequence currents, as will be pointed out in that same chapter.)

3-5. Find the $ABCD$ constants of a three-phase, 50-mile, 60-Hz, transmission line with a series impedance of $0.15 + j0.78$ ohm per mile and a shunt admittance of $j5.0 \times 10^{-6}$ mho per mile.

3-6. Determine both the π and T equivalents for the medium line of Prob. 3-5.

3-7. Determine the propagation constant m and the surge impedance (or characteristic impedance Z_0) for Prob. 3-5.

3-8. Work Prob. 3-5 for a 400-mile line. Find the $ABCD$ constants assuming (a) the medium line equations are sufficient, (b) the long-line equations are necessary. Note the difference between results of parts (a) and (b).

3-9. Find the series and shunt parameters for the equivalent π circuit for the 400-mile line of Prob. 3-8 (a) and (b).

3-10. Given the 400-mile line of Prob. 3-8. The load-end voltage is 220 kv. The load is 100 mw at 0.85 p.f. lagging. Find the sending-end voltage and current.

chapter **4**

POWER NETWORK REPRESENTATIONS

4-1. Introduction

The three-phase system was demonstrated on a per phase basis in Sec. 3-2. The line-to-neutral, single-phase representation is used throughout this text for both the balanced and the unbalanced systems. The unbalanced systems will be treated, with the aid of symmetrical components, in a later chapter. However, in this chapter it is assumed that conditions are balanced with respect to loading and/or short circuits.

4-2. Synchronous Generator Representation and Review

It is recalled from the study of a-c rotating machinery that numerous methods are available for determination of voltage regulation for various load conditions. Among these methods are the ASA method, the Doherty and Nickle two-reactance method, and various synchronous impedance methods. However, for representing the power network, the synchronous impedance method stands out from the others, not because of better accuracy, but rather because it offers as a by-product, an equivalent circuit for the generator. This equivalent circuit aids in the calculation of the network of which the generator is a part. Therefore the synchronous impedance method (unsaturated) will be singled out and a brief review of related generator concepts will be given.

A synchronous generator with cylindrical rotor is shown in Fig. 4-1. A simplified three-phase stator (armature) winding is demonstrated where the voltage induced into an individual phase coil is determined by Faraday's law, $e = Nd\phi/dt$. In this case, the rms value of voltage per phase is found to be $E_{\text{phase}} = 4.44 f N \phi_{\text{pole}} K_p K_d$ where f = frequency in Hz, N = series turns per phase, and ϕ_{pole} = pole flux in webers. The pitch factor, K_p corrects the voltage for windings which do not span a full 180 electrical degrees. K_d is the distribution factor, correcting the voltage for windings whose series turns are not concentrated into one pair of slots, but are distributed along the periphery of the stator. The voltages for

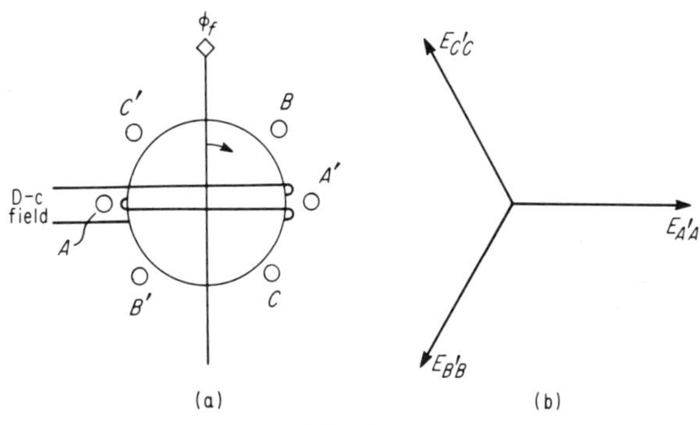

Fig. 4-1

phases A, B, and C are equal in magnitude since all voltages are dependent upon the same flux. However, the lag which is present corresponding to the time that ϕ_f(field flux) reaches phase A, B, and C (in that order) is responsible for the phase displacement of voltages as shown in Fig. 4-1b.

Now it is known that if a balanced load were added to the armature windings that three separate armature fluxes would be distributed with the peak density along the magnetic axis of individual phase windings. The resultant of these three pulsating fluxes is known to be a revolving field, rotating at the same speed and in same direction as the main field ϕ_f of the rotor. This armature reaction flux will be termed ϕ_{AR}. It was further learned from machine theory that ϕ_{AR} will either aid ϕ_f for zero power factor current (leading the internal voltage), oppose ϕ_f for zero power factor current (lagging), or travel in space quadrature with respect to ϕ_f under unity power factor conditions.

The so-called air-gap voltage is dependent upon all of the flux which crosses the air gap and therefore must include the combined effects of ϕ_f and ϕ_{AR}. A third flux is present termed *armature leakage flux*, ϕ_a, which does not link the main field poles. If saturation is neglected, flux is proportional to mmf. Also, voltage is proportional to flux. We say that the flux linking a coil is 90° displaced from the induced emf, since flux inside the turns is maximum when voltage is minimum. Therefore the voltage-mmf relationships may be clearly shown on the phasor diagram of Fig. 4-2a.

The mmf phasors F, A, and R are given in equivalent field amperes and the relative mmf positions also indicate the relative space arrangement of their corresponding fields. The terminal voltage V equals the air-gap voltage E_a minus the $I_a(r_e + jX_a)$ drop of the winding. The leakage reactance is represented by X_a. The resulting mmf (R) is leading the air-gap voltage by 90°. Armature reaction mmf A is in phase with I_a. If the armature circuit were opened, the only remaining mmf would be F, due to the field current, and this mmf is drawn 90° leading the open cir-

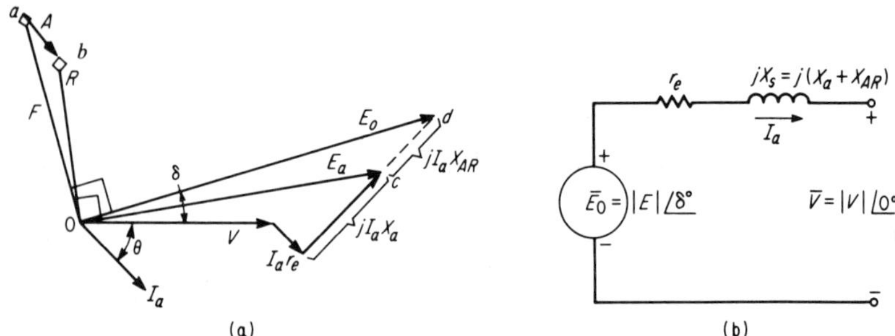

Fig. 4-2. (a) Phasor diagram for synchronous generator with cylindrical rotor. (b) Equivalent circuit for synchronous generator with cylindrical rotor.

cuit voltage E_0. With saturation neglected, $E_0/E_a = F/R$ and it is easily shown that in these two similar triangles (0*ab* and 0*cd*) that A is perpendicular to *cd*, which is an extension of the jI_aX_a line. The line *cd* is called jI_aX_{AR} where X_{AR} is the reactance of armature reaction. The phasors $jI_aX_a + jI_aX_{AR}$ are lumped together as $jI_a(X_a + X_{AR}) = jI_aX_s$, where X_s is termed the unsaturated synchronous reactance. The simple equivalent circuit of Fig. 4-2b represents the synchronous generator as the open circuit voltage (E_0) in series with the synchronous impedance of $r_e + JX_s$. Of course the value of X_s can be adjusted for effects of saturation. Any such correction will, theoretically, be good for only one given load condition, since the degree of saturation will vary with ϕ_{AR} and therefore with I_a. This text will not attempt coverage of the various saturated synchronous reactance methods.

The preceding discussion and review was based upon the cylindrical rotor synchronous generator. The salient-pole machine would appear to complicate the picture considerably, since the direct and quadrature axis of the rotor present such different reluctances to the mmf of armature reaction. I_q, the unity power factor amature current (with respect to the internal excitation voltage E_0) should be identified with a quadrature axis synchronous reactance (X_q), since this current component sets up flux in space quadrature with the main field poles. Similarly, lagging and leading zero power factor currents are identified with the direct axis reactance (X_d), since these currents set up flux in line with the axis of the main field poles. For review purposes, the phasor diagram is shown in Fig. 4-3a.

The two-reactance method is obviously more involved than the simple X_s of the cylindrical rotor where $X_s = X_d = X_q$. However, it happens that for the short-circuit studies, the salient-pole phasor diagram reduces to that of Fig. 4-3b. The reason for this is that short-circuit currents, in general, feed through lines or transformers where $X \gg r$, making the power factor approach zero as currents go lagging. The I_q component can often be neglected in such a case, resulting in the phasor

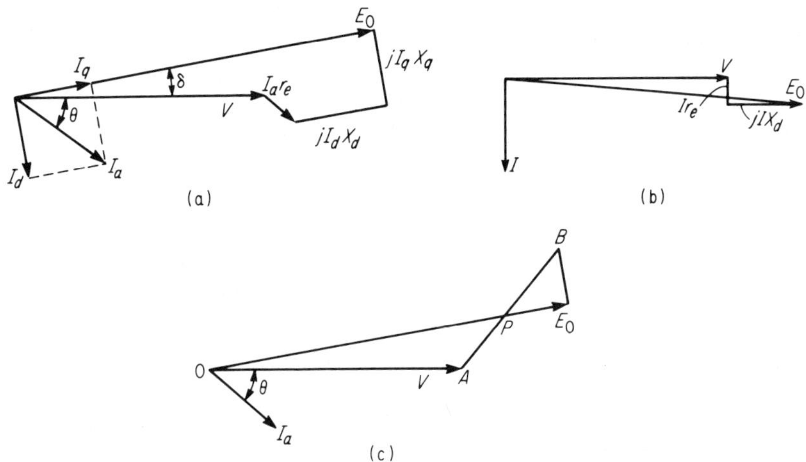

Fig. 4-3. (a) Phasor diagram for salient-pole synchronous generator. (b) Special case where power factor = 0. (c) Alternate diagram to Fig. 4-3a for construction purposes (r_e neglected).

diagram of Fig. 4-3b. This shows that for the short-circuited salient pole machine, X_d merely replaces the X_s used for the cylindrical rotor machine. Later in the text it will be shown that X_d takes on different values, depending upon the transient time following the short circuit.

While the phasor diagram of Fig. 4-3a is essentially correct, it would not be possible to construct as shown where V, I_a, r_e, and X_d are known and δ as yet is unknown. However, the phasor diagram can be drawn as per Fig. 4-3c by taking the following steps.

1. Draw V and $I\underline{/\theta^\circ}$.
2. Draw $I_a r_e$. Draw $jI_a X_d$ (line $\overline{AB}$) $\perp$ to I (r_e was neglected).
3. Establish point P where $|\overline{AP}| = |I_a X_q|$.
4. Draw a line through 0 and P for the E_0 line.
5. Drop a $\perp$ from B to line $\overline{OP}$ to establish E_0.

It will be recalled briefly that the power angle δ has both electrical and physical significance. It is not only the angle between phase terminal voltage and open circuit voltage (or excitation voltage) on the phasor diagram but also the approximate angle between the rotor axis and the synchronously revolving reference axis. The power angle δ will be studied in more detail in the coverage of power system stability. Refer to Sec. 14-2.

4-3. The Infinite Bus

It is often the case that a three-phase synchronous machine is paralleled through an equivalent system reactance (X_e) to a large system of high kva generating capacity relative to the single unit. When a change in shaft power or field excitation to the single unit does not cause the system frequency or terminal voltage to vary, the system is referred to as an infinite bus. Sometimes V is taken as Thévenin's voltage, while X_e is the Thévenin's reactance looking from the machine terminals. A diagram of this arrangement is seen in Fig. 4-4a. Note that the value of r_e has been neglected for the generator. This may or may not be the case, de-

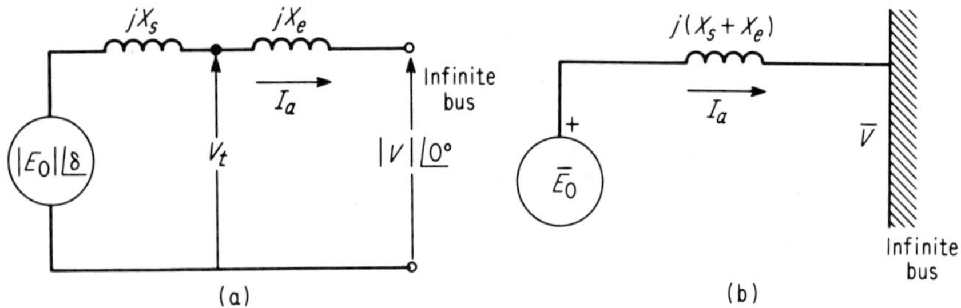

Fig. 4-4. (a) Per phase equivalent of generator and infinite bus in parallel. (b) One-line diagram representation of Fig. 4-4a.

pending upon the ratio of X_s to r_e and upon the nature of the problem to be solved. The impedance X_e between E_0 and V may include that of lines and/or transformers between the infinite bus and the generator terminals. Figure 4-4b shows a common one-line schematic used for the infinite bus situation. It is recalled from machine theory that (for the cylindrical rotor machine) the equation for power transferred from generator to bus is

$$P_{\text{phase}} = \frac{|E||V|}{X_s + X_e} \sin \delta \tag{4-1}$$

where δ is positive for generator action and negative for motor action. This equation is derived in Sec. 14-3.

The same generator-infinite bus relationships can be used for an entire plant and system, provided the system is large with respect to the plant. The separate units of the plant could be lumped together as one source voltage in series with the equivalent parallel impedance of the units.

4-4. Representing Transformers

The power systems engineer should be familiar with a number of equivalent circuits for transformers. Several of the most useful types will be summarized briefly in this section.

The most common transformer type is the two-winding transformer, either single-phase or the three-phase transformer represented on a per

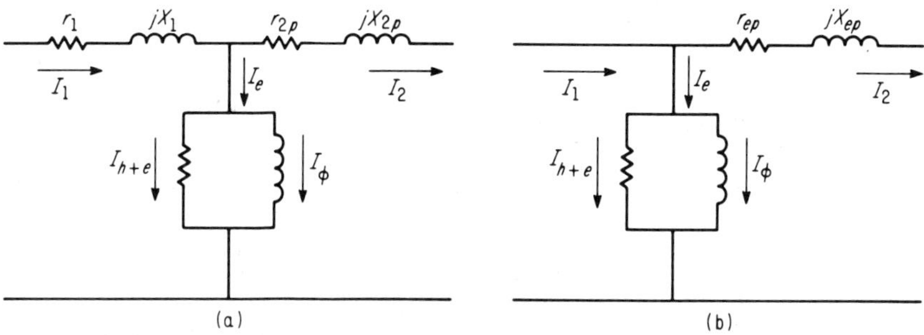

Fig. 4-5. (a) "Exact" equivalent circuit of a two-winding transformer. (b) "Approximate" equivalent circuit of the two-winding transformer.

phase basis. Two equivalent circuits are shown in Fig. 4-5. The "exact" equivalent circuit of Fig. 4-5a has all impedances referred to the primary side. The apparent impedance of the secondary winding as viewed from the primary side is $Z_{2p} = Z_2 \times \left(\frac{N_p}{N_s}\right)^2$. Figure 4-5b is an approximate equivalent circuit which moves the shunt branch to the left. This approximation is based upon the fact that exciting current I_e of the shunt branch is small with respect to load current, and moving this branch to the left will have little effect upon the IZ drop of the series branch. The series elements Z_e are lumped together in Fig. 4-5b as

$$Z_{ep} = Z_1 + [N_p/N_s]^2 Z_2$$

Further approximations can be made with the transformer. Since the shunt branch draws small exciting current, it may at times be omitted altogether as per Fig. 4-6a. Now with large power transformers where

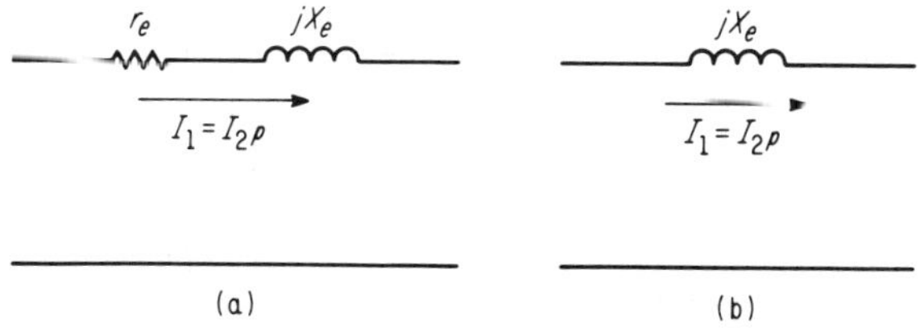

Fig. 4-6. Other approximations for the two-winding transformer.

$X_e \gg r_e$, one last approximation could be made in eliminating r_e. This leaves the transformer represented by a simple series reactance X_e. A typical application of this form of equivalent is made for short-circuit studies.

In load-flow studies another transformer representation is often most useful. This one involves the transformer to be used in a system study where the per unit values are based upon nominal system voltages but where the transformer itself has an *off-nominal tap setting*. In this case the effect of the additional buck or boost tap setting may superimpose a "circulating current" upon the network. Approximate methods for determination of such circulating currents are laid out in Secs. 11-2 and 11-6. However, a more precise π model of such a transformer (for load studies) is treated in Sec. 11-9 with the resulting equivalent of Fig. 11-20.

Three-winding transformers may take such forms as a single-phase transformer with two loaded secondaries or possibly a three-phase Y-Y transformer with Δ tertiary. The latter case will be used as an example. However, the resulting per phase equivalent circuit of Fig. 4-7b will apply to three-winding transformers in general.

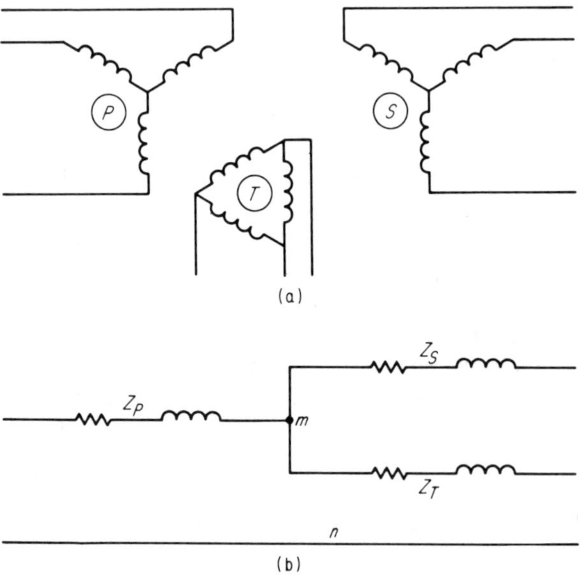

Fig. 4-7. (a) Schematic diagram of a three-winding transformer. (b) Approximate equivalent circuit of three-winding transformer.

Three short-circuit tests will determine the values of Z_P, Z_S and Z_T. The three tests are similar in that in each case one winding is open, one shorted, and reduced voltage is applied to the remaining winding. Impedance is measured on the side to which the voltage is applied. As an example of one test, Fig. 4-8 shows the secondary shorted, the tertiary open, with applied voltage (and meters) on the primary side. Readings of

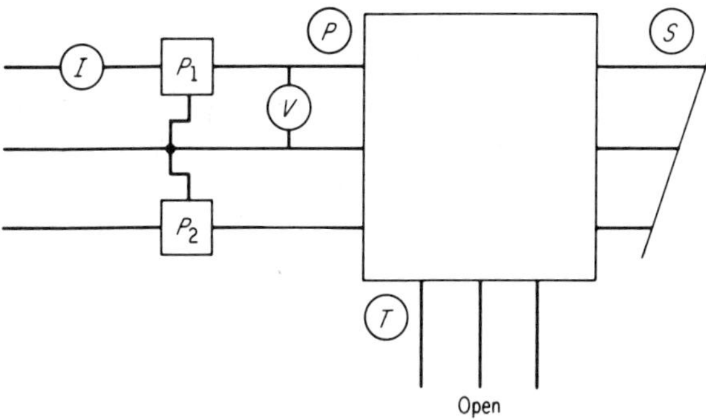

Fig. 4-8. Short-circuit test on a three-winding transformer.

P_1, P_2, V, and I are taken. Per phase values of Z_{PS}, r_{PS}, and X_{PS} are

$$|Z_{PS}| = \frac{V}{\sqrt{3}I}; \qquad r_{PS} = \frac{P_1 + P_2}{3I^2}$$

$$X_{PS} = \sqrt{Z_{PS}^2 - r_{PS}^2}$$

Results of three similar tests are:

$$\left.\begin{aligned} Z_{PS} &= Z_P + Z_S \\ Z_{PT} &= Z_P + Z_T \\ Z_{ST} &= Z_S + Z_T \end{aligned}\right\} \tag{4-2}$$

Assume the complex impedances Z_{PS}, Z_{PT} and Z_{ST} have been referred to the same side. Solving Eq. 4-2 for Z_P, Z_S and Z_T yields

$$\left.\begin{aligned} Z_P &= \frac{1}{2}(Z_{PS} + Z_{PT} - Z_{ST}) \\ Z_S &= \frac{1}{2}(Z_{PS} + Z_{ST} - Z_{PT}) \\ Z_T &= \frac{1}{2}(Z_{ST} + Z_{PT} - Z_{PS}) \end{aligned}\right\} \tag{4-3}$$

These results for the approximate equivalent circuit of Figure 4-7b have neglected the shunt branch of exciting current. If the exciting current is considered appreciable, the shunt branch could be included between n and m.

The *autotransformer* of Fig. 4-9a may be represented by a simple series impedance (Z_e) referred to the primary side. Z_e can be found by means of a short-circuit test or can be written in terms of the impedances of the common and series windings. Again, from machine theory it was determined that

$$Z_e = Z_S + (a - 1)^2 Z_C \tag{4-4}$$

In Eq. 4-4, Z_S is the a-c winding resistance and leakage reactance of the series winding or $Z_S = r_S + jx_S$. Likewise Z_C is the common winding resistance and leakage reactance, while a is the ratio of N_1/N_2. Note that Fig. 4-9b again has neglected the excitation branch.

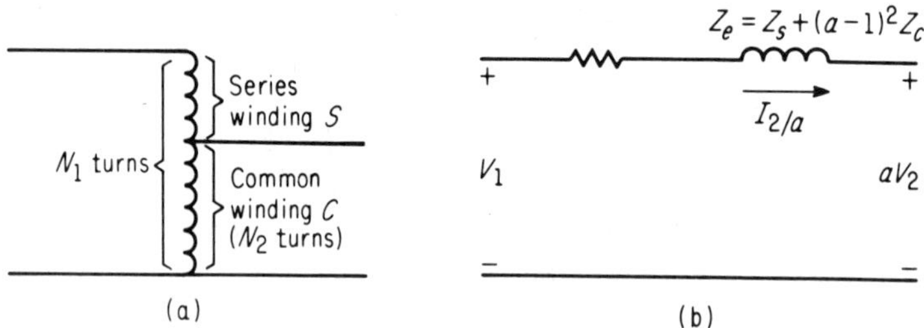

Fig. 4-9 (a) Autotransformer. (b) Autotransformer equivalent circuit.

4-5. Representing Lines

This subject has been treated in Chapter 3, where various line equivalents are dealt with in detail. Again it should be stated that emphasis in this text has not been placed upon the time proven calculations of the line parameters themselves. Rather, it is felt that for the purposes of the network solution, a knowledge of the use of these parameters would be of primary importance.

In π-equivalent line representations, shunt capacitance is divided into two equal parts at the sending and receiving ends, while series line impedance is lumped in between. Hyperbolic functions are used to calculate π equivalents for long lines, as covered in Chapter 3.

Three of the variations of line equivalents are represented in Fig. 4-10 in order of decreasing accuracy. Figure 4-10b has neglected shunt capacitance, while Fig. 4-10c is often considered to be of sufficient accuracy to use in short-circuit studies, although it might be completely unsatisfactory for certain load-flow studies or economic considerations.

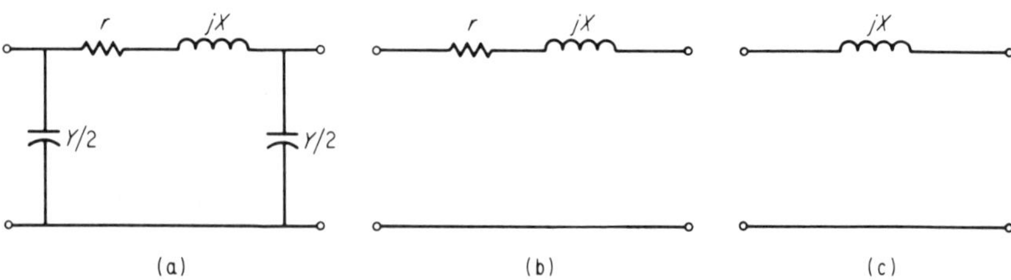

Fig. 4-10. Representing a three-phase line.

4-6. Ohmic Representation of Components in the Network

Previous sections have discussed the representations of lines, transformers, and generators when treated as a unit. For network representation, these compenents may be shown together in an equivalent circuit, often termed the *impedance* diagram. However, before network calculations may proceed, some attention must be given to the ohmic values placed on the impedance diagram. Refer to Fig. 4-11. Here the simple network progresses from generator to load through transformer T_1, line, and T_2 in that order. It must first be decided as to which side of T_1 (or T_2) will serve as reference. The primary of T_1 will be chosen in this example. This means that all actual ohmic values to the right of point a must be referred to side one of T_1 by appropriate factors. For example,

$$Z_2 \times \left(\frac{N_1}{N_2}\right)^2 = Z_{2(1)}$$

$$Z_{\text{line}(1)} = Z_{\text{line}}\left(\frac{N_1}{N_2}\right)^2, \qquad Z_{3(1)} = Z_3\left(\frac{N_1}{N_2}\right)^2, \qquad Z_{4(3)} = Z_4\left(\frac{N_3}{N_4}\right)^2$$

$$Z_{4(1)} = Z_{4(3)}\left(\frac{N_1}{N_2}\right)^2$$

or

$$Z_{4(1)} = Z_4\left(\frac{N_3}{N_4}\right)^2 \times \left(\frac{N_1}{N_2}\right)^2$$

Also,

$$Z_{\text{load}(1)} = Z_{\text{load}}\left(\frac{N_3}{N_4}\right)^2 \times \left(\frac{N_1}{N_2}\right)^2$$

In each instance the bracketed subscript refers the impedance to winding one, two, three, or four of T_1 or T_2. The network may now be treated as though all transformer turns ratios were unity.

The impedance diagram of Fig. 4-11 may be more exact than necessary for some applications. Approximate forms mentioned in the earlier sections of this chapter may serve well enough and with considerable saving in the complexity of the solution. Two variations are seen in Figs. 4-12a and 4-12b. The first case neglects shunt capacitance of the line and exciting current of the transformer. A further approximation, useful in the short-circuit problems, may eliminate the resistance values where $X \gg R$.

Voltage regulation was covered briefly in Sec. 3-4 and Eq. 3-12 was used for the consideration of short line regulation. If Fig. 4-12a is

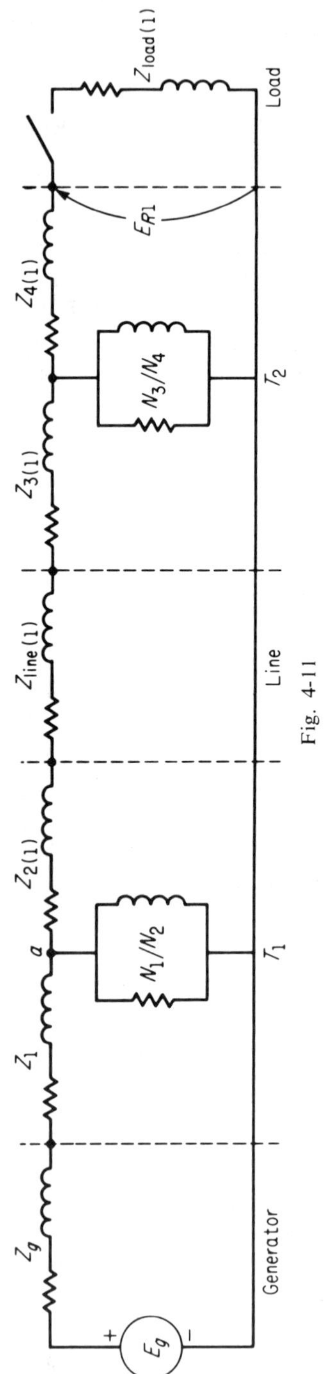

Fig. 4-11

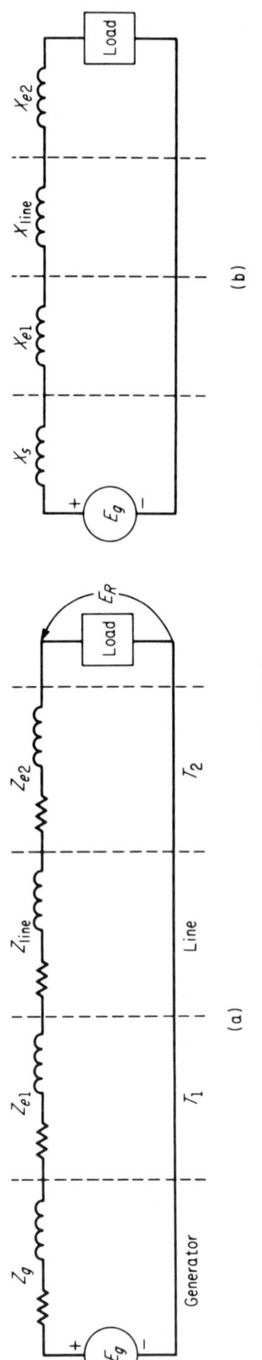

Fig. 4-12

used as our network equivalent, the equation becomes

$$\text{Percent voltage regulation} = \frac{|E_g| - |E_R|}{|E_R|} \times 100$$

It is obvious that the no load voltage (with load switch open) is equal to E_g, the open-circuit voltage of the generator.

4-7. Per Unit Representation

While network calculations could be made in terms of actual ohms or amperes, either the per unit (pu) or percent method will in general do the job in a much handier fashion. Percent quantities (of V, I, P, Z, etc.) differ from per unit quantities by a factor of 100. While the percent and pu methods are similar, the per unit method is preferred and will be used exclusively in this section. The tools of circuit analysis may be applied directly to a per unit system of voltage sources, pu impedances, etc., exactly as they would be applied to a system of actual values. Phasor values of pu quantities are used for steady-state analysis, and Ohm's law and Kirchhoff's laws still hold true. However if one were to place values in percent terms, some care must be taken in multiplying percent values times one another. For example, 5 percent ohms × 100 percent amperes = 5 percent volts, not 500 percent volts. Still, with the per unit method, 0.05 pu ohms × 1.0 pu ampere = 0.05 pu volts. Various advantages of the per unit method versus the method of actual values will be more evident after discussion of the method itself.

To express one quantity as a decimal fraction of another would indicate that some base value is inferred. Consider first the single phase system. Some volt-ampere base (VA_{base}) is first chosen which is the base value throughout the system. Next the base voltage is established. When transformers are in use, the base voltage on one side of the transformer will differ from the base voltage of another side, the base voltage on any particular winding being proportional to the number of turns on that winding. Of course, if one component (generator, transformer is considered apart from the system, the volt-ampere base and voltage base will normally be taken as the rating of that particular piece of equipment in question. For example, the manufacture of power transformers will stamp the transformer nameplate with percent or pu impedance values, based on the transformers own rating. The pu values will need to be corrected to system base values when integrating the transformer into the system.

Upon determining the VA_{base} and V_{base}, the other base values are fixed. Again, for the single-phase system,

$$VA_{\text{base}} = P_{\text{base}} = Q_{\text{base}} = V_{\text{base}} I_{\text{base}}$$

$$I_{base} = \frac{VA_{base}}{V_{base}} \tag{4-5}$$

$$Z_{base} = X_{base} = R_{base} = \frac{V_{base}}{I_{base}} \tag{4-6}$$

$$Y_{base} = B_{base} = G_{base} = \frac{I_{base}}{V_{base}} \tag{4-7}$$

A form for Z_{base} which is often more convenient is found below and given as Eq. 4-8.

$$Z_{base} = \frac{V_{base}}{I_{base}} = \frac{V_{base}}{VA_{base}/V_{base}} = \frac{V_{base}^2}{VA_{base}}$$

or

$$Z_{base} = \left(\frac{(KV_{base})^2}{MVA_{base}}\right) \tag{4-8}$$

Per unit quantities are determined in terms of actual and base values as follows:

$$\text{Value in per unit} = \frac{\text{actual value}}{\text{base value}} \tag{4-9}$$

For example,

$$\text{pu } Z = \frac{Z \text{ actual}}{Z \text{ base}}; \qquad \text{pu } I = \frac{I \text{ actual}}{I \text{ base}}$$

Another expression for pu Z may again prove the more convenient form. Substituting results of Eq. 4-8 into Eq. 4-9,

$$\text{pu } Z = \frac{Z_{actual}}{(KV_{base})^2 \div MVA_{base}}$$

$$\text{pu } Z = Z \text{ actual} \times \frac{MVA_{base}}{(KV_{base})^2} \tag{4-10}$$

For the three-phase system, Eqs. 4-5–4-10 would still hold if one were careful to use per phase values consistently. However, three phase voltage and current values are normally expressed as line quantities, and volt-amperes as a total three-phase quantity, so the per unit quantities for three phase systems will be treated separately, in terms of line-to-line base voltages, line currents, and total volt amperes. Again it is convenient to deal with the three-phase system on a per phase basis, isolating one phase of a Y arrangement. Slight revisions of Eqs. 4-4 through 4-10 are necessary:

$$VA_{base} = P_{base} = Q_{base} = \sqrt{3}\, V_{base}\, I_{base} \tag{4-11}$$

$$I_{base} = \frac{VA_{base}}{\sqrt{3}\,V_{base}} \tag{4-12}$$

$$Z_{base} = X_{base} = R_{base} = \frac{V_{base}}{\sqrt{3}\,I_{base}} \text{ ohms per phase} \tag{4-13}$$

$$Y_{base} = B_{base} = G_{base} = \frac{\sqrt{3}\,I_{base}}{V_{base}} \tag{4-14}$$

Also,

$$Z_{base} = \frac{(KV_{base})^2}{MVA_{base}} \tag{4-15}$$

obtained by multiplying Eq. 4-13 by $\frac{MV_{base}}{MV_{base}}$ and

$$\text{pu } Z = Z_{actual} \frac{MVA_{base}}{KV^2_{base}} \tag{4-16}$$

Note that Eqs. 4-15 and 4-16 for the three-phase case are identical with the single-phase forms of Eqs. 4-8 and 4-10 respectively.

Generally the pu values of devices are given in terms of their own kva and voltage ratings. Yet these devices are to be included together within a network. It is necessary to refer all of the given pu values to the system base values. Assume that an impedance (Z actual) is referred to base (1) as a per unit quantity (pu Z_1) and it is desired to refer pu Z, to the new base (2) as pu Z_2. This is easily accomplished by first writing the pu impedances in terms of actual and based quantities.

$$\text{pu } Z_1 = Z_{actual} \frac{MVA_1}{(KV_1)^2} \tag{4-17}$$

$$\text{pu } Z_2 = Z_{actual} \frac{MVA_2}{(KV_2)^2} \tag{4-18}$$

Dividing Eq. 4-17 by Eq. 4-18,

$$\frac{\text{pu } Z_1}{\text{pu } Z_2} = \frac{MVA_1}{MVA_2} \frac{(KV_2)^2}{(KV_1)^2}$$

or

$$\text{pu } Z_2 = \text{pu } Z_1 \times \frac{MVA_2}{MVA_1} \cdot \left(\frac{KV_1}{KV_2}\right)^2 \tag{4-19}$$

One should not be tempted to use the voltage-ratio correction of Eq. 4-19 because of a mere voltage transformation. This voltage correction is primarily of use where the system base voltage at the location of some particular device differs from the base voltage from which the pu value of the device was calculated. It should be said that the pu Z_e of a two-winding transformer is the same whether the calculation was made from the h-v or the l-v side. To demonstrate this, it will be shown that the pu impedance calculated from the l-v winding (pu Z_{LV}) is equal to the value of pu Z_{HV}:

$$\text{pu } Z_{LV} = \frac{Z_{LV}}{\text{base } Z_{LV}} = \frac{Z_{HV} \times (N_2/N_1)^2}{\text{base } V_{LV}/\text{base } I_{LV}} = \frac{Z_{hV} \times (N_2/N_1)^2}{\dfrac{\text{base } V_{HV}(N_2/N_1)}{\text{base } I_{HV}(N_1/N_2)}}$$

$$= \frac{Z_{HV}}{\text{base } Z_{HV}} = \text{pu } Z_{HV}$$

4-8. Expressing Mutual Impedance as a Per Unit Quantity Between Lines of Different Voltage Levels

Assume that two three-phase lines of different voltage levels are running together, with mutual reactance present as shown in Fig. 4-13

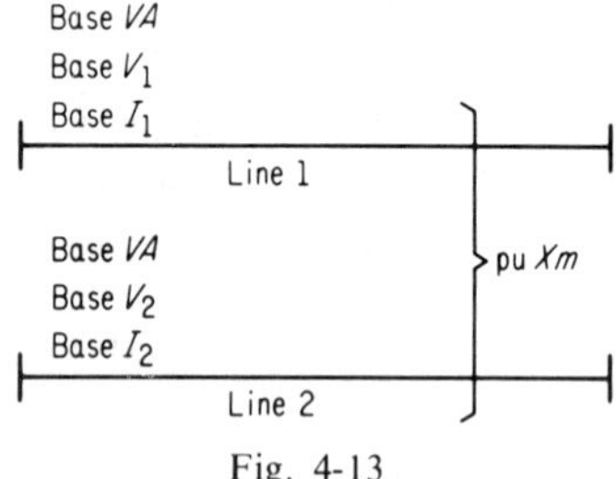

Fig. 4-13

line 1. It is often necessary to be able to express this mutual reactance in per unit terms. There is one value of pu X_m which will serve both lines. In terms of line 2,

$$\text{pu } X_{m2} = \frac{X_{m(\text{actual})}}{\text{base } X} = \frac{X_m}{\text{base } V_2/\text{base } I_1} \tag{4-20}$$

We are using (in Eq. 4-20) the base current of line 1_1 since I_1 is the current that induces mutual voltage into line 2. Then

$$\text{pu } X_{m2} = \frac{X_m}{\text{base } V_2} \text{base } I_1 \times \frac{\text{base } V_1}{\text{base } V_1}$$

$$\text{pu } X_m = \frac{X_m \text{ base VA}}{\text{base } V_2 \cdot \text{base } V_1}$$

or

$$\text{pu } X_m = \frac{X_m \cdot \text{base MVA}}{\text{base } KV_2 \times \text{base } KV_1} \tag{4-21}$$

It is interesting to note the similarities of Eq. 4-16 and Eq. 4-21.

4-9. Pu vs. Ohmic Representation

When a transformer is stamped with a 5 percent impedance value ($\% Z_e$), this figure should have real significance to the power engineer. For example it is recognized that 1 pu ampere flowing through this series equivalent impedance will cause a 5 percent voltage drop between input and output. Of course, this voltage drop is a phasor quantity and does not, in general, equal the voltage regulation except for the special case where the ratio of R to X of the load is equal to the r_e/X_e of the transformer. The impedance concept is thus simplified by the fact that $\% Z_e$ is closely related to percent voltage drop. The statement that a transformer series impedance is 15 ohms or that 200 volts is dropped across the unit has little meaning unless related to transformer ratings.

The pu impedance values for machines of like ratings fall within a narrow range which is recognizable to the power engineer. The National Electrical Manufacturers Association (NEMA) publishes acceptable limits for percent impedance values of various devices. For example, an oil-filled 60-cycle 100-kva single-phase transformer in the 25-kv class will be expected to have an impedance of about 5 percent. This value rises with the kva and/or kv rating. Standard tolerance of variation from the standard values for the two-winding transformer is $\pm 7\frac{1}{2}$ percent of the specified values. In general, the ratio of X_e/r_e is greater for the large kva ratings. Values of percent synchronous reactance for generators likewise fall within certain acceptable limits. In a case where equipment impedance is unknown or difficult to determine, fair estimates may be made without too much loss of accuracy. Even where pu values are given, gross errors in these values are easily detected.

If ohmic values were to be used in a network problem, it would be necessary to refer all values of one part of the circuit to that of another whenever the two circuits are joined together by transformer connection. For a Y-Y or Δ-Δ transformation this is accomplished by an appropriate factor of the square of the turns ratio applied to the ohmic value. If a Δ-Y transformation is considered, the ohmic value on the Y side would be referred to the Δ side by the square of the line to line-voltage ratio. Now, in comparison, the pu impedance of a circuit does not require correction for voltage transformation. Naturally, the pu values must be expressed in terms of system base values as per Eq. 4-19.

Stamped values of pu impedances of transformers are based upon the coil ratings. These coil ratings do not change with a mere change in connection (from Y-Y to Δ-Y). Therefore, the *pu impedance is not affected* by such a reconnection of windings. This is not to say that line to line base voltage for the transformer is unaffected. System base voltage must again be considered as per Eq. 4-19 when the transformer is integrated into the system.

4-10. One-Line Diagrams

Per phase impedance diagrams may represent a system on either the ohmic basis as per Sec. 4-6 or on a pu basis. However the one-line diagram shows the essentials of the system in a most simplified form. An example of this form is shown in Fig. 4-14, where the simple circle repre-

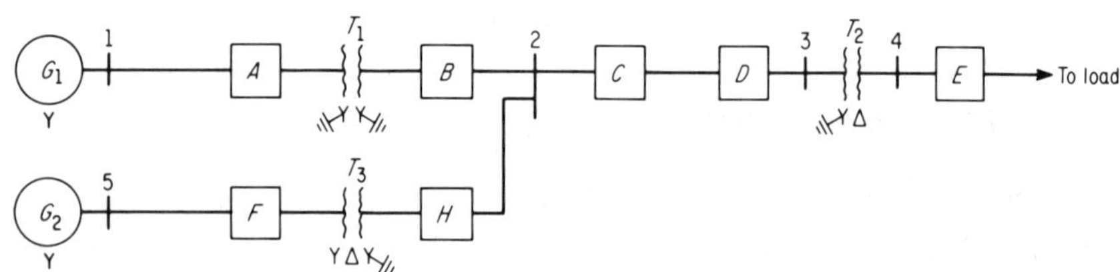

Fig. 4-14. Example of a one-line diagram.

sents one phase of a rotating machine (in this case the synchronous generators G_1 and G_2). Only one line is carried through for each three-phase branch, even the neutral connection being excluded. Several of the standard symbols are indicated on the drawing such as two-winding transformers T_1 and T_2, three-winding transformer T_3, and breakers A through H. Several bus or node positions are indicated, numbered 1–5. These positions may be actual junction points or may be simply defined as buses, where the bus is to represent any position of particular interest in the system. The one-line diagram is easily converted to an impedance diagram, or vice versa. Machine and equipment ratings are given on the one line diagram in either actual or per unit values. Note that Fig. 4-14 includes the type of transformer and generator connections. Grounded Y's are shown along with any current-limiting resistance or reactance which might be present in the neutral or ground connection.

One application of the one-line diagram will exclude breakers (load-flow application) while another may include breakers as well as relay positions (stability studies). Still another possibility would be to use three separate diagrams to represent the positive-, negative-, and zero-sequence networks separately. The latter use is found in short-circuit studies of unsymmetrical faults, where all impedances are expressed in per unit to the same kva base. No set rule applies with regard to the exact form of the one-line diagram.

4-11. A-C and D-C Analyzer Boards

Chapter one makes brief mention of the a-c and d-c analyzer boards (also called *calculating boards*) for power system calculations. When viewed against the backstop of hand-calculated solutions, these boards were for years indispensable for the problems of short circuits, load flow, and stability. The boards are mere replicas of the system on a per phase per unit basis. Generator sources and variable impedance boxes are available for representing the actual system. Meters for V, I, P, and Q can easily be switched in for metering various points in the network.

The a-c boards have provision for varying both the magnitude and phase angle of generator voltages. Variable inductance, capacitance, and resistance parameters are available, whereas the d-c boards use only resistance boxes to represent reactance or scalar impedance parameters of the network. Frequencies of 60 to 10,000 Hz are in use for the a-c boards. The higher frequencies allow a size reduction for the variable-inductance components. The d-c board is not sufficiently representative of the system to justify its use in load flow or stability problems, however it does serve well for most fault studies.

The digital computer is rapidly replacing the calculating boards today. The computer is, in general, faster and more accurate. It also will provide a neat and complete set of output data for the record. At the same time, analyzers are still being put to good use in many instances. The physical concept of the system on the board is much clearer to most minds than a deck of cards, a roll of magnetic tape, or even the printed output data. There are computer programs, however, that are capable of printing out a one-line diagram of the system, properly connected according to instructions. This diagram may then be compared with the original one-line diagram from which the data was prepared. An analyzer board can be left connected for a system and proposed modifications can be easily made on the board. With the modification, new information, such as redistribution of currents, can be read from the meters on the board. The engineer is thus left free to tinker with any number of possible changes before making a decision on the change. While the d-c and a-c boards are generally thought of as occupying the better part of a room or small building, they are also found in much smaller sizes. A portable d-c board would be battery-operated and could be small enough to fold and carry as a suitcase.

Rather than consider both boards in detail a brief coverage of the d-c board will be given. In order to use this board for short-circuit studies, the following assumptions will be made.

1. Neglect transformer exciting current.
2. Neglect resistance of lines, transformers, generators, etc., or represent all impedances as scalar values.
3. Neglect all loads, assuming load currents are small with respect to fault currents. One exception to this rule involves the synchronous motor load, since the motor emf will contribute to short-circuit currents.
4. Assume generator voltages equal in magnitude.
5. Assume generator phase angles are equal.
6. Omit shunt capacitance of lines.

For this brief discussion of the d-c board and its related short-circuit solution, only the symmetrical three-phase fault will be considered. The unsymmetrical fault will be treated in more detail following a coverage of the tools of symmetrical components.

Example 4-1. Figures 4-15 through 4-18 illustrate a simple three-phase system with a symmetrical three-phase fault at position x. Figure 4-15 is the one-line diagram for the system. Prefault loading in this prob-

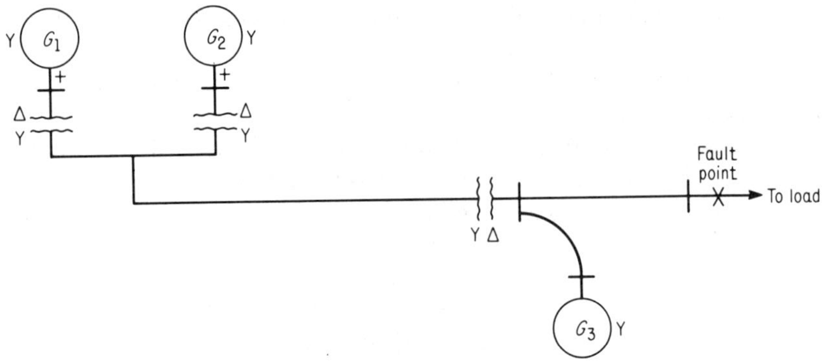

Fig. 4-15. One-line diagram with three-phase fault at x.

lem is considered negligible with respect to the fault current. Figure 4-16 is the reactance diagram which includes the given pu reactance values (resistance neglected) for the transformers, lines and generators. All reactance values have been referred to the common system base kva. Figure 4-17 is the diagram corresponding to the d-c board connection. Note that Figs. 4-16 and 4-17 differ in two respects. First of all, the reactance values of Fig. 4-16 have been replaced with resistances which will produce the same effect of impedance to current magnitudes in the d-c circuit. Of

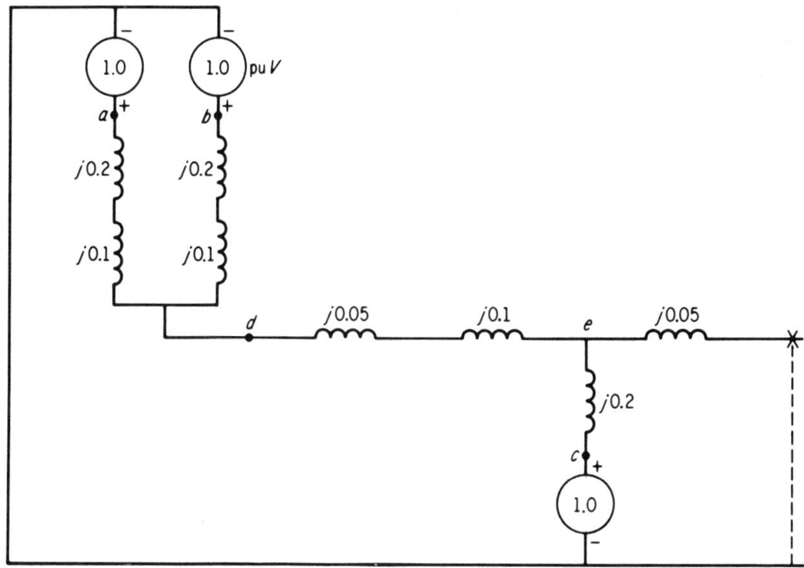

Fig. 4-16. Reactance diagram with fault at x.

course all resulting d-c board currents will in reality be considered as lagging the source voltage by 90°.

The other difference between the two figures is seen in the joining together of three voltage sources into one. This move will be obvious when it is realized that for this study the magnitudes and phase angles of these voltages are assumed equal. Therefore the potentials at a, b, and c are the same. These points may then be joined together and all constant voltage sources replaced with one source. If current at the fault were the only unknown in question, then Fig. 4-17 would not be the simplest form for the board connection. However, it is likely that individual branch cur-

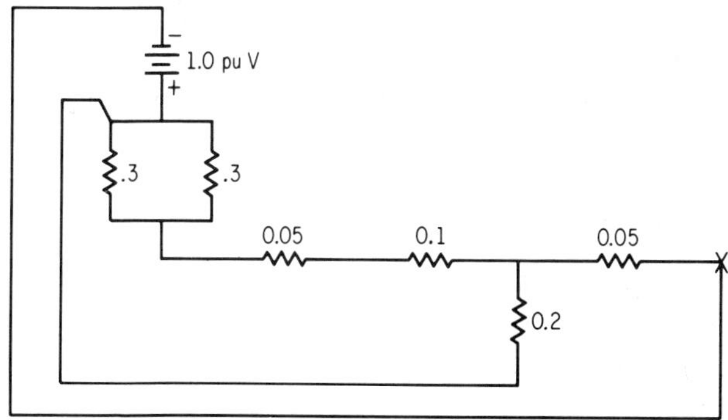

Fig. 4-17. Diagram for d-c board connection.

rents and voltages will be needed, so the identity of these branches is retained for metering purposes.

For three-phase fault consideration Fig. 4-17 would also be referred to as the positive sequence network for the board.

Figures 4-18a, b, and c are used merely to demonstrate the results of a hand calculation. The final series impedance of Fig. 4-18c between

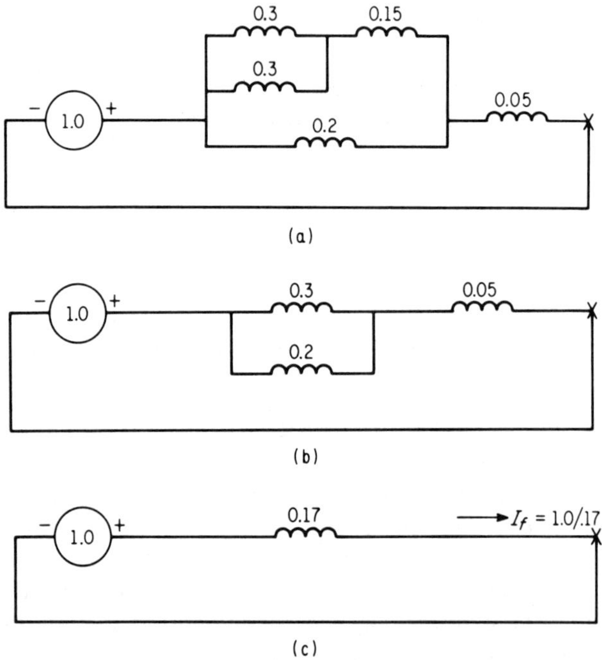

Fig. 4-18. The simplified network for longhand calculation.

generation and fault is 0.17 pu ohm and $I_{\text{fault}} = \dfrac{1.0}{0.17} = 5.88$ pu amp. The circuits may be worked in reverse to determine the fault contributions from the various branches. For example, the current of branch *c-e* of Fig. 4-16 is found from Fig. 4-18b and the current-divider principle as $I_{ce} = I_f \times \dfrac{0.3}{0.3 + 0.2} = 5.88 \times \dfrac{0.3}{0.5} = 3.53$ pu amp. This procedure of working the equivalents Figs. 4-18c, b, and a in reverse may be continued until all needed currents and voltages are obtained.

Since the connected board is of itself a miniature system, base values must be established for the board. For example, a d-c board may read current in milliamps and have a voltage range of 0–150 volts. Assume board base values of 10 volt-amperes and 100 volts are arbitrarily established. Then base amperes $= \dfrac{10 \text{ va}}{100 \text{ v}} = 0.1$ amp $= 100$ ma. Base ohms would be fixed at $\dfrac{100 \text{ v}}{0.1 \text{ amp}} = 1{,}000$ ohms. If a system pu impedance of 0.05 were to be represented, the board value would then be $0.05 \times 1000 = 50$ ohms. A current reading of 300 ma would correspond to 3.0 pu amperes.

4-12. Representing the "Outside" Systems on the D-C Board

Few systems today are isolated. Most system studies, whether total or partial, must include the effect of tie lines to neighboring systems. Information is necessary for areas on the other side of these tie points (or tie buses) and which are outside of the system under study. If only one tie is involved it is merely a matter of replacing the "outside area" with Thévenin's equivalent E and Z. For the d-c board study, with loads neglected and all generation assumed equal to 1.0 pu volt, then generation is again lumped together.

Consider the example demonstrated in Fig. 4-19. System generation has been lumped together in Fig. 4-19a and the passive network is as

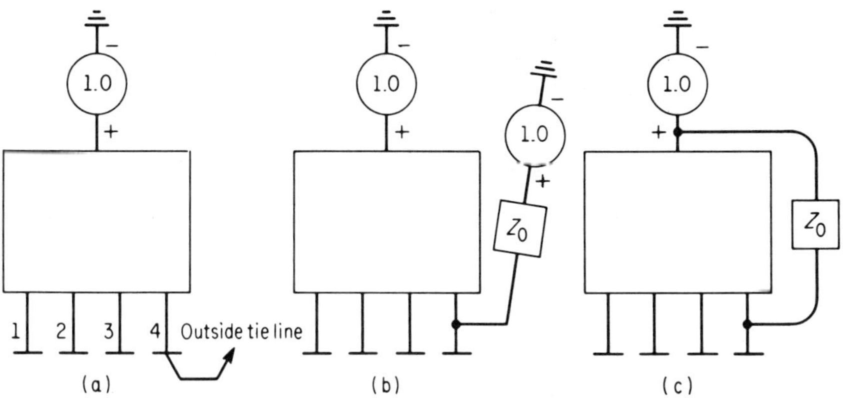

Fig. 4-19. System with outside tie line.

shown in the box. Points or buses of interest are brought out as buses 1–4. Notice that bus 4 is tied to an outside system. The effect of this tie must be included in the network in order to study bus faults. It is known that the outside tie will contribute to these faults. (A three-phase fault on bus 1 could be shown by connecting bus 1 to the generator neutral represented by a ground symbol.) Now the neighboring power company would be requested to furnish the values of Thévenin's Z_0 in one form or another. The actual value of Z_0 could be given as the reciprocal of I_{sc}, where I_{sc} is the current flowing through this tie to bus 4 with bus 4 faulted. This short-circuit current could be determined from a simple board study of the neighboring company's system.

Another more common way to furnish this information about Z_0 is to give either the value of I_{sc} or the so-called short-circuit mva. Short-circuit mva may be expressed as

$$\text{S.C. mva} = \sqrt{3}\,V_{\text{base}} \times I_{sc} \times 10^{-6} \tag{4-22}$$

or

$$I_{sc} = \frac{\text{S.C. mva}}{\sqrt{3}\,V_{\text{base}} \times 10^{-6}}$$

Thévenin's series reactance X_0 (neglecting R_0) is found as

$$\begin{aligned} X_0 &= \frac{V_{\text{base}}/\sqrt{3}}{I_{sc}} = \frac{V_{\text{base}}}{\sqrt{3}} \frac{1}{\text{S.C. mva}/(\sqrt{3}\,V_{\text{base}} \times 10^{-6})} \\ &= (KV_{\text{base}})^2/\text{S.C. mva} \end{aligned} \tag{4-23}$$

and

$$\text{pu } X_0 = \frac{X_0}{\text{base } X} = \frac{(KV_{\text{base}})^2/\text{S.C. mva}}{(KV_{\text{base}})^2/\text{base mva}}$$

$$\text{pu } X_0 = \frac{\text{base mva}}{\text{S.C. mva}} \tag{4-24}$$

or

$$\text{pu } X_0 = \frac{\text{base } I}{I_{sc}} \tag{4-25}$$

It is seen from Eqs. 4-24 and 4-25 that Thévenin's pu X_0 is readily found in terms of either I_{sc} or S.C. mva.

The case just described served for only one tie line, but not for plural interconnections. Take the case where two tie lines move out toward the same outside area (see Fig. 4-20). This requires that the outside passive network be replaced by an equivalent circuit before incorporating into the d-c board study. The equivalent is a simple delta mesh with three nodes including the source node 0.

Again, a simple board study would obtain this equivalent, or the equivalent could be produced by computer elimination of all nodes in the neighboring system except 0, 3, and 4. Upon finding the values of pu X_{04}, pu X_{03}, and pu X_{34}, the values are included on the d-c board connection.

It should be mentioned that the Δ equivalent is not the only form which could be used for this three bus system. The "rake equivalent" is another possibility, as discussed in Chapter 2.

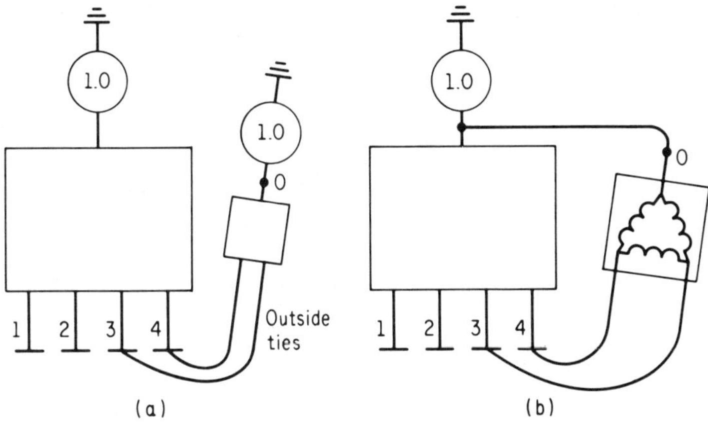

Fig. 4-20. System with two interconnections.

4-13. Obtaining a Rake Equivalent from the Analyzer Board

A per phase representation of a passive network of n nodes plus reference node may be given in equivalent form by either a mesh equivalent or a rake equivalent as explained in Sec. 2-5. A complex network may be connected in miniature on the analyzer board and certain tests made to find an equivalent containing a reduced number of nodes. Recall that the rake equivalent is identified with the bus impedance matrix. For fault studies it is sometimes termed the short-circuit equivalent.

First, the procedure for finding a three-bus (plus reference bus)

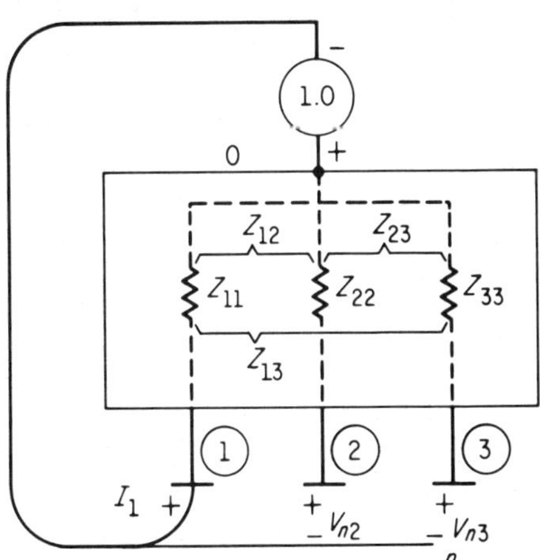

Fig. 4-21. Network test circuit for analyzer board to determine equivalent (dotted).

equivalent will be given. Values would normally be expressed as per unit quantities.

1. Connect the n bus network on the board as it exists for the actual system represented by the per phase, per unit impedance diagram.
2. Place a voltage source (1.0 pu volts is used here) at reference node 0, shorting bus 1 back to the source as seen in Fig. 4-21. Take meter readings of I_1, V_{20} and V_{30}.
3. Calculate Z_{11}, Z_{12}, and Z_{13}. A look at Fig. 4-21 reveals that

$$Z_{11} = 1.0/I_1 \tag{4-26}$$

Also

$$V_{20} = I_1 Z_{12}$$

$$Z_{12} = \frac{V_{20}}{I_1} \tag{4-27}$$

and

$$Z_{13} = \frac{V_{30}}{I_1} \tag{4-28}$$

4. Z_{22}, Z_{33}, and Z_{23} may be calculated using the general procedure of steps 2 and 3, only with buses 2 and 3 shorted one at a time.

4-14. Obtaining the Mesh Equivalent from the Analyzer Board

In order to determine the impedances of the mesh equivalent, set up the network on the board and place a voltage source (1.0 pu volts) on bus 0. Buses which are to be identified in the equivalent are shorted through ammeters back to the neutral bus as shown in Fig. 4-22. Since buses 1, 2, and 3 are effectively shorted and at equal potential, no current will pass through Z_{12}, Z_{23}, or Z_{13}. The metered currents I_1, I_2, and I_3 pass through Z_{01}, Z_{02}, and Z_{03} respectively. Therefore

$$Z_{01} = \frac{1.0}{I_1}, \quad Z_{02} = \frac{1.0}{I_2}, \quad Z_{03} = \frac{1.0}{I_3} \tag{4-29}$$

Other similar tests are necessary to determine the remaining impedances. For example, Z_{21}, Z_{20}, and Z_{23} can be found by applying the source voltage to bus 2 with buses 0, 1, and 3 shorted through ammeters back to the neutral bus. Four such tests are possible for a three-bus system (plus reference bus 0), each test yielding three impedances, or a total of twelve results. Since there are only six unknown impedances, this added information serves as a check on results.

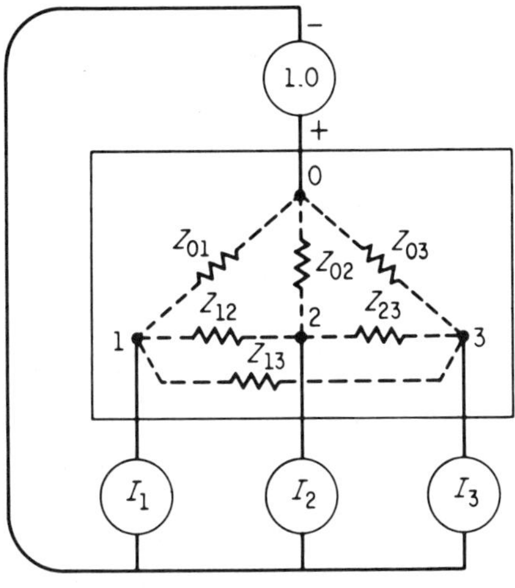

Fig. 4-22

Equivalents of the type covered in both Secs. 4-13 and 4-14 are often requested from one company by a neighboring company, especially where adjoining companies are linked together with one or more tie buses. Section 4-12 dealt with this problem where not more than two tie buses were involved. However, there is theoretically no limit to the number of buses that could be retained in either a mesh or short-circuit equivalent.

4-15. Representing Mutual Impedances on the D-C Board

Figure 3-11a shows one method for representing mutual impedances between two lines where the lines have one end in common. The more general case presents more of a problem however, where end points terminate on four separate buses.

Figure 4-23a shows the flux linkages due to the currents of two lines running alongside one another. Note that the mutual flux linkages (ϕ_{m1} and ϕ_{m2}) add to the leakage flux linkages (ϕ_{L1} and ϕ_{L2}) in the case where I_1 and I_2 are flowing in the same direction. This means that a mutual voltage induced into one line from another will be in phase with a self-induced voltage, assuming currents are in phase and that both self and mutual impedances are considered purely reactive.

Also note that total self-linkage flux (ϕ_{11}) due to I_1 equals ϕ_{m1} + ϕ_{L1} and $\phi_{22} = \phi_{m2} + \phi_{L2}$. It is likewise true that

$$\text{Total self-reactance} = \text{leakage reactance} + \text{mutual reactance} \tag{4-30}$$

or

$$X_{11} = X_{L1} + X_{m1} \tag{4-31}$$

and

$$X_{22} = X_{L2} + X_{m2} \tag{4-32}$$

where

$$X_{m1} = X_{m2}$$

Refer to the one-line diagram of Fig. 4-23b.

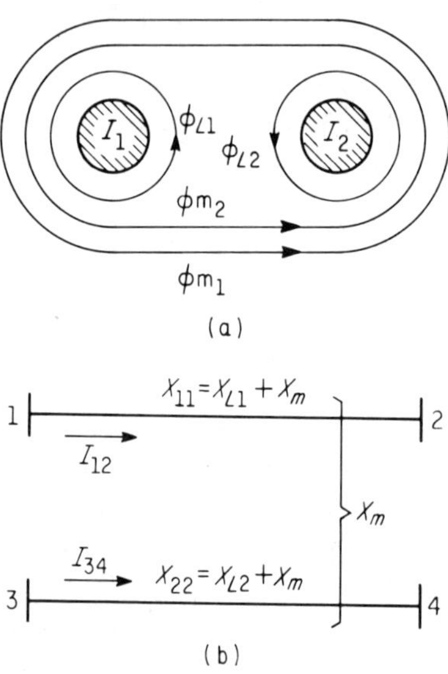

Fig. 4-23. (a) Separate flux linkages of two current-carrying conductors. (b) Mutually coupled lines without common terminals.

One solution for this problem of representing mutuals on the d-c board involves the use of what is called the *insertion unit*. This method will be better understood by writing the equation for line-voltage drops as follows:

$$\begin{aligned} V_{12} &= I_{12}Z_{11} + I_{34}Z_m \\ &= I_{12}jX_{11} + I_{34}jX_m \\ &= I_{12}j(X_{L1} + X_m) + I_{34}jX_m \\ &= I_{12}jX_{L1} + I_{12}jX_m + (I_{34}jX_m) \end{aligned} \tag{4-33}$$

E_x

Similarly,

$$V_{34} = I_{34}jX_{L2} + (I_{34}jX_m) + I_{12}jX_m \tag{4-34}$$

E_y

Notice there are component voltages of Eq. 4-33 which are identical to two components of Eq. 4-34. It is this simple fact upon which the theory of the insertion unit is based. See Fig. 4-24. If a d-c voltage of E_x (representing the mutual voltage due to I_{34}) is inserted into line 1–2, this voltage

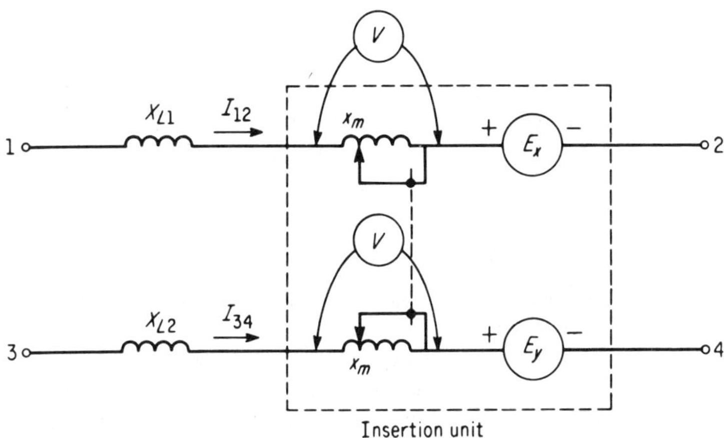

Fig. 4-24. Representing mutuals with an insertion unit.

should equal the $I_{34}X_m$ drop of line 3–4. Likewise the mutual voltage E_y due to the current I_{12} should equal the $I_{12}X_m$ drop of line 1–2. This calls for an adjustment on the part of the inserted values of E_x and E_y in order to meet the condition that

$$E_x = I_{34}X_m, \quad \text{and} \quad E_y = I_{12}X_m \tag{4-35}$$

Naturally, any adjustment on the part of these inserted voltages will affect the currents I_{12} and I_{34}, as well as every other current in the network. A bit of readjustment will insure that the conditions of Eq. 4-35 are met. The insertion units in use at Detroit Edison Company are complete with four voltmeters (for metering E_x, E_y, $I_{12}X_m$, and $I_{34}X_m$) as well as a ganged variable resistance for setting the correct value of X_m into each line. (It is recalled that resistance is used to represent reactance on the d-c board.)

Problems

4-1. A two-winding, three-phase, Δ-Y, 120,000-volt to 24,000-volt 30-mva transformer has a series per unit impedance of pu r_e = 0.0047 and pu X_e = 0.0862. The core loss at rated voltage is 65,800 watts.

(a) If the transformer low voltage secondary winding is loaded to rated mva at 0.8 p.f. lagging, what is the percent voltage regulation and efficiency? Use the approximate circuit of Fig. 4-5b.

(b) Repeat (a) with a load of 0.8 p.f. leading.

(c) If the 24,000-volt terminals are short-circuited, what high-side voltage would be required to drive rated current through the transformer?

(d) Assuming the per unit series impedance is split evenly between high and low windings, find the actual ohmic values per phase for r_1, X_1, r_2, and X_2.

4-2. A three-winding, three-phase, 68-mva transformer is connected as follows:

Winding 1: 120KV, Y-connected, rating of 68 mva
Winding 2: 13.2KV, Δ-connected, rating of 34 mva
Winding 3: 13.2KV, Δ-connected, rating of 34 mva

Three short-circuit tests revealed the following percent reactances (R neglected). All values are on a 34 mva base.

$$\%X_{12} = 16.1, \quad \%X_{13} = 16.1, \quad \%X_{23} = 34.6$$

(a) Find the per unit impedance of the three-winding equivalent circuit on a 34-mva base.

(b) Two generators are feeding the two 13.2-kv windings equally with the 120-kv winding loaded to 68 mva at 0.9 p.f. lagging and rated voltage. What is the voltage regulation of the transformer if the generator terminals are held at constant voltage from no load to full load?

4-3. A three-phase, 60-Hz, 300-mva, synchronous generator feeds into a large system through a step-up transformer. The generator synchronous impedance is $j1.30$ pu ohms with a transformer impedance of $j0.10$ pu ohms, both values on a 300-mva base. The system impedance as seen from the transformer high-voltage terminals is $j0.03$ pu ohms on a 100-mva base. Assume no base-voltage corrections necessary on the impedance values.

(a) A current of 1.0 per unit (300-mva base) flows at a 0.9 p.f. lagging with respect to the generator terminal voltage E_t. Consider the infinite bus voltage as 1.0 pu. Find the pu voltage E_0 behind the synchronous reactance, the pu generator terminal voltage E_t and the power angle δ between E_0 and V of the infinite bus.

(b) According to the power-transfer equation, what is the maximum pu power which can be transferred from this generator to the system? Assume the generator field is held constant.

4-4. Given the machine of Prob. 4-3. Rework part (a) with the same current but where $|E_t| = 1.0$ and $|V|$ is unknown, find E_0, V, and δ.

4-5. A three-phase, 130-mva, 13.8-kv synchronous generator has a synchronous reactance of 1.87 pu on its own base. This machine is tied to a 130-mva transformer which is rated 13.2 kv to 126 kv. The nominal system voltage is 120 kv. The transformer impedance is $j9.7$ percent on its own base. The system Thévenin's impedance as seen from the high-voltage terminals of the transformer is $j15$ percent on a 100-mva, 120-kv base.

(a) What total value of series reactance would be used in the power-transfer equation? Use the 120-kv system voltage as a base and base mva = 130.

(b) Repeat part (a) using the generator voltage rating as base.

4-6. The machine of Prob. 4-5 is delivering rated current from its terminals at 13,800 volts. The infinite bus voltage is 120 kv.

(a) What is the p.f. of the generator measured at its terminals?

(b) What is the power angle δ and voltage E_0 behind synchronous reactance?

4-7. A d-c analyzer board has a 10-va, 100-volt base. A system is to be represented (per phase) on the board on a 50-mva base. All resistor dials have been marked in percent ohms to correspond to the 50-mva system base. One hundred volts on the source gives 100 volts on the voltmeter and corresponds to 100 percent system volts. A reading of 1 ampere on the board meter corresponds to 1 pu ampere on the 50-mva system base. A 120-kv to 24-kv, Y-Y, three-phase transformer has a series impedance (X_e) of $j.08$ pu ohm on its own base of 30 mva.

(a) What is the percent X of the transformer on a 50-mva base? (The resistor dial for this element would be set to this value.)

(b) What actual ohms does this represent in the transformer as seen from the h-v side?

(c) What actual ohms does the dial setting of part (a) require in the board resistor?

4-8. Given the d-c analyzer and transformer of Prob. 4-7. Assume the 100-volt source is applied directly to the transformer impedance and a fault is applied to the transformer secondary.

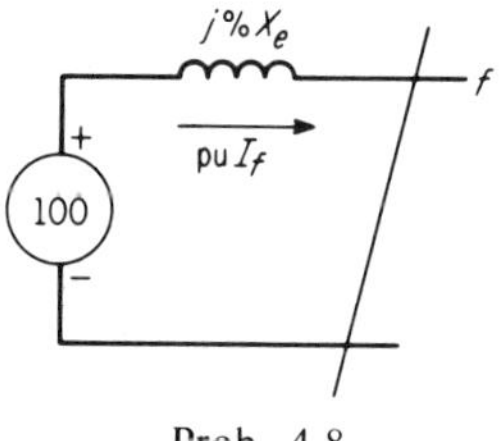

Prob. 4-8

(a) What type of fault would this represent?

(b) What per unit current would you expect?

(c) What reading would you expect on the d-c ammeter?

(d) What actual board current flows?

(e) What actual transformer current would flow (in both primary and secondary wingings) for this type of fault?

4-9. Refer to Fig. 4-15. Remove the fault from the load circuit and place a symmetrical three-phase fault on the terminals of generator G_3. Find the per unit fault current and the individual fault contributions of G_1, G_2 and G_3. Consider the load branch as an infinite impedance. Also sketch the diagram for a d-c board connection.

4-10. Work Prob. 4-9, but with the fault removed from G_3 terminals and placed at the junction of the two transformers which are fed by G_1 and G_2.

4-11. A four-bus system (plus reference bus) is set up on a d-c analyzer on a 100-mva base. Various tests are made to determine elements of the rake

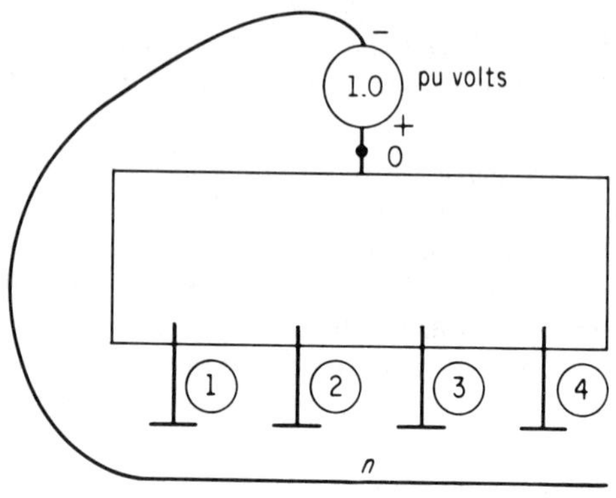

Prob. 4-11

equivalent. Assume all elements here are purely reactive. With bus 1 faulted:

$$I_{1f} = 15.0 \text{ pu}; \quad V_{n2} = 0.7 \text{ pu}; \quad V_{n3} = 0.8 \text{ pu}; \quad V_{n4} = 0.9 \text{ pu}$$

With bus 2 faulted:

$$I_{2f} = 18.0 \text{ pu}; \quad V_{n3} = 0.75; \quad V_{n4} = 0.85$$

With bus 3 faulted:

$$I_{3f} = 20 \text{ pu}; \quad V_{n4} = 0.80$$

With bus 4 faulted:

$$I_{4f} = 22 \text{ pu}$$

(a) Determine all elements of the rake equivalent and show the equivalent.

(b) With bus 4 faulted, what readings would you expect for V_{n1}, V_{n2}, and V_{n3}?

4-12. Determine the mesh equivalent for a three-bus (plus reference bus 0) system with four tests as follows. Consider all elements are reactive.

(1) 1.0 pu voltage on bus 0, metered buses 1, 2, and 3 faulted:

$$I_1 = 10.0 \text{ pu}, \quad I_2 = 12.5 \text{ pu}, \quad I_3 = 11.1 \text{ pu}$$

(2) 1.0 pu voltage on bus 1, metered buses 0, 2, and 3 faulted:

$$I_0 = 10.0, \quad I_2 = 3.33, \quad I_3 = 2.0$$

(3) 1.0 pu voltage on bus 2, metered buses 0, 1, and 3 faulted:

$$I_1 = 3.33, \quad I_3 = 2.5, \quad I_0 = 12.5$$

(4) 1.0 pu voltage on bus 3, metered buses 0, 1, and 2 faulted:

$$I_0 = 11.1, \quad I_1 = 2.0, \quad I_2 = 2.5$$

chapter 5

BASIC NETWORK SOLUTIONS

5-1. Introduction

It would seem profitable to look ahead at Chapters 5 and 6 in order to classify certain methods of network solutions. The following listing is offered for this purpose.

I. (a) *The General Loop-Impedance Method.* Here Kirchhoff's voltage equations are set up around the loops where loop currents are normally unknown. In determining the proper number of independent loops, the tree-link approach is taken.

(b) *Mesh-Impedance Method.* This method is a special case of I(a) where only open loops (meshes) are chosen. Both methods I(a) and I(b) are treated together in Sec. 5-2.

II. (a) *The General Cut-Set Admittance Method.* Where current sources are known, the unknown independent node-pair voltages are determined. The network is broken or divided into pieces or cut-sets and Kirchhoff's current law must be satisfied about each cut-set line. No reference node is given. Again the proper number of independent current equations is determined by the tree-link approach. The method is treated in Sec. 5-7.

(b) *The Nodal-Admittance Method.* This method is a special case of II(a) where independent node pairs are taken from a reference node and current equations are written about the nodes themselves. This method is more common than that of II(a) and is covered in Sec. 5-4.

III. *Branch-Impedance Transformation Method.* Refer to Sec. 6-2.

IV. *Branch-Admittance Transformation Method.* Refer to Sec. 6-3. Table 6-1 of Sec. 6-3 also shows a comparison of equations for the foregoing methods. Dual expressions exist for I(a) and II(a); I(b) and II(b); III and IV.

It will be assumed that the reader is familiar with basic matrix methods and solutions of linear equations by determinants. If such is not the case it is suggested that Appendix C be covered first.

5-2. The Loop-Impedance Method

This most familiar method utilizes Kirchhoff's voltage law which states that the sum of the voltages around any closed loop equals zero. A potential rise is considered positive and a potential drop is negative. Both the symbols E and V are used (sometimes interchangeably) in denoting voltage, no attempt being made to identify them as drops or rises.

Most of the reference works dealing with basic circuit theory treat the loop method in fairly complete fashion. Even the simple two and three loop examples serve fairly well to demonstrate the loop method. Yet in a large network one must be assured that the proper number of independent loop equations has been written. In order to illustrate the possibility of trouble when attempting to "fly blind" in the writing of loop equations, refer to Fig. 5-1. Two correct choices of loop currents are shown in Figs. 5-1a and 5-1b; Fig. 5-1c illustrates an incorrect choice of loop currents, a choice which does not yield a sufficient number of independent voltage equations. True, the loop currents have included all branches but in the general case the current through Z_a will not equal the current through Z_b, even though Fig. 5-1c shows them to be equal.

One systematic method to assure the proper number of independent loop equations is the *tree-link* approach. A *tree* of the network is formed which includes as many branches as possible without closing a loop. Now each branch addition (termed a *link*) to this tree will close a new loop and a corresponding loop equation is written. When all the links have been closed in turn, the resulting loop equations are complete. Each loop will

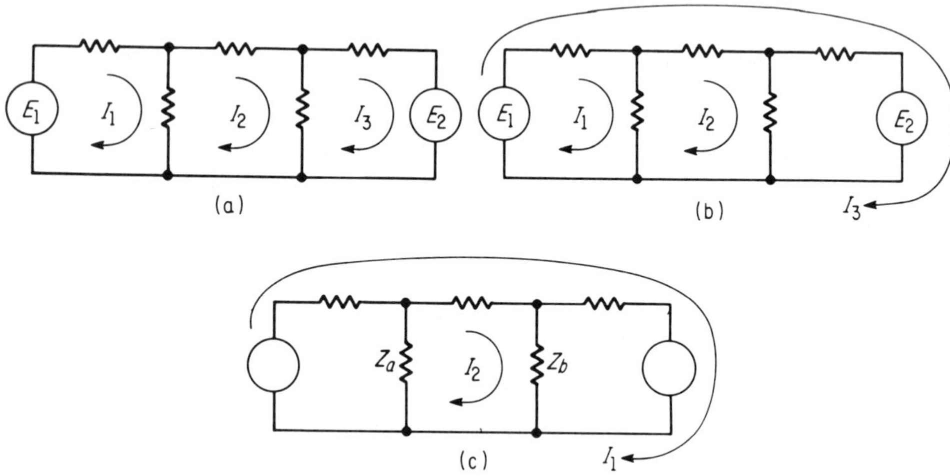

Fig. 5-1. (a) One correct choice of loop currents. (b) Another correct choice of loop currents. (c) Incorrect choice of loop currents.

contain one link and a part of the original tree. One could, of course, replace the network with a simple graph representation, a *graph* being a diagram where each element is replaced by a line. Refer to "topology" of Sec. 5-6. The term *loop* is purposely used instead of the word *mesh*. A mesh may be defined as only the open spaces of a planar network, where a planar network is one that can be drawn on a plane surface without the crossing of lines. The loop is not always the equivalent of a mesh by this definition, since a loop is any closed path of the network.

The number of independent loop currents in a network equals the number of tree links. Another way to quickly determine the number of independent loops L without even the formation of a tree is to express L in terms of the number of branches B and nodes N. This relationship is

$$L = B - N + 1 \tag{5-1}$$

This equation is true where all branches are metallically coupled. When the network consists of several subnetworks (S) coupled only magnetically, then

$$L = B - N + S \tag{5-2}$$

Example 5-1. Given the circuit of Fig. 5-2. Indicate the independent loop currents, using the tree-link concept. Write the voltage matrix equations for the network.

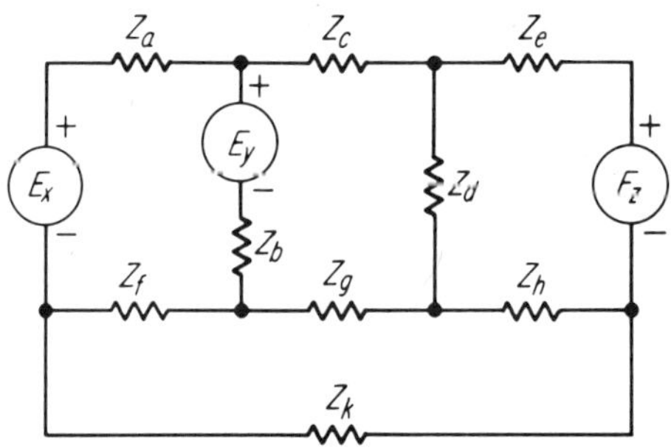

Fig. 5-2

Several tree formations are possible. Two possibilities are indicated in Fig. 5-3. The link branches are dotted in. The loop currents are formed in each case by using the original tree and one new link. A set of branches which makes up such a loop is referred to as a *tie set*.

Now that independent loop currents have been established, the voltage equations are written. Using the tree of Fig. 5-3a, the sum of the voltage sources around each closed loop is set equal to the IZ drops. The four loop equations for this example are shown on page 78. The bars shown above the elements of Eq. 5-4 are used to denote matrix quantities.

Loop 1

$$E_x - E_y = I_1(Z_a + Z_b + Z_f) + I_2(-Z_b) + I_3(0) + I_4 Z_a$$

Loop 2

$$E_y = I_1(-Z_b) + I_2(Z_b + Z_c + Z_d + Z_g) + I_3(-Z_d) + I_4 Z_c$$

Loop 3

$$-E_z = I_1(0) + I_2(-Z_d) + I_3(Z_d + Z_e + Z_h) + I_4 Z_e$$

Loop 4

$$E_x - E_z = I_1 Z_a + I_2 Z_c + I_3 Z_e + I_4(Z_a + Z_c + Z_e + Z_k)$$

The equations in matrix form are

$$\begin{bmatrix} E_x - E_y \\ E_y \\ -E_z \\ E_x - E_z \end{bmatrix} = \begin{bmatrix} (Z_a + Z_b + Z_f) & -Z_b & 0 & Z_a \\ -Z_b & (Z_b + Z_c + Z_d + Z_g) & -Z_d & Z_c \\ 0 & -Z_d & (Z_d + Z_e + Z_h) & Z_e \\ Z_a & Z_c & Z_e & (Z_a + Z_c + Z_e + Z_k) \end{bmatrix} \begin{bmatrix} I_1 \\ I_2 \\ I_3 \\ I_4 \end{bmatrix} \tag{5-3}$$

or

$$\overline{E} = \overline{Z}\overline{I} \tag{5-4}$$

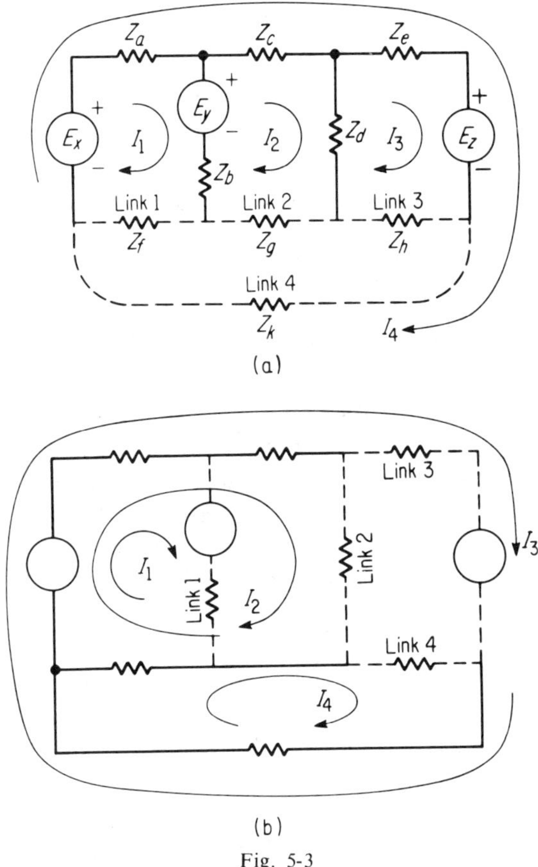

Fig. 5-3

The matrix, Eq. 5-3, may be written directly by inspection, without writing out the individual voltage equations. The diagonal elements of the Z matrix are equal to the self-impedances around each loop, while off diagonal elements are the impedances common to two conductively coupled loops. An off diagonal element (Z_{KL}) is negative with respect to the diagonal element (Z_{KK}) of the same row when this impedance carries the current of loop K in the opposite direction to the loop L current.

Mutual coupling may be present between two loops and may be represented on the off-diagonal terms in the same manner that impedances of conductively coupled loops are represented. Assume for example that there is mutual coupling between loops K and L. The diagonal term Z must again include all self-impedance of loop K, while Z_{KL} will contain the mutual impedance between loops K and L. The sign on Z_{KL} will be of the same sign as Z_{KK} if the direction of flux set up by the two loop currents

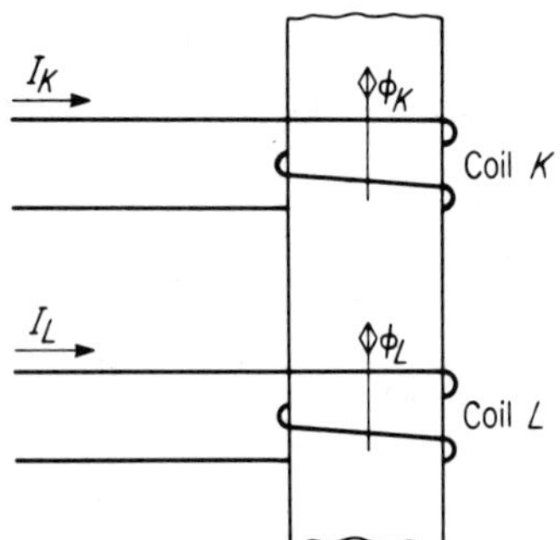

Fig. 5-4. Example of loops with mutual coupling.

(I_K and I_L) is additive as in Fig. 5-4. When such is the case, that part of ϕ_L which couples coil K would naturally induce a mutual voltage into loop K in the same direction as the self-induced voltage caused by I_K.

Refer again to Example 5-1. Since $\overline{E}$ and $\overline{Z}$ of Eq. 5-4 are the known matrices, the job remains to find the array of unknown currents ($\overline{I}$). This may be accomplished by inverting $\overline{Z}$ and multiplying both sides of Eq. 5-4 by $\overline{Z}^{-1}$. Then

$$\begin{bmatrix} I_1 \\ I_2 \\ I_3 \\ I_4 \end{bmatrix} = \overline{Z}^{-1} \cdot \overline{E} \tag{5-5}$$

The matrix Eq. 5-3 may be rewritten in more general terms as

$$\begin{bmatrix} E_1 \\ E_2 \\ E_3 \\ E_4 \end{bmatrix} = \begin{bmatrix} Z_{11} & Z_{12} & Z_{13} & Z_{14} \\ Z_{21} & Z_{22} & Z_{23} & Z_{24} \\ Z_{31} & Z_{32} & Z_{33} & Z_{34} \\ Z_{41} & Z_{42} & Z_{43} & Z_{44} \end{bmatrix} \begin{bmatrix} I_1 \\ I_2 \\ I_3 \\ I_4 \end{bmatrix} \tag{5-6}$$

It is of interest to realize that an equivalent circuit can be drawn which will satisfy Eq. 5-6. This information will be helpful in Chapter 10 where the short-circuit equivalent of Sec. 2-4 is justified. This equivalent form of Fig. 5-5 will not necessarily have any resemblance to the actual original network, yet currents flowing in this loop-impedance equivalent are identical to those loop currents flowing in the original network. In Fig. 5-5, note the diagonal elements of the impedance matrix of Eq. 5-6 are represented as branch self-impedances (Z_{11}, Z_{22}, Z_{33}, Z_{44}). Note that the impedances of this matrix are of the same nature as those in the bus impedance matrix and rake equivalent of Sec. 2-4. Off diagonal elements of Eq. 5-4 are shown as mutual coupling between branches. In the equivalent of Fig. 5-5, the identity of various nodes and branches in the original network has been lost, but the loop currents have remained unchanged.

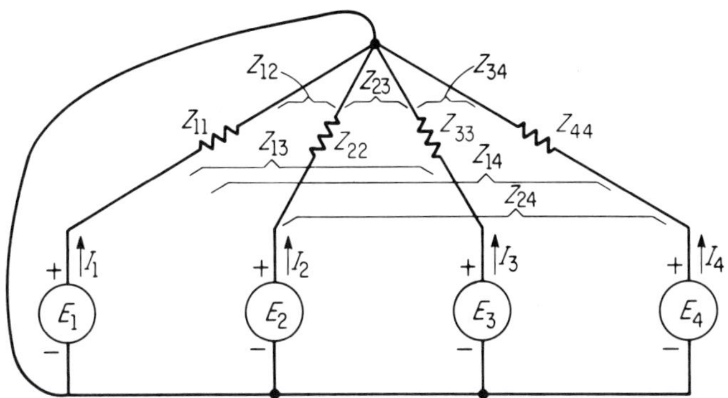

Fig. 5-5. The loop-impedance equivalent for the network of Fig. 5-3a.

5-3. Elimination of Loop Currents by Matrix Partitioning

There are times when the complete array of unknown loop currents is not needed. In such a case it is possible to eliminate unwanted loop currents from the matrix equation by partitioning and reduction of the matrix. This is possible provided the unwanted loops contain no source voltages. Assume a 5-loop matrix equation where only I_1, I_2, and I_3 are needed. The matrix may be considered a compound matrix and partitioned as follows.

$$\begin{bmatrix} E_1 \\ \bar{E}_x \\ E_2 \\ E_3 \\ \hline E_4 \\ \bar{E}_y \\ E_5 \end{bmatrix} = \left[\begin{array}{ccc|cc} Z_{11} & Z_{12} & Z_{13} & Z_{14} & Z_{15} \\ & \bar{Z}_1 & & & \bar{Z}_2 \\ Z_{21} & Z_{22} & Z_{23} & Z_{24} & Z_{25} \\ Z_{31} & Z_{32} & Z_{33} & Z_{34} & Z_{35} \\ \hline Z_{41} & Z_{42} & Z_{43} & Z_{44} & Z_{45} \\ & \bar{Z}_3 & & & \bar{Z}_4 \\ Z_{51} & Z_{52} & Z_{53} & Z_{54} & Z_{55} \end{array}\right] \begin{bmatrix} I_1 \\ \bar{I}_x \\ I_2 \\ I_3 \\ \hline I_4 \\ \bar{I}_y \\ I_5 \end{bmatrix} \tag{5-7}$$

Rewriting in terms of the above submatrices,

$$\begin{bmatrix} \bar{E}_x \\ \bar{E}_y \end{bmatrix} = \begin{bmatrix} \bar{Z}_1 & \bar{Z}_2 \\ \bar{Z}_3 & \bar{Z}_4 \end{bmatrix} \begin{bmatrix} \bar{I}_x \\ \bar{I}_y \end{bmatrix} \tag{5-8}$$

Now expanding Eq. 5-8 into two voltage equations and solving for $\bar{I}_x$,

$$\bar{E}_x = \bar{Z}_1 \bar{I}_x + \bar{Z}_2 \bar{I}_y \tag{5-9}$$

$$\bar{E}_y = \bar{Z}_3 \bar{I}_x + \bar{Z}_4 \bar{I}_y \tag{5-10}$$

From Eq. 5-10,

$$\bar{Z}_4 \bar{I}_y = \bar{E}_y - \bar{Z}_3 \bar{I}_x$$

$$\bar{I}_y = \bar{Z}_4^{-1}(\bar{E}_y - \bar{Z}_3 \bar{I}_x)$$

Substituting this value of $\bar{I}_y$ into Eq. 5-9,

$$\bar{E}_x = \bar{Z}_1 \bar{I}_x + \bar{Z}_2 [\bar{Z}_4^{-1}(\bar{E}_y - \bar{Z}_3 \bar{I}_x)]$$

$$\bar{E}_x = (\bar{Z}_1 - \bar{Z}_2 \bar{Z}_4^{-1} \bar{Z}_3)\bar{I}_x + \bar{Z}_2 \bar{Z}_4^{-1} \bar{E}_y \tag{5-11}$$

Now the source voltages (E_4 and E_5) around loops 4 and 5 are zero. Therefore, $\bar{E}_y$ is a zero matrix and Eq. 5-11 becomes

$$\bar{E}_x = (\bar{Z}_1 - \bar{Z}_2 \bar{Z}_4^{-1} \bar{Z}_3)\bar{I}_x \tag{5-12}$$

Solving now for the current array, represented by I_x,

$$\begin{bmatrix} I_1 \\ I_2 \\ I_3 \end{bmatrix} = \bar{I}_x = [\bar{Z}_1 - \bar{Z}_2 \bar{Z}_4^{-1} \bar{Z}_3]^{-1} \bar{E}_x \tag{5-13}$$

5-4. The Nodal-Admittance Method

In the loop-impedance method of Sec. 5-2, voltage equations were written in terms of known constant voltage sources, known impedances, and unknown loop currents. In the nodal method (the dual of the mesh method) current equations are written in terms of known admittances and unknown node-pair voltages. In the loop method each branch was expressed in the form of Thévenin's E_0 in series with some impedance, Z_0. Likewise in the nodal-admittance method, each branch may be expressed as a simple Norton's equivalent of a constant-current source in parallel with an admittance.

As review, the Norton's equivalent circuit will be developed in terms of the Thévenin's equivalent.

$$V = E_0 - I_L Z_0; \qquad I_L = \frac{E_0 - V}{Z_0} = \frac{E_0}{Z_0} - \frac{V}{Z_0}$$

$$I_L = E_0 Y_0 - V Y_0 \tag{5-14}$$

$E_0 Y_0$ → Constant-current source; $V Y_0$ → Shunt admittance current

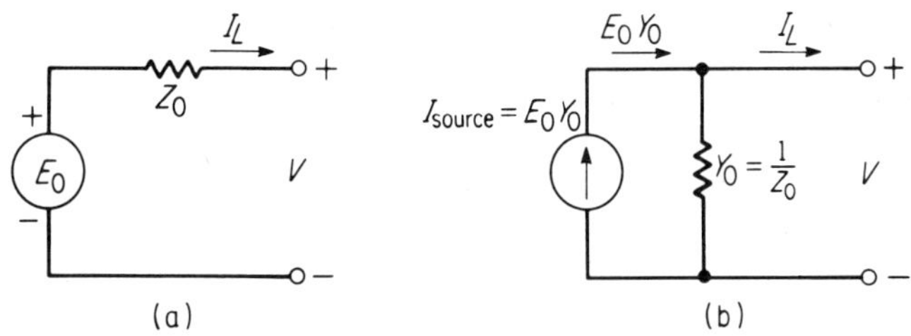

Fig. 5-6. (a) Thévenin's branch equivalent. (b) Norton's branch equivalent.

Equation 5-14 divided I_L into two components, one being a constant-current source ($E_0 Y_0$) and the other being a current dependent upon the terminal voltage. Eq. 5-14 is satisfied by Fig. 5-6b, the Norton's branch equivalent. It is reasoned that if Fig. 5-6b is the equivalent of Fig. 5-6a, then terminal conditions should be the same. This condition is met for both circuits, since the same terminal voltage V and load current I_L are present for both.

The nodal solution is based upon Kirchhoff's current law, which states

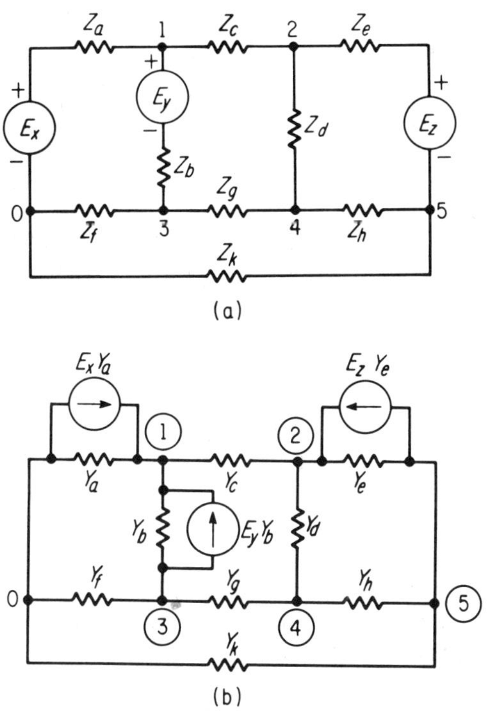

Fig. 5-7

$$\Sigma \text{ currents entering node} = \Sigma \text{ currents leaving node} \tag{5-15}$$

The constant-current sources of Norton's branch equivalents will be considered as feeding the nodes, and will normally be written on the left-hand side of the equal sign. Of course, if a source current arrow is directed away from the node in question, a minus sign is associated with this term. The shunt admittance of each branch will be considered as drawing current away from the node. To demonstrate this method, the circuit of Fig. 5-7a has been redrawn in Fig. 5-7b, substituting the Norton's equivalent for the Thévenin's equivalent.

Node 0 will be chosen as the reference node. Five independent node-pair voltages are possible, taken from node 0 out to each of the other nodes in the circuit. There is one less independent node-pair voltage than total nodes. This is always the case where all branches are metallically coupled. When the total network is divided into magnetically coupled subnetworks, then the number of independent node pairs P equals the number of nodes N minus the number of subnetworks S, or

$$P = N - S \tag{5-16}$$

A current equation must be written at all nodes except the reference node for Fig. 5-7b. (See page 85.) A more general form of the nodal matrix equation is

$$\begin{bmatrix} I_1 \\ I_2 \\ I_3 \\ I_4 \\ I_5 \end{bmatrix} = \begin{bmatrix} Y_{11} & Y_{12} & Y_{13} & Y_{14} & Y_{15} \\ Y_{21} & Y_{22} & Y_{23} & Y_{24} & Y_{25} \\ Y_{31} & Y_{32} & Y_{33} & Y_{34} & Y_{35} \\ Y_{41} & Y_{42} & Y_{43} & Y_{44} & Y_{45} \\ Y_{51} & Y_{52} & Y_{53} & Y_{54} & Y_{55} \end{bmatrix} \begin{bmatrix} V_{01} \\ V_{02} \\ V_{03} \\ V_{04} \\ V_{05} \end{bmatrix} \tag{5-19}$$

or

$$\overline{I}_s = \overline{Y}\,\overline{V} \tag{5-20}$$

Notice in comparing Eq. 5-18 with Eq. 5-19 that the diagonal element (Y_{kk}) equals the summation of all admittances surrounding node k. The off-diagonal admittances ($Y_{kL} = Y_{Lk}$) will contain the negative of the admittance joining node L with node k.

Keep in mind that the $\overline{I}_s$ of Eq. 5-20 is the matrix array of known constant-current sources feeding the nodes and $\overline{V}$ is the unknown array of voltages. To complete the solution, $\overline{V}$ may be found by inverting the Y matrix and premultiplying both sides of Eq. 5-20 by $\overline{Y}^{-1}$, or

$$\overline{V} = \overline{Y}^{-1}\overline{I}_s \tag{5-21}$$

Node 1:

$$E_x Y_a + E_y Y_b = V_{01} Y_a + (V_{01} - V_{03}) Y_b + (V_{01} - V_{02}) Y_c$$

Collecting terms,

$$\left.\begin{aligned} E_x Y_a + E_y Y_b &= V_{01}(Y_a + Y_b + Y_c) - V_{02} Y_c - V_{03} Y_b \\ &\text{Node 2:} \\ E_z Y_e &= V_{02}(Y_c + Y_d + Y_e) - V_{01} Y_c - V_{04} Y_d - V_{05} Y_e \\ &\text{Node 3:} \\ -E_y Y_b &= V_{03}(Y_b + Y_f + Y_g) - V_{01} Y_b - V_{04} Y_g \\ &\text{Node 4:} \\ 0 &= V_{04}(Y_d + Y_g + Y_h) - V_{02} Y_d - V_{03} Y_g - V_{05} Y_h \\ &\text{Node 5:} \\ -E_z Y_e &= V_{05}(Y_h + Y_e + Y_k) - V_{02} Y_e - V_{04} Y_h \end{aligned}\right\} \tag{5-17}$$

The equations of 5-17 could be written in matrix form by inspection. The result will be

$$\begin{bmatrix} E_x Y_a + E_y Y_b \\ E_z Y_e \\ -E_y Y_b \\ 0 \\ -E_z Y_e \end{bmatrix} = \begin{bmatrix} (Y_a + Y_b + Y_c) & -Y_c & -Y_b & 0 & 0 \\ -Y_c & (Y_c + Y_d + Y_e) & 0 & -Y_d & -Y_e \\ -Y_b & 0 & (Y_b + Y_f + Y_g) & -Y_g & 0 \\ 0 & -Y_d & -Y_g & (Y_d + Y_g + Y_h) & -Y_h \\ 0 & -Y_e & 0 & -Y_h & (Y_h + Y_e + Y_k) \end{bmatrix} \begin{bmatrix} V_{01} \\ V_{02} \\ V_{03} \\ V_{04} \\ V_{05} \end{bmatrix} \tag{5-18}$$

It is understood that conventional matrix inversion is not ordinarily the most convenient means for solving the unknown vector $\overline{V}$. Refer to Chapter 12, Sec. 1–8.

Example 5-2. Given the circuit of Fig. 5-8a. Use the nodal-admittance method to find the independent node pair voltages, V_{01} and V_{02}.

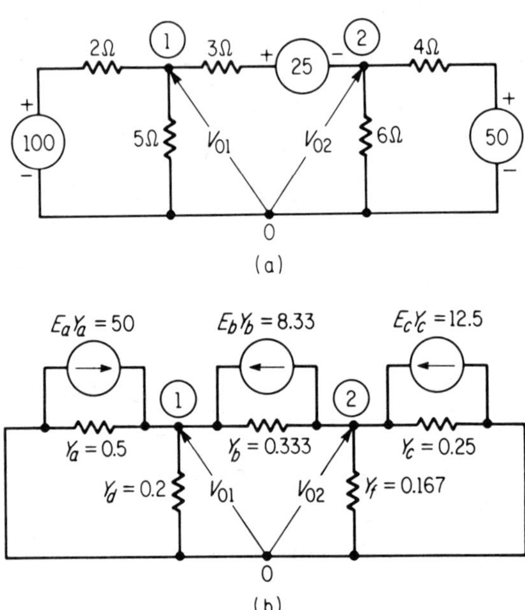

Fig. 5-8

The given circuit is first replaced with Fig. 5-8b where each branch is replaced with its Norton's equivalent. Node 0 is the reference node. Note that the number of independent voltages equals one less than the number of nodes.

Consider the current sources as feeding the nodes and branch admittances as drawing current from the nodes. The matrix equation is normally written by inspection, but for this example the separate nodal equations will first be written out.

$$\Sigma \text{ currents feeding node } = \Sigma \text{ currents leaving node}$$

Node 1:

$$e_a Y_a + e_b Y_b + e_d Y_d = V_{01}(Y_a + Y_b + Y_d) - V_{02} Y_b$$

or

$$50 + 8.33 = V_{01}(0.5 + 0.333 + 0.2) - V_{02}(0.333)$$

Node 2:

$$12.5 - 8.33 = V_{02}(0.333 + 0.25 + 167) - V_{01}(0.333)$$

Writing the matrix equation,

$$\bar{I}_s = \bar{Y}\,\bar{V} \qquad [5\text{-}20]$$

or

$$\begin{bmatrix} I_{s1} \\ I_{s2} \end{bmatrix} = \begin{bmatrix} Y_{11} & Y_{12} \\ Y_{21} & Y_{22} \end{bmatrix} \begin{bmatrix} V_{01} \\ V_{02} \end{bmatrix}$$

$$\begin{bmatrix} 50 + 8.33 \\ 12.5 - 8.33 \end{bmatrix} = \begin{bmatrix} (0.5 + 0.333 + 0.2) & -0.333 \\ -0.333 & (0.333 + 0.25 + 0.167) \end{bmatrix} \begin{bmatrix} V_{01} \\ V_{02} \end{bmatrix}$$

$$\begin{bmatrix} 58.33 \\ 4.17 \end{bmatrix} = \begin{bmatrix} 1.033 & -0.333 \\ -0.333 & 0.750 \end{bmatrix} \begin{bmatrix} V_{01} \\ V_{02} \end{bmatrix}$$

$$\bar{V} = \bar{Y}^{-1}\bar{I}_s$$

$$\bar{Y}^{-1} = \begin{bmatrix} 1.033 & -0.333 \\ -0.333 & 0.750 \end{bmatrix}^{-1} = \frac{1}{\Delta} \begin{bmatrix} 0.750 & 0.333 \\ 0.333 & 1.033 \end{bmatrix}$$

where $\Delta = 0.750 \times 1.033 - 0.333 \times 0.333 = 0.664$

$$\bar{Y}^{-1} = \begin{bmatrix} 1.13 & 0.502 \\ 0.502 & 1.56 \end{bmatrix}$$

$$\begin{bmatrix} V_{01} \\ V_{02} \end{bmatrix} = \bar{Y}^{-1}\bar{I} = \begin{bmatrix} 1.13 & 0.502 \\ 0.502 & 1.56 \end{bmatrix} \begin{bmatrix} 58.33 \\ 4.17 \end{bmatrix} = \begin{bmatrix} 68.1 \\ 35.7 \end{bmatrix}$$

or

$$V_{01} = \underline{\underline{68.1}} \quad \text{and} \quad V_{02} = \underline{\underline{35.7}}$$

5-5. Node Elimination by Matrix Partitioning

Section 2-2 treated one method of node reduction using the star-mesh conversion and eliminating one node at a time. However, it is possible to eliminate a whole grouping of node-pair voltages from the nodal matrix equation. This is done in the same manner that unwanted loop currents of Sec. 5-3 were eliminated from the loop matrix equation. The

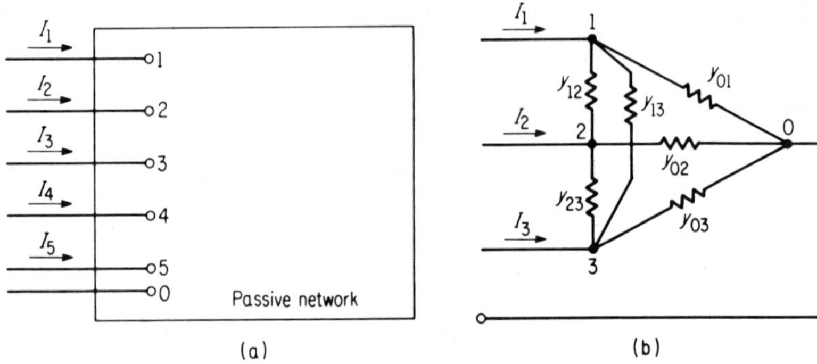

Fig. 5-9. (a) Five-node system plus reference node. (b) Reduced three-node equivalent system plus reference node.

elimination procedure will, for clarity, be repeated in this section as it applies to the nodal method.

As an example, assume a small system containing 5 nodes plus the reference node 0 (Fig. 5-9). The boxed portion of the network is passive with no current sources within. If current sources exist, they will enter from outside the boxed region. Now assume that nodes 4 and 5 are to be eliminated. This will be possible if no current sources are feeding these nodes. The matrix equation is

$$\begin{bmatrix} I_1 \\ I_2 \\ I_3 \\ \hline I_4 \\ I_5 \end{bmatrix} = \left[\begin{array}{ccc|cc} Y_{11} & Y_{12} & Y_{13} & Y_{14} & Y_{15} \\ Y_{21} & Y_{22} & Y_{23} & Y_{24} & Y_{25} \\ Y_{31} & Y_{32} & Y_{33} & Y_{34} & Y_{35} \\ \hline Y_{41} & Y_{42} & Y_{43} & Y_{44} & Y_{45} \\ Y_{51} & Y_{52} & Y_{53} & Y_{54} & Y_{55} \end{array}\right] \begin{bmatrix} V_{01} \\ V_{02} \\ V_{03} \\ \hline V_{04} \\ V_{05} \end{bmatrix} \tag{5-22}$$

or, in terms of the submatrices,

$$\begin{bmatrix} \overline{I}_x \\ \overline{I}_y \end{bmatrix} = \begin{bmatrix} \overline{Y}_1 & \overline{Y}_2 \\ \overline{Y}_3 & \overline{Y}_4 \end{bmatrix} \begin{bmatrix} \overline{V}_x \\ \overline{V}_y \end{bmatrix} \tag{5-22a}$$

The unknowns V_{04} and V_{05} of submatrix $\overline{V}_y$ may be eliminated by substitution as follows:

$$\overline{I}_x = \overline{Y}_1 \overline{V}_x + \overline{Y}_2 \overline{V}_y \tag{5-23}$$

$$\cancelto{0}{\overline{I}_y} = \overline{Y}_3 \overline{V}_x + \overline{Y}_4 \overline{V}_y \tag{5-24}$$

from Eq. 5-24,

$$\overline{V}_y = -\overline{Y}_4^{-1} \overline{Y}_3 \overline{V}_x$$

now substituting into Eq. 5-23,

$$\bar{I}_x = \bar{Y}_1 \bar{V}_x - \bar{Y}_2 \bar{Y}_4^{-1} \bar{Y}_3 V_x$$

$$\bar{I}_x = [\bar{Y}_1 - \bar{Y}_2 \bar{Y}_4^{-1} \bar{Y}_3] \bar{V}_x \qquad (5\text{-}25)$$

The reduced admittance matrix is

$$\bar{Y}_{\text{reduced}} = \bar{Y}_1 - \bar{Y}_2 \bar{Y}_4^{-1} \bar{Y}_3 \qquad (5\text{-}26)$$

The three-node network of Eq. 5-26 has replaced the five-node network of Eq. 5-22.

Actually two objects of importance are obtained from our network reduction.

1. A network solution for the node-pair voltages of $\bar{V}_x$ is possible when the current sources of $\bar{I}_x$ are known. This may be accomplished by inverting the reduced $\bar{Y}$ matrix and premultiplying both sides of Eq. 5-25 by Y_{reduced}^{-1}, or

$$\bar{V}_x = [\bar{Y}_1 - \bar{Y}_2 \bar{Y}_4^{-1} \bar{Y}_3]^{-1} \cdot \bar{I}_x \qquad (5\text{-}27)$$

2. A reduced mesh equivalent is represented by the reduced Y matrix. As indicated in Sec. 5-4, the off-diagonal element (Y_{kL}) is the negative of the admittance between node k and node L, while Y_{kk} is the sum of all admittances surrounding node k. This information is sufficient to determine all values of Fig. 5-9b. Admittance from the reference node to node k in the reduced network will exist only when Y_{kk} of the admittance matrix is not equal to the negative of the sum of the off-diagonal elements of the kth row.

There is often the temptation on the part of the student to include admittance branches that are outside of the area under consideration. For this reason it is sometimes helpful in setting up the original Y matrix, to isolate the passive network of the diagram from the remaining circuit by means of dotted lines. Any admittance outside of these lines may be excluded from consideration in forming the reduced Y matrix. External tie

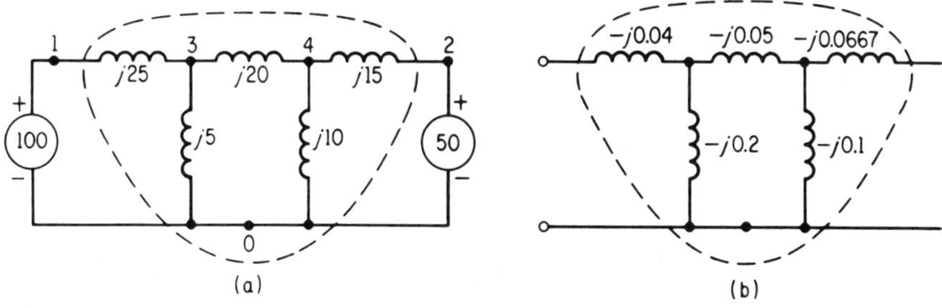

Fig. 5-10. (a) Impedance diagram for Example 5-3. (b) Admittance network to be reduced.

lines will be retained with the outside area and later reconnected to the reduced passive network. Example 5-3 should serve to demonstrate this procedure.

Example 5-3. Given the circuit of Fig. 5-10. Eliminate nodes 3 and 4 and show the resulting impedance diagram. Use the method of nodal elimination by matrix partitioning. The complete nodal matrix equation is

$$\begin{bmatrix} I_1 \\ I_2 \\ I_3 = 0 \\ I_4 = 0 \end{bmatrix}$$

$$= \begin{bmatrix} -j0.04 & 0 & j0.04 & 0 \\ 0 & -j0.0667 & 0 & j0.0667 \\ j0.04 & 0 & -j(0.04 + 0.05 + 0.2) & j0.05 \\ 0 & 0.0667 & j0.05 & -j(0.05 + 0.0667 + 0.1) \end{bmatrix}$$

$$\times \begin{bmatrix} V_{01} \\ V_{02} \\ V_{03} \\ V_{04} \end{bmatrix}$$

Actually, only the admittance matrix is needed for the reduction, but the complete matrix equation is shown for clarity. Notice that source currents identified with nodes 3 and 4 are zero. Outside currents feeding into nodes 1 and 2 may be written simply as I_1 and I_2, whether or not the actual values are known. The reduction is accomplished as follows.

$$\overline{Y}_{\text{reduced}} = \overline{Y}_1 - \overline{Y}_2\overline{Y}_4^{-1}\overline{Y}_3$$

$$\overline{Y}_4^{-1} = \frac{1}{\Delta}\begin{bmatrix} -j0.217 & -j0.05 \\ -J0.05 & -j0.29 \end{bmatrix} \qquad \text{where } \Delta = -0.0628 + 0.0025 = -0.0603$$

$$= \begin{bmatrix} j3.6 & j0.83 \\ j0.83 & j4.82 \end{bmatrix}$$

$$\overline{Y}_4^{-1}\overline{Y}_3 = \begin{bmatrix} j3.6 & j0.83 \\ j0.83 & j4.82 \end{bmatrix}\begin{bmatrix} j0.04 & j0 \\ j0 & j0.0667 \end{bmatrix} = \begin{bmatrix} -0.144 & -0.0553 \\ -0.0332 & -0.321 \end{bmatrix}$$

$$\overline{Y}_2\overline{Y}_4^{-1}\overline{Y}_3 = \begin{bmatrix} j0.04 & 0 \\ 0 & j0.0667 \end{bmatrix} \begin{bmatrix} -0.144 & -0.0553 \\ -0.0332 & -0.321 \end{bmatrix}$$

$$= \begin{bmatrix} -j0.00576 & -j0.00221 \\ -j0.00221 & -j0.0214 \end{bmatrix}$$

$$\overline{Y}_1 - \overline{Y}_2\overline{Y}_4^{-1}\overline{Y}_3 = \begin{bmatrix} -j0.04 & 0 \\ 0 & -j0.0667 \end{bmatrix} - \begin{bmatrix} -j0.00576 & -j0.00221 \\ -j0.00221 & -j0.0214 \end{bmatrix}$$

$$\overline{Y}_{\text{reduced}} = \begin{bmatrix} -j0.0342 & j0.00221 \\ j0.00221 & -j0.0453 \end{bmatrix} = \begin{bmatrix} Y_{11} & Y_{12} \\ Y_{21} & Y_{22} \end{bmatrix}$$

Now with a bit of interpretation, the above Y matrix will yield the equivalent circuit of Fig. 5-11.

$$y_{12} = -Y_{12} = -j\underline{\underline{0.00221}}$$

$$y_{01} = Y_{11} - y_{12} = -j0.0342 - (-j0.00221) = -j\underline{\underline{0.0320}}$$

$$y_{02} = Y_{22} - y_{12} = -j0.0453 - (-j0.00221) = -j\underline{\underline{0.0431}}$$

Taking reciprocals of the Y values of Fig. 5-10a gives the resulting impedance diagram of Fig. 5-11b. The student may easily verify the results of Fig. 5-11b by the alternate method of reduction using Y-Δ conversion.

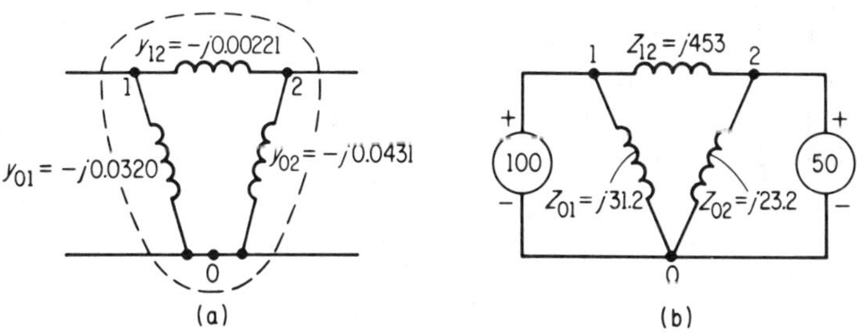

Fig. 5-11. (a) Reduced equivalent. (b) Resulting equivalent impedance diagram.

5-6. Additional Network Topology

The term *topology* may be thought of as a "rubber geometry," or geometry of distortion, helpful in translating graphical information to systematic algebraic forms, in which the figure may be distorted while other properties remain unchanged. For example, a network drawn upon a thin sheet of rubber may be stretched into various shapes. Still, the parts of the network (branches, meshes) in which we are interested are not

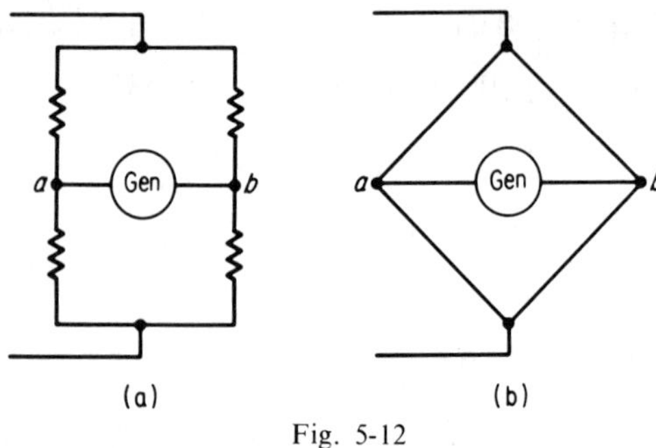

Fig. 5-12

affected. Take the example of a Wheatstone bridge drawn upon a rubber sheet in Fig. 5-12a. If the sheet is stretched horizontally from points *a* and *b* we have a more familiar form of the bridge. A topological *graph* or *map* of an electrical network may represent the actual network as simple lines. These lines are derived from the original network by shorting branch emf's, opening branch current sources and considering all branch impedances as zero. Refer to Fig. 5-13b as an example of a topological graph representation of the network of Fig. 5-13a. Note that six line segments replace branch elements between nodes where the nodes are usually taken as the intersections or junctions of two or more branches. Whenever branches are connected in such a way as to include all nodes, yet failing to close a loop, this set of branches is called a *tree*, as indicated in Sec. 5-2. Branches included in this tree are called *tree branches*. Various combinations can be found for forming such a tree. Once the tree has been established, each of the remaining branches (or links) will close a new loop, each loop containing only one *link*.

The *tree-link* concept is especially helpful in large networks where there may be some question as to the proper choice and number of independent equations. An example of a faulty choice was demonstrated in Fig. 5-1c of Sec. 5-2. The tree-link method provides a consistent and dependable method for determining the proper independent loop equations. First a tree is chosen, then links are inserted which identify the

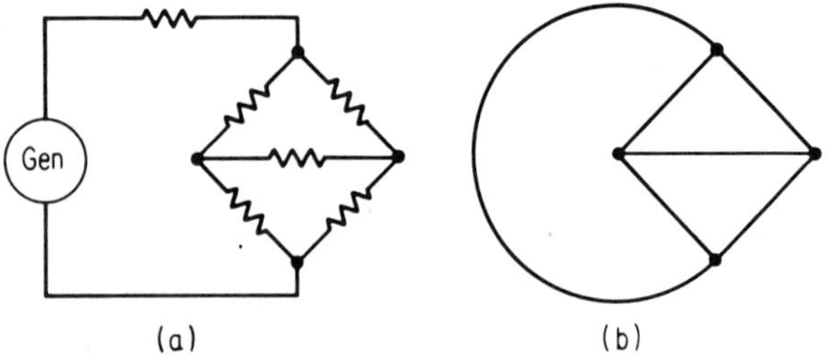

Fig. 5-13. (a) Original network. (b) Graph of network.

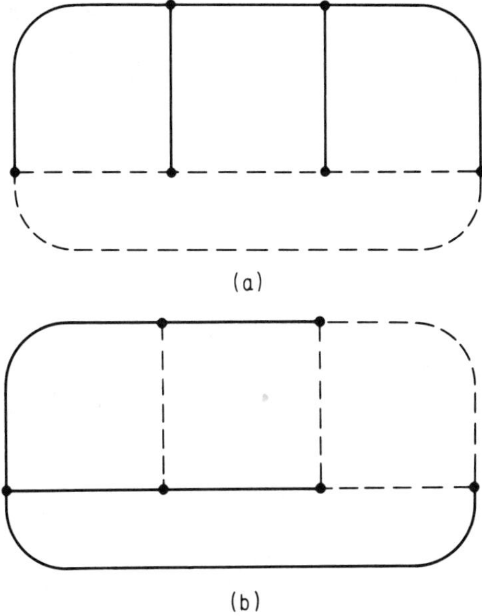

Fig. 5-14. Tree-link graphs for Fig. 5-2.

loop currents. The resulting loop equations form an independent set which are correct and complete. Whenever a link closes a loop, a so called *tie set* has been formed which is a loop containing one link and one or more branches of the established tree. Refer to Fig. 5-3 for examples of the formation of a tree. Of course the graph of straight line elements would be simpler than the actual branch representation of Fig. 5-3. This is done in Fig. 5-14. Links are dotted and tree branches are shown with solid lines.

It is of importance to realize that the number of links equals the number of independent loops and it is also true that the number of independent node pairs equals the number of tree branches. Establishing a tree will not only establish the independent loop currents but will at the same time establish an independent set of node-pair voltages. This fact is especially important in the subsequent section which covers a nodal method without a reference node. The number of tree branches will equal the number of nodes minus one, which is also a criterion for the number of independent node pairs. Some such useful relationships are summarized as

$$\text{Loops } (L) = \text{Links } (l)$$

and from Eq. 5-1,

$$L = B - N + 1 \qquad [5\text{-}1]$$

Number of independent Node Pairs (P) = Number of tree Branches (B_t)

or

$$P = B_t \tag{5-28}$$

also

$$P = N - 1 \tag{5-29}$$

When the network contains several subnetworks, Eqs. 5-1 and 5-29 become

$$L = B - N + S \tag{5-30}$$

$$P = N - S \tag{5-31}$$

combining Eqs. 5-30 and 5-31,

$$L = B - P \tag{5-32}$$

To demonstrate the use of Eqs. 5-30 and 5-31, refer to Fig. 5-15 which includes three subnetworks, magnetically coupled. $B = 13$ and $N = 11$.

$$L = B - N + S = 13 - 11 + 3 = 5$$

The result is obviously correct since there are five open meshes in the network:

$$P = N - S = 11 - 3 = 8$$

One way to check the value of P by inspection is to treat node 1 as a reference and direct node pair voltages from node 1 to the other nodes. To do this it is convenient to establish a reference between subnetworks by tying them together as shown in the dotted lines. This puts node two at the potential of node 1 and 5 at the potential of 4. From this it can be seen that only 8 independent voltages can be directed from node 1.

It has been pointed out that the tree branches could be used to establish independent node-pair voltages (which are identical with these tree-branch voltages). In order to form an independent set of nodal equations

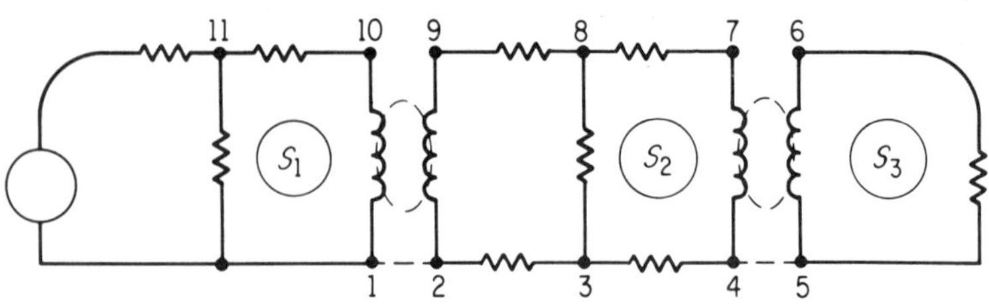

Fig. 5-15

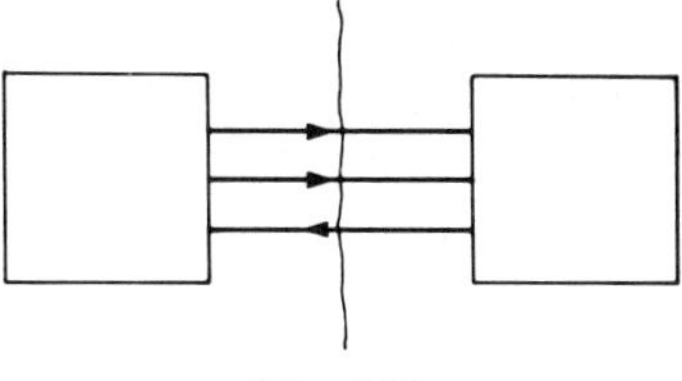

Fig. 5-16

(where no reference node is used) it will be seen (Sec. 5-7) that the concept of cut-sets is necessary. A *cut-set* is a set of elements or edges of a connected graph such that the removal of these edges separate the graph into two subgraphs. Each cut-set contains one tree branch, the other elements being links. The reader should keep in mind that nodal solutions involve the use of Kirchhoff's current law, either written about a node or about a cut-set. Suppose a network is cut in two parts as per Fig. 5-16. If we look at one side of the dividing line, we know that the sum of currents flowing toward the line equals the sum of the currents flowing away from the division. Just as Kirchhoff's current equations were written at all nodes (but reference node) for the method of Sec. 5-4, they could also be written about each cut-set, as will be shown in Sec. 5-7.

Given the graph of Fig. 5-17 with the chosen tree (solid) and links (dotted). This is a directed graph in that arbitrary directions are chosen for each branch. The various cut-sets for this chosen tree are shown in Fig. 5-17 where there are as many independent cut-sets as there are tree branches. Each cut-set is formed by passing a line of separation through one tree branch and continuing on through appropriate links until the graph is divided in two parts. Elements are marked with an X at a cut-set line to indicate elements belonging to the cut-set. For example in the a cut-set, the elements of a, f, and k make up the set. It is also obvious that

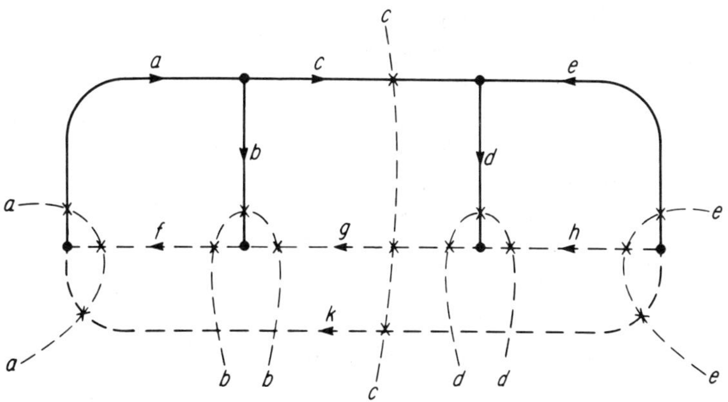

Fig. 5-17

$i_a - i_f - i_k = 0$. Notice also in cut-set b that the branch k is not included in the cut set, since the separation line cuts element k twice. The graph could have been separated without piercing k at all. Another way to visualize this fact is to sum the currents entering the cut-set b as $i_b - i_f + i_g + i_k - i_k = 0$, and observe that the current of element k drops out.

A cut-set matrix may be formed which will relate cut-set currents to branch currents. This will be of special importance in Chapter 6, where a connection matrix (or transformation matrix) $\bar{A}$ is necessary to accomplish the branch-admittance transformation method of solution.

5-7. The General Cut-Set Admittance Method

In this section we will revise the nodal method of Sec. 5-4 to include the general case *without* reference node. This method is perfectly general in that no reference node is necessary and arbitrary tree selection (and selection of independent node pairs) is possible. Refer to the preceding section for discussion of the cut-set line. The procedure for this method is as follows:

1. Choose a tree for the network with corresponding links. Assign a direction to each element of the graph.
2. Sketch in the cut-set lines, one for each tree branch. If current sources entering the cut-set lines are to be considered *positive*, then the assigned directions of the tree branch voltages should be directed *toward* these lines; if *negative*, then direct *away* from these lines.
3. Write a current equation about each cut-set line, setting source currents on the left-hand side of the equal sign and admittance currents of the cut-set elements on the right. The $Y \cdot V$ admittance current terms contain the unknown tree-branch voltages which are to be found.

In practice, step 3 will be written as the matrix form $\bar{I}_s = \bar{Y}\,\bar{V}$. The forms of the three matrices ($\bar{I}_s$, $\bar{Y}$, and $\bar{V}$) are easily recognizable by inspection of the graph. Possibly the best way to demonstrate this procedure is by means of an Example. Necessary equations will be determined first by an analytical breakdown of the separate cut-sets. Later it will be indicated how the matrix form may be written out by inspection.

Example 5-4. Find the independent node-pair voltages of Fig. 5-18a by the cut-set admittance method. To better understand the cut-set approach, we will first write out the separate current equations from the circuit of Fig. 5-18b. Of course the cut-set lines were determined from the tree of Fig. 5-18c. Cut-set lines *b-b*, *d-d*, and *g-g* pierce tree branches b, d, and g respectively, continuing on through appropriate links until each cut-set line divides the graph into two parts.

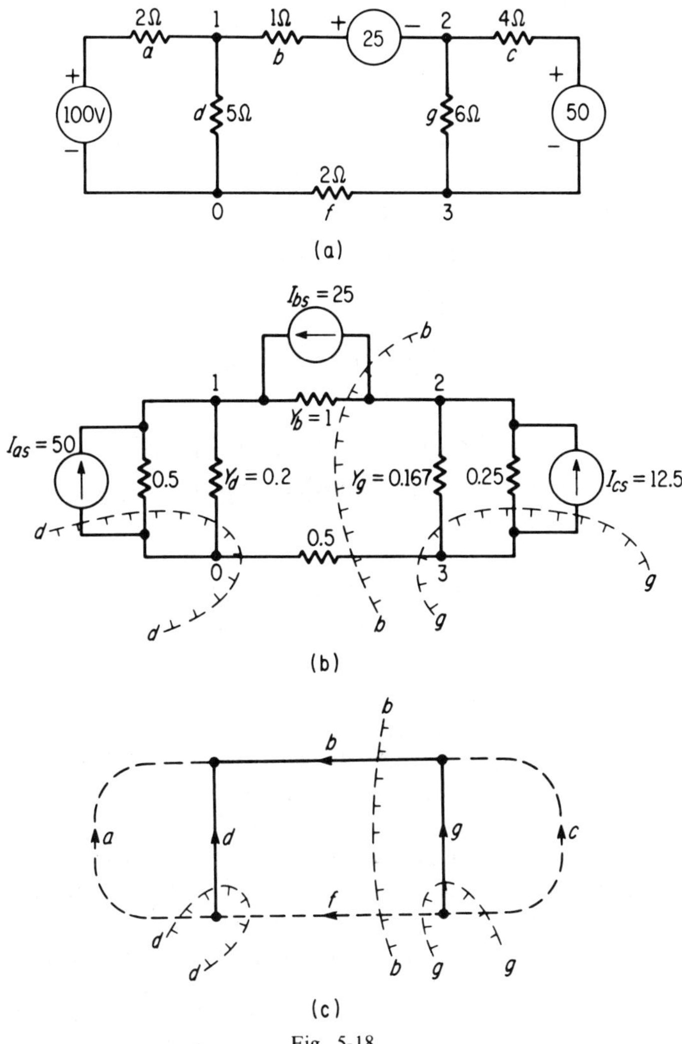

Fig. 5-18

In this problem we have chosen to set the summation of source currents leaving the cut-set line equal to the currents entering the cut-set line through the admittance branches.

At *b-b:*

$$E_b Y_b + \overset{0}{\cancel{E_f Y_f}} = Y_b V_{21} + Y_f V_{30}$$

where

$$V_{30} = V_{32} + V_{21} + V_{10}$$

$$E_b Y_b = Y_b V_{21} + Y_f(V_{32} + V_{21} + V_{10})$$

$$E_b Y_b = (Y_b + Y_f) V_{21} - Y_f V_{01} + Y_f V_{32} \qquad (1)$$

At *d-d:*

$$E_a Y_a + \overset{0}{\cancel{E_d Y_d}} - \overset{0}{\cancel{E_f Y_f}} = (Y_a + Y_d) V_{01} + Y_f V_{03}$$

Again, putting V_{03} in terms of tree-branch voltages,

$$E_a Y_a = (Y_a + Y_d) V_{01} + Y_f (V_{23} + V_{12} + V_{01})$$

$$E_a Y_a = -Y_f V_{32} + (Y_a + Y_d + Y_f) V_{01} - Y_f V_{21} \qquad (2)$$

At *g-g:*

$$\overset{0}{\cancel{E_f Y_f}} + \overset{0}{\cancel{E_g Y_g}} + E_c Y_c = (Y_g + Y_c) V_{32} + V_{30} Y_f$$

$$E_c Y_c = (Y_g + Y_c) V_{32} + Y_f (V_{32} + V_{21} + V_{10})$$

$$E_c Y_c = Y_f V_{21} - Y_f V_{01} + (Y_c + Y_g + Y_f) V_{32} \qquad (3)$$

Equations 1, 2, and 3 can be solved for V_{21}, V_{01}, and V_{32}. Placing the equations in matrix form,

$$\begin{bmatrix} E_b Y_b \\ E_a Y_a \\ E_c Y_c \end{bmatrix} = \begin{matrix} b \\ d \\ g \end{matrix} \begin{bmatrix} Y_b + Y_f & -Y_f & Y_f \\ -Y_f & Y_a + Y_d + Y_f & -Y_f \\ Y_f & -Y_f & Y_c + Y_g + Y_f \end{bmatrix} \begin{bmatrix} V_{21} \\ V_{01} \\ V_{32} \end{bmatrix}$$

It is well to investigate the preceding matrix equation in order to be able to write the matrix equation directly by inspection of Fig. 5-18c. The voltage array contains simply the independent branch voltages of the chosen tree branches *b*, *d*, and *g*. As for the *Y* matrix, the diagonal element of Y_{dd}, for example, is the sum of all admittances associated with the *d-d* cut set. The off-diagonal element (Y_{bd} for example) contains the admittance common to both cut-sets *b* and *d*. Notice Y_{bd} includes a minus sign. The sign of the off-diagonal elements may be determined by inspection of the directed graph, and where element *f* is directed toward cut-set line *d-d* and away from cut-set line *b-b*, the minus sign is added. Y_{bg} is positive since the common element (f) is directed away from both cut-set lines. Substituting numerical values into the matrix equation yields

$$\begin{bmatrix} 25 \\ 50 \\ 12.5 \end{bmatrix} = \begin{bmatrix} 1.5 & -0.5 & 0.5 \\ -0.5 & 1.2 & -0.5 \\ 0.5 & -0.5 & 0.917 \end{bmatrix} \begin{bmatrix} V_{21} \\ V_{01} \\ V_{32} \end{bmatrix}$$

The voltage matrix will be determined by inverting the *Y* matrix and premultiplying both sides by $\overline{Y}^{-1}$. The reader may verify the results of the inverse as

$$\bar{Y}^{-1} = \begin{bmatrix} 0.853 & 0.21 & -0.352 \\ 0.21 & 1.13 & 0.502 \\ -0.352 & 0.502 & 1.56 \end{bmatrix}$$

then

$$\begin{bmatrix} V_{21} \\ V_{01} \\ V_{32} \end{bmatrix} = \begin{bmatrix} 0.853 & 0.21 & -0.352 \\ 0.21 & 1.13 & 0.502 \\ -0.352 & 0.502 & 1.56 \end{bmatrix} \begin{bmatrix} 25 \\ 50 \\ 12.5 \end{bmatrix}$$

$$= \begin{bmatrix} (0.853)(25) + (0.21)(50) + (-0.352)(12.5) \\ (0.21)(25) + (1.13)(50) + (0.502)(12.5) \\ (-0.352)(25) + (0.502)(50) + (1.56)(12.5) \end{bmatrix} = \begin{bmatrix} 27.4 \\ 68.0 \\ 35.8 \end{bmatrix}$$

or

$$V_{21} = \underline{\underline{27.4}}, \quad V_{01} = \underline{\underline{68.0}}, \quad V_{32} = \underline{\underline{35.8}}$$

Problems

5-1. Given the circuit of Fig. 5-2. Consider all impedances as pure resistance, with values as follows:

$$\begin{aligned} &Z_a = 3.0, \quad Z_b = 2.0, \quad Z_c = 2.5, \quad Z_d = 3.8 \\ &Z_e = 4.0, \quad Z_f = 4.4, \quad Z_g = 6.0, \quad Z_h = 2.4, \quad Z_k = 6.5 \\ &E_x = 100, \quad E_y = 25, \quad E_z = 50 \end{aligned}$$

(a) Sketch a new tree differing from that of Fig. 5-3. Exclude the links containing Z_a, Z_c, Z_e, and Z_h.

(b) Now write the voltage matrix equation for this network using this tree. Show your assumed directions for all loop currents.

5-2. Solve Prob. 5-1 for all loop currents using the tree suggested in that problem. Use the method of matrix inversion as outlined in Appendix C-2.

5-3. Repeat Prob. 5-1 choosing a tree which excludes links containing Z_g, Z_b, Z_c, and Z_d.

5-4. In Prob. 5-1, eliminate the loop current (by matrix partitioning) which passes through the link containing Z_h. Show the resulting 3×3 impedance matrix.

5-5. In Prob. 5-3 eliminate the loop current (by matrix partitioning) which passes through the link containing Z_g. Show the resulting 3×3 impedance matrix.

5-6. Set up the voltage equations for Prob. 5-1 using the mesh-impedance method, where only open loops of the given network are used. Sketch the tree which would correspond to such a selection of open-loop currents.

5-7. Refer to Fig. 5-7b and the nodal matrix equation of 5-18. Substituting the voltage and impedance numerical data of Prob. 5-1, rewrite Eq. 5-18.

5-8. Eliminate node 4 in the matrix equation of Prob. 5-7 and re-write the new four-node matrix equation without solving.

5-9. Solve for the four unknown voltages of Prob. 5-8 (V_{01}, V_{02}, V_{03}, V_{05}). Use conventional matrix inversion as outlined in Appendix C-2.

5-10. (Optional) Solve Prob. 5-9 using the Gauss reduction method of Sec. 12-3.

5-11. (Optional) Solve Prob. 5-9 using the Crout reduction method of Sec. 12-4.

5-12. (Optional) Solve Prob. 5-9 using the method of Sec. 12-5 involving elementary row operations.

5-13. Refer to Fig. 5-18a of Example 5-4. Choose a new tree with branches which include $Z_a(2\Omega)$, $Z_f(2\Omega)$, and $Z_g(6\Omega)$. Now set up the current matrix equation in terms of the independent node-pair voltages, using the cut-set admittance method. Sketch your figure identifying cut-set lines.

5-14. Solve the matrix equation of Prob. 5-13 for the independent node-pair voltages.

5-15. Again, as in Prob. 5-13, set up the current matrix equation using the cut-set admittance method, for Fig. 5-18a. However, in this case, choose a tree with branches which include $Z_a(2\Omega)$, $Z_b(1\Omega)$, and $Z_c(4\Omega)$.

5-16. Given the circuit of Fig. 5-2 with voltage and impedance data of Prob. 5-1. Choosing the same tree as Prob. 5-1a, set up a current matrix equation using the cut-set admittance method. Sketch your figure identifying all cut-set lines.

5-17. Repeat Prob. 5-16, choosing a tree which excludes links containing Z_g, Z_b, Z_c, and Z_d.

chapter 6

NETWORK SOLUTIONS WITH MATRIX TRANSFORMATIONS

6-1. Transformation Concepts

Two basic methods of network solution were dealt with early in Chapter 5; the loop-impedance method and the nodal-admittance method. These were dual methods with regard to the form of the matrix equations. Chapter 6 will also center around two methods of solution. They likewise are dual methods and both involve the use of *connection* matrices, which transform a column matrix of voltages or currents from one set of coordinates to another. By coordinates we refer to the elements or variables of the column matrix. A transformation may be thought of here as a substitution of variables. As an example of a linear transformation, two equations are given.

$$y_1 = c_1x_1 + c_2x_2 \tag{6-1}$$

$$y_2 = c_3x_1 + c_4x_2 \tag{6-2}$$

Rewriting Eqs. 6-1 and 6-2 in matrix form,

$$\begin{bmatrix} y_1 \\ y_2 \end{bmatrix} = \begin{bmatrix} c_1 & c_2 \\ c_3 & c_4 \end{bmatrix} \begin{bmatrix} x_1 \\ x_2 \end{bmatrix} \tag{6-3}$$

or

$$\overline{y} = \overline{c}\,\overline{x} \tag{6-4}$$

It is said that the x variables of Eq. 6-3 are transformed into the y variables through the transformation matrix $\overline{c}$. Taken in this light, even an impedance matrix ($\overline{Z}$) would fall into the category of a transformation matrix, transforming a column array (or vector array) of currents into a voltage array, where $\overline{V} = \overline{Z}\,\overline{I}$.

Sometimes it is helpful, in gaining a better insight of a procedure, to change the order of development. By this we mean to first investigate the resulting equations of a transformation method before actually becoming involved with the details of proof. Four transformation equations are shown which will be used in the branch-impedance transformation method. In Sec. 6-2 these expressions will be verified and exemplified. Two reference frames are used in this method, one involving the chosen loop quantities and the other the branch quantities. First assume that the branch-current variables in the matrix $\bar{I}_B$ are found in terms of the loop-current variables of $\bar{I}_L$. We may write

$$\bar{I}_B = \bar{C}\bar{I}_L \tag{6-5}$$

For our second relationships, it will be proven in Sec. 6-2 that loop and branch source emf's are related by the equation

$$\bar{E}_{\text{Loop}} = \bar{C}_t^* \bar{E}_{\text{Branch}} \tag{6-6}$$

The asterisk indicated on the transposed $\bar{C}$ matrix indicates the complex conjungate of the matrix. Actually for this application, the $\bar{C}$ matrix is made up of ± 1 and 0 elements and the asterisk could be excluded.

The loop impedance matrix $\bar{Z}_L$, which was written in Chapter 5 by inspection of the network loops, may also be written in terms of the simple branch impedances matrix Z_B and the connection matrix $\bar{C}$ as

$$\bar{Z}_L = \bar{C}_t^* \bar{Z}_B \bar{C} \tag{6-7}$$

The matrix $\bar{Z}_B$ is more clearly defined in Sec. 6-2.

An equation for the branch currents in terms of the simply $\bar{E}_B$, $\bar{Z}_B$, and $\bar{C}$ matrices is finally determined by substitution of Eqs. 6-6 and 6-7 into Eq. 6-5:

$$\begin{aligned} \bar{I}_B &= \bar{C}\bar{I}_L \\ &= \bar{C}(\bar{Z}_L^{-1}\bar{E}_L) \end{aligned}$$

$$\bar{I}_B = \bar{C}[\bar{C}_t^* \bar{Z}_B \bar{C}]^{-1} \cdot \bar{C}_t^* \bar{E}_B \tag{6-8}$$

The fact that the branch currents of the matrix array ($\bar{I}_B$) can be found in such a direct manner does not imply that this approach will be simpler than the loop-impedance method. In fact, the use of Eq. 6-8 may prove to be tedious. The Eqs. 6-5 through 6-8 were applied to a system with branch and loop reference frames. Had a different set of loop currents been chosen, (that is a different loop-current reference frame) the equations would still hold true, as long as a connection matrix can be found to relate the elements or coordinates of one reference frame to that of the other.

6-2. The Branch-Impedance Transformation Method

The method briefly outlined in the foregoing section requires some verification. Refer to Fig. 6-1. The connection matrix $\overline{C}$ which relates branch currents to chosen loop currents can be written by inspection.

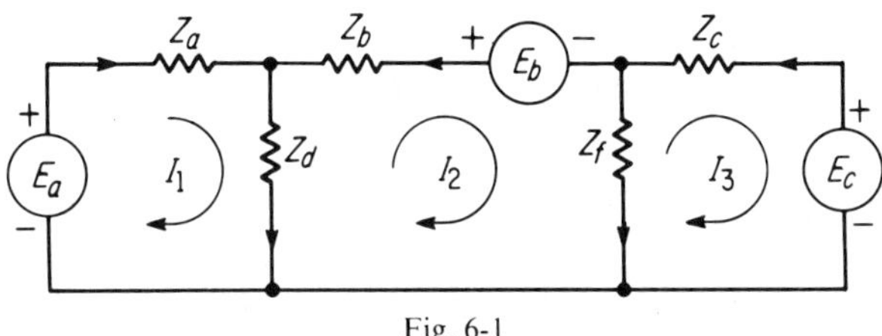

Fig. 6-1

Notice that assumed directions of both branch and loop currents are indicated on the diagram. Equations are as follows:

$$\left.\begin{aligned} I_a &= I_1 \\ I_b &= -I_2 \\ I_c &= -I_3 \\ I_d &= I_1 - I_2 \\ I_f &= I_2 - I_3 \end{aligned}\right\} \tag{6-9}$$

Expressed in matrix form,

$$\begin{bmatrix} I_a \\ I_b \\ I_c \\ I_d \\ I_f \end{bmatrix} = \begin{bmatrix} 1 & 0 & 0 \\ 0 & -1 & 0 \\ 0 & 0 & -1 \\ 1 & -1 & 0 \\ 0 & 1 & -1 \end{bmatrix} \begin{bmatrix} I_1 \\ I_2 \\ I_3 \end{bmatrix} \tag{6-10}$$

or

$$\overline{I}_B = \overline{C}\,\overline{I}_L \qquad [6\text{-}5]$$

Equation 6-6 must be verified which states that $\overline{E}_L = \overline{C}_t^* \overline{E}_B$. First we will test its validity on the special case at hand, followed by a more general proof. Writing the loop-voltage sources in terms of branch-voltage sources will yield

Loop 1:

$$E_1 = E_a + E_d$$

Loop 2:

$$E_2 = -E_d - E_b + E_f \tag{6-11}$$

Loop 3:

$$E_3 = -E_f - E_c$$

or

$$\begin{bmatrix} E_1 \\ E_2 \\ E_3 \end{bmatrix} = \underbrace{\begin{bmatrix} 1 & 0 & 0 & 1 & 0 \\ 0 & -1 & 0 & -1 & 1 \\ 0 & 0 & -1 & 0 & -1 \end{bmatrix}}_{\bar{C}_t} \begin{bmatrix} E_a \\ E_b \\ E_c \\ \cancel{E_d}^{\,0} \\ \cancel{E_f}^{\,0} \end{bmatrix} \tag{6-12}$$

or

$$\bar{E}_L = \bar{C}_t \bar{E}_B$$

Notice that Eq. 6-6 is verified in our sample case.

A more general proof of Eq. 6-6 may be given by first realizing that the total network power due to the loop source emf's must be equal to that power due to all of the branch emf's. Stated another way, the network power is invariant, and will not change with the choice of reference frames (loop or branch coordinates). The same applies to reactive vars.

$$\text{Total VA due to loop source emf's} = \text{Total VA due to branch source emf's} \tag{6-13}$$

The $P + jQ$ of a given branch (or loop) is the product of the complex conjugate of the branch (or loop) current times the source emf of that branch of loop:

$$I_1^* E_1 + I_2^* E_2 + I_3^* E_3 = I_a^* E_a + I_b^* E_b + I_c^* E_c + I_d^* E_d + I_f^* E_f \tag{6-14}$$

Writing Eq. 6-14 in matrix form,

$$[I_1 I_2 I_3]^* \begin{bmatrix} E_1 \\ E_2 \\ E_3 \end{bmatrix} = [I_a I_b I_c I_d I_f]^* \begin{bmatrix} E_a \\ E_b \\ E_c \\ E_d \\ E_f \end{bmatrix} \tag{6-15}$$

or

$$\bar{I}^*_{L_t}\bar{E}_L = \bar{I}^*_{B_t}\bar{E}_B \tag{6-16}$$

Substituting the value of I_B from Eq. 6-5 into Eq. 6-16,

$$\bar{I}^*_{L_t}\bar{E}_L = [\bar{C}\bar{I}_L]^*_t\,\bar{E}_B$$

Applying the reversal rule to a transposed product,

$$\bar{I}^*_{L_t}\bar{E}_L = \bar{I}^*_{L_t}\bar{C}^*_t\,\bar{E}_B$$

and

$$\bar{E}_L = \bar{C}^*_t\,\bar{E}_B \qquad [6\text{-}6]$$

Next, Eq. 6-7 will be verified. The voltage equations for the branches of Fig. 6-1 will be written by referring to Fig. 6-2. At this point,

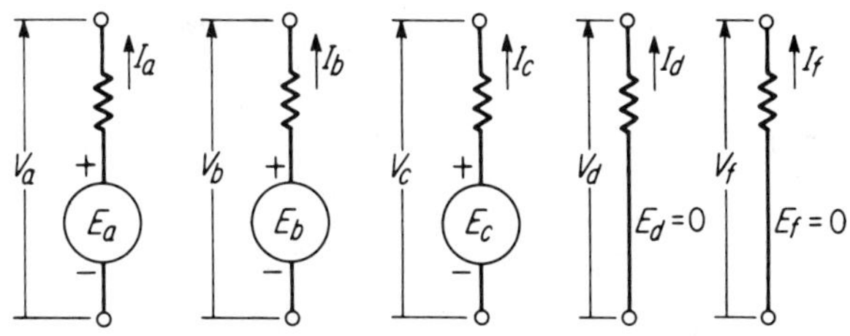

Fig. 6-2

we will think of the network in terms of its simple branches. Regardless of their particular interconnection, the following branch-voltage equations hold true:

$$\left.\begin{aligned} V_a &= E_a \quad Z_a I_a \\ V_b &= E_b - Z_b I_b \\ V_c &= E_c - Z_c I_c \\ V_d &= E_d - Z_d I_d \\ V_f &= E_f - Z_f I_f \end{aligned}\right\} \tag{6-17}$$

In matrix form,

$$\begin{bmatrix} V_a \\ V_b \\ V_c \\ V_d \\ V_f \end{bmatrix} = \begin{bmatrix} E_a \\ E_b \\ E_c \\ E_d \\ E_f \end{bmatrix} - \begin{bmatrix} Z_a & 0 & 0 & 0 & 0 \\ 0 & Z_b & 0 & 0 & 0 \\ 0 & 0 & Z_c & 0 & 0 \\ 0 & 0 & 0 & Z_d & 0 \\ 0 & 0 & 0 & 0 & Z_f \end{bmatrix} \begin{bmatrix} I_a \\ I_b \\ I_c \\ I_d \\ I_f \end{bmatrix} \tag{6-18}$$

or

$$\overline{V}_B = \overline{E}_B - \overline{Z}_B \overline{I}_B \tag{6-19}$$

Notice that the so-called branch impedance matrix ($\overline{Z}_B$) is a simple matrix containing only branch impedances for diagonal elements and (for this example) zeros for the off-diagonal elements. In general, however, the off-diagonal element (Z_{ij}, where $i \neq j$) equals any mutual impedance which might exist between branch i and branch j. Now multiply Eq. 6-19 by $\overline{C}_t^*$:

$$\overline{C}_t^* \overline{V}_B = \overline{C}_t^* \overline{E}_B - C_t^* \overline{Z}_B \overline{I}_B \tag{6-20}$$

The left-hand side of Eq. 6-20 vanishes, since $\overline{C}_t^* \overline{V}_B = \overline{V}_{\text{Loop}}$ and $\overline{V}_{\text{Loop}}$ is a zero matrix, since the sum of all branch voltages around any closed loop is zero. Equation 6-20 then becomes

$$\overline{C}_t^* \overline{E}_B = \overline{C}_t^* \overline{Z}_B \overline{I}_B \tag{6-21}$$

Substituting loop values in place of branch values (as per Eqs. 6-5 and 6-6) into Eq. 6-21,

$$\overline{E}_{\text{Loop}} = [\underbrace{\overline{C}_t^* \overline{Z}_B \overline{C}}_{Z_{\text{Loop}}}]\, \overline{I}_{\text{Loop}} \tag{6-22}$$

Equation 6-22 is now a loop matrix equation where

$$Z_{\text{Loop}} = \overline{C}_t^* \overline{Z}_B \overline{C} \qquad [6\text{-}7]$$

Example 6-1. Find all branch currents for the circuit of Fig. 6-3 using the branch-impedance transformation method.

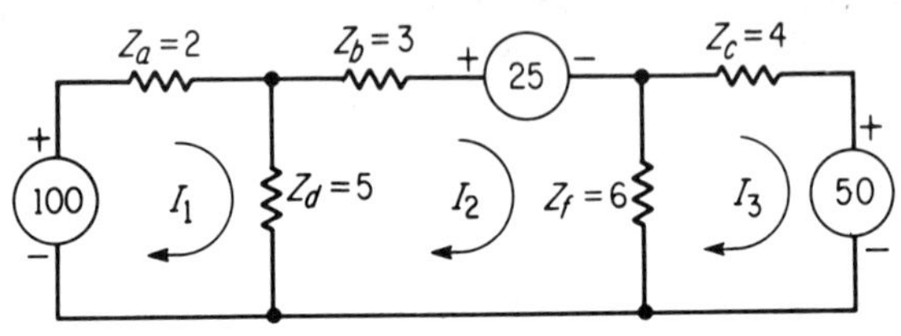

Fig. 6-3

From Eq. 6-8,

$$\overline{I}_B = \overline{C}[\overline{C}_t \overline{Z}_B \overline{C}]^{-1} \overline{C}_t \overline{E}_B$$

Because of the real nature of $\overline{C}$, the conjugations on Eq. 6-8 were dropped. The connection matrix $\overline{C}$ is first determined, from the relation $\overline{I}_B = \overline{C}\,\overline{I}_L$

$$\overline{C} = \text{Branch}\overset{\text{Loop}}{\begin{bmatrix} 1 & 0 & 0 \\ 0 & -1 & 0 \\ 0 & 0 & -1 \\ 1 & -1 & 0 \\ 0 & 1 & -1 \end{bmatrix}}; \qquad \overline{C}_t = \begin{bmatrix} 1 & 0 & 0 & 1 & 0 \\ 0 & -1 & 0 & -1 & 1 \\ 0 & 0 & -1 & 0 & -1 \end{bmatrix}$$

$$\overline{E}_B = \begin{bmatrix} 100 \\ 25 \\ 50 \\ 0 \\ 0 \end{bmatrix}; \qquad \overline{Z}_B = \begin{bmatrix} 2 & 0 & 0 & 0 & 0 \\ 0 & 3 & 0 & 0 & 0 \\ 0 & 0 & 4 & 0 & 0 \\ 0 & 0 & 0 & 5 & 0 \\ 0 & 0 & 0 & 0 & 6 \end{bmatrix}$$

First, form $\overline{Z}_B \overline{C} = \begin{bmatrix} 2 & 0 & 0 & 0 & 0 \\ 0 & 3 & 0 & 0 & 0 \\ 0 & 0 & 4 & 0 & 0 \\ 0 & 0 & 0 & 5 & 0 \\ 0 & 0 & 0 & 0 & 6 \end{bmatrix} \cdot \begin{bmatrix} 1 & 0 & 0 \\ 0 & -1 & 0 \\ 0 & 0 & -1 \\ 1 & -1 & 0 \\ 0 & 1 & -1 \end{bmatrix}$

$$= \begin{bmatrix} 2 & 0 & 0 \\ 0 & -3 & 0 \\ 0 & 0 & -4 \\ 5 & -5 & 0 \\ 0 & 6 & -6 \end{bmatrix}$$

Next $\overline{C}_t \overline{Z}_B \overline{C} = \begin{bmatrix} 1 & 0 & 0 & 1 & 0 \\ 0 & -1 & 0 & -1 & 1 \\ 0 & 0 & -1 & 0 & -1 \end{bmatrix} \cdot \begin{bmatrix} 2 & 0 & 0 \\ 0 & -3 & 0 \\ 0 & 0 & -4 \\ 5 & -5 & 0 \\ 0 & 6 & -6 \end{bmatrix}$

$$= \begin{bmatrix} 7 & -5 & 0 \\ -5 & 14 & -6 \\ 0 & -6 & 10 \end{bmatrix}$$

At this point it is well to recall that $\overline{C}_t \overline{Z}_B \overline{C} = \overline{Z}_L$, and the results of the $\overline{C}_t \overline{Z}_B \overline{C}$ product may be checked by forming $\overline{Z}_L$ by inspection. For example the element in the 1,1 position equals 7, which is also the sum of impedances around loop one.

$$[\overline{C}_t \overline{Z}_B \overline{C}]^{-1} = \begin{bmatrix} 0.218 & 0.1045 & 0.0627 \\ 0.1045 & 0.1485 & 0.0878 \\ 0.0627 & 0.0878 & 0.153 \end{bmatrix}$$

Steps for taking the foregoing inverse are not shown but should be verified by the student.

$$C(\overline{C}_t \overline{Z}_B \overline{C})^{-1} = \begin{bmatrix} 1 & 0 & 0 \\ 0 & -1 & 0 \\ 0 & 0 & -1 \\ 1 & -1 & 0 \\ 0 & 1 & -1 \end{bmatrix} \begin{bmatrix} 0.218 & 0.1045 & 0.0627 \\ 0.1045 & 0.1485 & 0.0878 \\ 0.0627 & 0.0878 & 0.153 \end{bmatrix}$$

$$= \begin{bmatrix} 0.218 & 0.1045 & 0.0627 \\ -0.1045 & -0.1485 & -0.0878 \\ -0.0627 & -0.0878 & -0.153 \\ 0.113 & -0.044 & -0.0251 \\ 0.0418 & 0.0607 & -0.0652 \end{bmatrix}$$

$$\overline{C}(\overline{C}_t \overline{Z}_B \overline{C})^{-1} \overline{C}_t = \begin{bmatrix} 0.218 & -0.1045 & -0.0627 & 0.1135 & 0.0418 \\ -0.1045 & 0.1485 & 0.0878 & 0.044 & -0.0607 \\ -0.0627 & 0.0878 & 0.153 & 0.0251 & 0.0652 \\ 0.1135 & 0.044 & 0.0251 & 0.157 & -0.0189 \\ 0.0418 & -0.0607 & 0.0652 & -0.0189 & 0.1259 \end{bmatrix}$$

Finally, $\overline{I}_B$ equals the above matrix times $\overline{E}_B$:

$$\overline{I}_B = \overline{C}(\overline{C}_t \overline{Z}_B \overline{C})^{-1} \overline{C}_t \overline{E}_B$$

$$\overline{I}_B = \begin{bmatrix} 0.218 & -0.1045 & -0.0627 & 0.1135 & 0.0418 \\ -0.1045 & 0.1485 & 0.0878 & 0.044 & -0.607 \\ -0.0627 & 0.0878 & 0.153 & 0.0251 & 0.0652 \\ 0.1135 & 0.044 & 0.0251 & 0.157 & -0.0189 \\ 0.0418 & -0.607 & 0.0652 & -0.0189 & 0.1259 \end{bmatrix} \begin{bmatrix} 100 \\ 25 \\ 50 \\ 0 \\ 0 \end{bmatrix}$$

$$\bar{I}_B = \begin{bmatrix} I_a \\ I_B \\ I_c \\ I_d \\ I_f \end{bmatrix} = \begin{bmatrix} 16.0 \\ -2.35 \\ 3.58 \\ 13.7 \\ 5.92 \end{bmatrix}$$

6-3. The Branch-Admittance Transformation Method

A nodal method exists which is completely analogous to the method of Sec. 6-2. We will refer to this as the branch-admittance transformation method. This differs from the nodal-admittance method of Chapter 5 in that it is required to determine directly all branch voltages rather than just the independent node pair voltages. These branch voltages may be found by a matrix equation which relates the branch voltage matrix ($\bar{V}_B$) to the known branch admittance matrix $\bar{Y}_B$, the known branch source-current matrix ($\bar{I}_{BS}$) and a known connection matrix $\bar{A}$. The final matrix equation for this method is later proved to be

$$\bar{V}_B = \bar{A}\,[\bar{A}_t \bar{Y}_B \bar{A}]^{-1} \bar{A}_t \bar{I}_{BS} \qquad (6\text{-}23)$$

Henceforth conjugations on $\bar{A}_t$ and $\bar{C}_t$ will be dropped due to the real nature of $\bar{A}$ and $\bar{C}$. Note the similarity of Eq. 6-23 with that of Eq. 6-8:

$$\bar{I}_B = \bar{C}\,[\bar{C}_t \bar{Z}_B \bar{C}]^{-1} \bar{C}_t \bar{E}_B \qquad [6\text{-}8]$$

The two dual methods of Chapter 5 and those of Secs. 6-2 and 6-3 are all compared in Table 6-1.

We are concerned in this section with the four equations in Table 6-1 numbered 6-23 through 6-26. We will not accept the complete set of equa-

Table 6-1

	I. Loop-Impedance Method	II. Nodal-Admittance Method or the more general Cut-Set Admittance Method	
Known:	$\bar{E}_L, \bar{Z}_L$	$\bar{I}_s, \bar{Y}$	
Find:	$\bar{I}_L$	$\bar{V}$	
Equation:	$\bar{I}_L = \bar{Z}_L^{-1} \bar{E}_L$	$\bar{V} = (\bar{Y})^{-1} \bar{I}_s$	
	III. Branch-Impedance Transformation Method	IV. Branch-Admittance Transformation Method (Method of this section)	
Known:	$\bar{E}_B, \bar{Z}_B, \bar{C}$	$\bar{I}_{BS}, \bar{Y}_B, \bar{A}$	
Find:	$\bar{I}_B$ (Branch Current Matrix)	$\bar{V}_B$ (Branch Voltage Matrix)	
Equations:	$\bar{I}_B = \bar{C}\bar{I}_L$	$\bar{V}_B = \bar{A}\bar{V}$	(6-24)
	$\bar{E}_L = \bar{C}_t \bar{E}_B$	$\bar{I}_s = \bar{A}_t \bar{I}_{BS}$	(6-25)
	$\bar{Z}_L = \bar{C}_t \bar{Z}_B \bar{C}$	$\bar{Y} = \bar{A}_t \bar{Y}_B \bar{A}$	(6-26)
	$\bar{I}_B = \bar{C}(\bar{C}_t \bar{Z}_B \bar{C}) \bar{C}_t \bar{E}_B$	$\bar{V}_B = \bar{A}(\bar{A}_t \bar{Y}_B \bar{A})^{-1} \bar{A}_T \bar{I}_{BS}$	[6-23]

tions merely on the grounds that they are or should be the duals of their counterpart forms arrived at in the foregoing section. Rather, verification of the four equations will be given in much the same way that Eqs. 6-5 through 6-8 were verified.

The connection matrix $\bar{A}$ will be here defined as that matrix which transforms the independent node pair voltage matrix $\bar{V}$ into the branch voltage matrix $\bar{V}_B$, or

$$\bar{V}_B = \bar{A}\,\bar{V} \qquad [6\text{-}24]$$

The independent node pair matrix $\bar{V}$ is easily determined where a reference node 0 has been chosen. In this case $\bar{V}$ contains the voltage array of $\bar{V}_{01}$, $\bar{V}_{02}$, V_{03} etc., voltages being directed from node 0 to every other node. However, if independent node pairs are arbitrarily chosen with no reference node, node pairs should be associated with tree-branch voltages only.

Next it will be verified that if Eq. 6-24 is true then

$$\bar{I}_s = \bar{A}_t \bar{I}_{BS} \qquad [6\text{-}25]$$

where $\bar{I}_{BS}$ is the branch source current matrix. Again, if independent node pairs are taken from a reference node 0, $\bar{I}_s$ is merely the matrix array $(I_1, I_2 \ldots,$ etc.) of total source currents entering the individual nodes $1, 2, 3 \ldots,$ etc. However, when no reference node is chosen and independent node pairs are identified with tree branches, we resort to the use of cut-sets.

Elements of $\bar{I}_s$ are then no longer identified with source currents entering a particular node, but rather the source currents of a particular cut-set. For this more general case, the A matrix may be found by an inspection of the graph (tree and links) of the network and its corresponding cut-sets, using either the relationships of Eqs. 6-24 or 6-25. Determination of $\bar{A}$ where cut-sets are required (with no reference node) will be covered in more detail in Sec. 6-5. The *verification* of Eq. 6-25 is given as follows. Again the total power delivered to the network by all of the individual branch-source currents must be equal to that power which would be supplied to the network through all of the source currents of $\bar{I}_s$. Setting these equal,

$$(\bar{I}'_s)^*_t\,\bar{V} = (\bar{I}_{BS})^*_t\,\bar{V}_B \qquad (6\text{-}27)$$

Substituting for $\bar{V}_B$ the value of (Eq. 6-24),

$$(\bar{I}_s)^*_t\,\bar{V} = (\bar{I}_{BS})^*\,\bar{A}\,\bar{V}$$

or

$$(\bar{I}_s)^*_t = (\bar{I}_{BS})^*_t\,\bar{A}$$

Taking the transpose of both sides with the aid of the reversal rule yields

$$\overline{I}_s = \overline{A}_t \overline{I}_{BS} \qquad [6\text{-}25]$$

Equation 6-26 relates the cut-set admittance matrix of Sec. 5-7 (or the nodal-admittance matrix of Sec. 5-4) to the simple branch-admittance matrix $\overline{Y}_B$. The value of the Y_{kk} element along the diagonal of $\overline{Y}_B$ is merely the admittance for branch k. We will refer to the separate branch current equations for the branches of Fig. 6-4 in verifying Eq. 6-26.

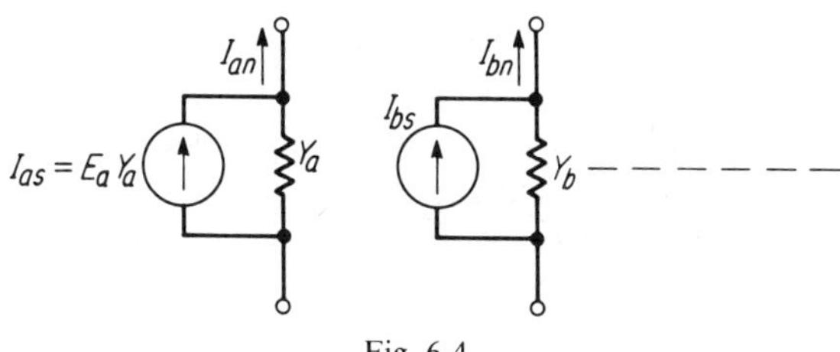

Fig. 6-4

The following relations hold:

$$I_{an} = I_{as} - Y_a V_a$$

$$I_{bn} = I_{bs} - Y_b V_b$$

or

$$\overline{I}_n = \overline{I}_{BS} - \overline{Y}_B \overline{V}_B \qquad (6\text{-}28)$$

Multiplying (Eq. 6-28) by $\overline{A}_t$,

$$\overline{A}_t \cancelto{0}{\overline{I}_n} = \overline{A}_t \overline{I}_{BS} - \overline{A}_t \overline{Y}_B \overline{V}_B \qquad (6\text{-}29)$$

The left-hand term of Eq. 6-29 is zero since it represents (when compared with Eq. 6-25) the array of total currents to the independent nodes (or cut-sets) just as the $\overline{I}_s$ array represents the total source currents to the independent nodes (or cut-sets). Then

$$\overline{A}_t \overline{I}_{BS} = \overline{A}_t \overline{Y}_B \overline{V}_B \qquad (6\text{-}30)$$

Substituting values from Eqs. 6-24 and 6-25 into Eq. 6-30,

$$\overline{I}_s = \Big[\underbrace{\overline{A}_t \overline{Y}_B \overline{A}}_{\overline{Y}}\Big] \overline{V} \qquad (6\text{-}31)$$

or

$$\overline{Y} = \overline{A}_t \overline{Y}_B \overline{A} \qquad [6\text{-}26]$$

The equation of 6-23 may now be found by direct substitution:

$$\overline{V}_B = \overline{A}\,\overline{V} \qquad [6\text{-}24]$$

$$\overline{V}_B = \overline{A}[\overline{Y}]^{-1}\cdot\overline{I}_s$$

Substituting the results of Eqs. 6-25 and 6-26,

$$\overline{V}_B = \overline{A}\,[\overline{A}_t\overline{Y}_B\overline{A}]^{-1}\cdot\overline{A}_t\overline{I}_{BS} \qquad [6\text{-}23]$$

Example 6-2. Determine all of the branch voltages for Fig. 6-5 using the transformation Eq. 6-23, or

$$\overline{V}_B = \overline{A}\,[\overline{A}_t\overline{Y}_B\overline{A}]^{-1}\overline{A}_t\overline{I}_{BS}$$

For the sake of comparison of methods, Fig. 6-5 is identical with the circuit of Example 5-4. The independent node pairs of the matrix $\overline{V}$ will be taken from the chosen tree of Fig. 6-5 (c). The connection matrix A

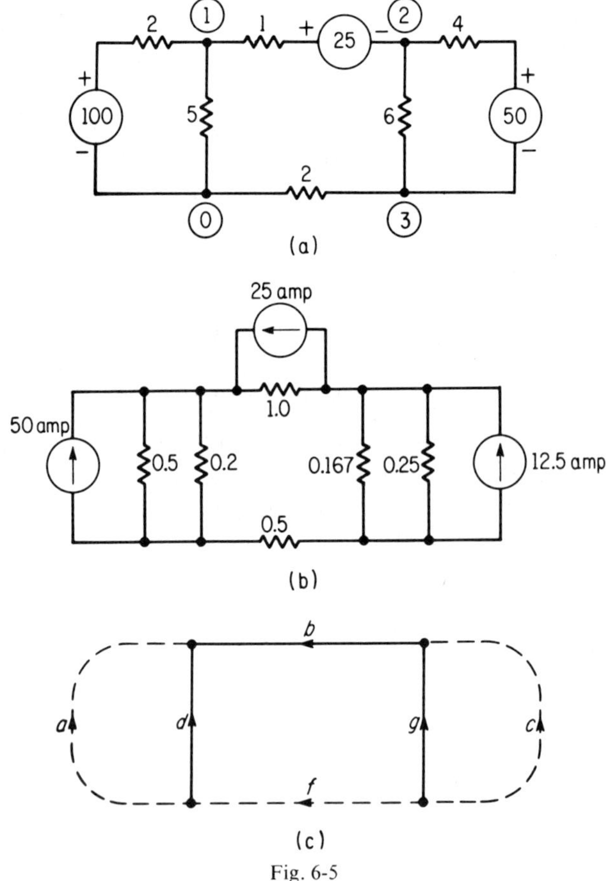

Fig. 6-5

will first be found by relating the branch voltage matrix $\overline{V}_B$ to the independent node pair voltage matrix $\overline{V}$, or $\overline{V}_B = \overline{A}\,\overline{V}$.

For the sake of clarity the relationships will be noted individually:

$$\left.\begin{aligned} V_a &= V_d \\ V_b &= V_b \\ V_c &= V_g \\ V_d &= V_d \\ V_f &= V_g + V_b - V_d \\ V_g &= V_g \end{aligned}\right\} \text{ or } \begin{bmatrix} V_a \\ V_b \\ V_c \\ V_d \\ V_f \\ V_g \end{bmatrix} = \underbrace{\begin{bmatrix} 0 & 1 & 0 \\ 1 & 0 & 0 \\ 0 & 0 & 1 \\ 0 & 1 & 0 \\ 1 & -1 & 1 \\ 0 & 0 & 1 \end{bmatrix}}_{\overline{A}} \begin{bmatrix} V_b \\ V_d \\ V_g \end{bmatrix}$$

The branch admittance matrix is

$$\overline{Y}_B = \begin{bmatrix} 0.5 & 0 & 0 & 0 & 0 & 0 \\ 0 & 1.0 & 0 & 0 & 0 & 0 \\ 0 & 0 & 0.25 & 0 & 0 & 0 \\ 0 & 0 & 0 & 0.2 & 0 & 0 \\ 0 & 0 & 0 & 0 & 0.5 & 0 \\ 0 & 0 & 0 & 0 & 0 & 0.167 \end{bmatrix}$$

$$\overline{Y}_B\overline{A} = \begin{bmatrix} 0 & 0.5 & 0 \\ 1 & 0 & 0 \\ 0 & 0 & 0.25 \\ 0 & 0.2 & 0 \\ 0.5 & -0.5 & 0.5 \\ 0 & 0 & 0.167 \end{bmatrix}; \quad \overline{A}_t = \begin{bmatrix} 0 & 1 & 0 & 0 & 1 & 0 \\ 1 & 0 & 0 & 1 & -1 & 0 \\ 0 & 0 & 1 & 0 & 1 & 1 \end{bmatrix}$$

$$\overline{A}_t\overline{Y}_B\overline{A} = \begin{bmatrix} 1.5 & -0.5 & 0.5 \\ -0.5 & 1.2 & -0.5 \\ 0.5 & -0.5 & 0.917 \end{bmatrix}$$

Notice from Eq. 6-26 that the preceding matrix should also be equal to the admittance matrix $\overline{Y}$ of either the nodal or cut set admittance method.

As a check on our results, it is seen that the matrix does indeed equal the $\overline{Y}$ matrix of Example 5-4.

Next taking the inverse yields

$$[\overline{A}_t\overline{Y}_b\overline{A}]^{-1} = \begin{bmatrix} 0.853 & 0.21 & -0.352 \\ 0.21 & 1.13 & 0.502 \\ -0.352 & 0.502 & 1.56 \end{bmatrix}$$

$$\overline{A}\,[\overline{A}_t\overline{Y}_B\overline{A}]^{-1} = \begin{bmatrix} 0.21 & 1.13 & 0.502 \\ 0.853 & 0.21 & -0.352 \\ -0.352 & 0.502 & 1.56 \\ 0.21 & 1.13 & 0.502 \\ 0.291 & -0.418 & 0.706 \\ -0.352 & 0.502 & 1.56 \end{bmatrix}$$

$$\overline{A}\,[\overline{A}_t\overline{Y}_B\overline{A}]^{-1}\overline{A}_t = \begin{bmatrix} 1.13 & 0.21 & 0.502 & 1.13 & -0.418 & 0.502 \\ 0.21 & 0.853 & -0.352 & 0.21 & 0.291 & -0.352 \\ 0.502 & -0.352 & 1.56 & 0.502 & 0.706 & 1.56 \\ 1.13 & 0.21 & 0.502 & 1.13 & -0.418 & 0.502 \\ -0.418 & 0.291 & 0.706 & -0.418 & 1.415 & 0.706 \\ 0.502 & -0.352 & 1.56 & 0.502 & 0.706 & 1.56 \end{bmatrix}$$

Finally, $\overline{V}_B = \overline{A}\,[\overline{A}_t\overline{Y}_B\overline{A}]^{-1}\overline{A}_t\overline{I}_{BS}$, where I_{BS} is the array of branch source currents:

$$\overline{I}_{BS} = \begin{bmatrix} I_a \\ I_b \\ I_c \\ I_d \\ I_f \\ I_g \end{bmatrix} = \begin{bmatrix} 50 \\ 25 \\ 12.5 \\ 0 \\ 0 \\ 0 \end{bmatrix}$$

and

$$\overline{V}_B = \begin{bmatrix} 1.13 \times 50 + 0.21 \times 25 + 0.502 \times 12.5 \\ 0.21 \times 50 + 0.853 \times 25 - 0.352 \times 12.5 \\ 0.502 \times 50 - 0.352 \times 25 + 1.56 \times 12.5 \\ 1.13 \times 50 + 0.21 \times 25 + 0.502 \times 12.5 \\ -0.418 \times 50 + 0.291 \times 25 + 0.706 \times 12.5 \\ 0.502 \times 50 - 0.352 \times 25 + 1.56 \times 12.5 \end{bmatrix}$$

or

$$\begin{bmatrix} V_a \\ V_b \\ V_c \\ V_d \\ V_f \\ V_g \end{bmatrix} = \begin{bmatrix} 68.0 \\ 27.4 \\ 35.8 \\ 68.0 \\ -4.71 \\ 35.8 \end{bmatrix}$$

The results check with those of Example 5-4.

6-4. Systematic Formation of the $\overline{C}$ Matrix

We learned from Sec. 6-2 that the connection matrix $\overline{C}$ will transform an array of loop currents into an array of branch currents, or $\overline{I}_B = \overline{C}\,\overline{I}_L$. Therefore, the $\overline{C}$ matrix may be set up by inspection of the branch and loop current relationships as was done in the example of Sec. 6-2. We also know that the transpose of $\overline{C}$ will transform a branch-voltage matrix into a loop voltage matrix, or $\overline{E}_L = \overline{C}_t\overline{E}_B$. It should be realized that this $\overline{C}$ matrix is akin to another matrix which is often referred to in the literature as the *tie-set matrix*. This so-called tie-set matrix is generally set up in such a manner as to be identical with $\overline{C}_t$. In other words, where $\overline{C}$ is a branch (row)-loop (column) matrix, the tie-set matrix may be set up as a loop-branch matrix.

A systematic procedure is in order at this point. This procedure is summarized as follows:

1. Select a tree from the network graph.
2. Include links (dotted) and show loop currents associated with

each link, keeping in mind that each loop current flows through a given link. The link and tree branches of one loop are said to constitute a tie-set.

3. Orient the branches and loop currents of the graph by assigning directions to all branches and loops. While branch current directions may be arbitrarily chosen, some care should be taken to choose loop currents in the same direction as corresponding link currents. Some identifying scheme is also necessary. For example, branches might be lettered and loops numbered in order to keep columns and rows of the matrix straight.
4. Prepare the $\overline{C}$ matrix, row by row, with +1, −1, and 0 elements by relating branch and loop currents according to the equation $\overline{I}_B = \overline{C}\,\overline{I}_L$.

As an alternative to the row-by-row formation, $\overline{C}$ might also be set up column-by-column from the knowledge that $\overline{E}_L = \overline{C}_t\overline{E}_B$. For example, column 1 would relate voltages of loop 1 to branches of loop 1. An example problem will serve to demonstrate the procedure.

Example 6-3. Find the $\overline{C}$ matrix for the circuit whose graph is shown in Fig. 6-6 (a).

First a tree is chosen as per Fig. 6-6b. Next, the loop currents are shown and arrows are added to indicate assumed current directions for all branches and loops. Again note that loops 1, 2, 3, and 4 are all assumed in the same direction as links *f*, *g*, *h* and *l*. For the sake of clarity the branch-loop current equations will be written out separately before writing out the *C* matrix.

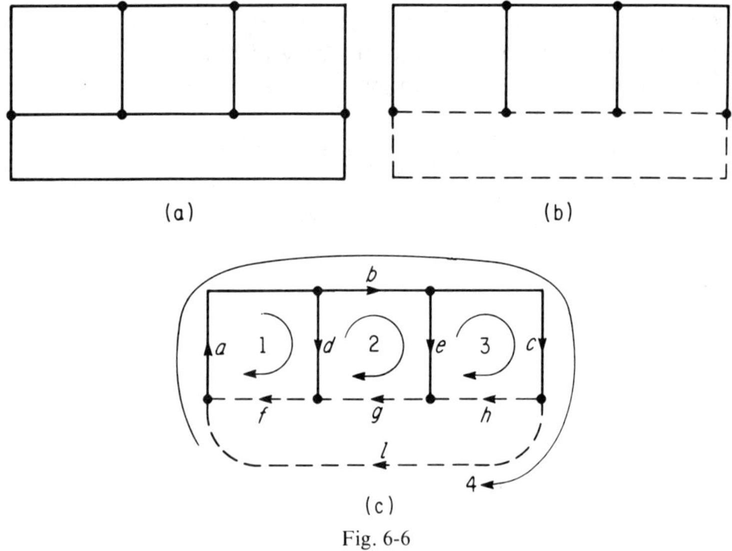

Fig. 6-6

$$I_a = I_1 + I_4 \qquad I_f = I_1$$
$$I_b = I_2 + I_4 \qquad I_g = I_2$$
$$I_c = I_3 + I_4 \qquad I_h = I_3$$
$$I_d = I_1 - I_2 \qquad I_l = I_4$$
$$I_e = I_2 - I_3$$

Since $\overline{I}_B = \overline{C}\,\overline{I}_L$, then

$\overline{C} =$

	1	2	3	4
a	1	0	0	1
b	0	1	0	1
c	0	0	1	1
d	1	−1	0	0
e	0	1	−1	0
f	1	0	0	0
g	0	1	0	0
h	0	0	1	0
l	0	0	0	1

The alternate means of forming the matrix by columns should be investigated. Notice that if one were to sum the scource voltages around loop 1 the result would be $E_1 = E_a + E_d + E_f$. In matrix form we say $\overline{E}_L = \overline{C}_t\overline{E}_B$. Notice that column 1 contains +1 in position *a*, *d*, and *f*. The $\overline{C}$ matrix can then be formed by rows from the branch-loop current relationships or by columns from the loop-branch voltage relationships. In either case, $\overline{C}$ would be easily written out by inspection.

6-5. Systematic Determination of the $\overline{A}$ Matrix

An example problem involving the $\overline{A}$ matrix was given in Sec. 6-3. This matrix can be formed by relating branch voltages to independent node-pair voltages or $\overline{V}_B = \overline{A}\,\overline{V}$. Another way to form the matrix is to realize that the source current array from various cut-sets are related to the branch source current array by the transpose of $\overline{A}$, current array by the transpose of $\overline{A}$, or $\overline{I}_s$, $= \overline{A}_t\overline{I}_{BS}$. In some literature, the transpose of $\overline{A}$ is termed the *cut-set matrix*.

A systematic procedure for finding $\overline{A}$ is offered.

1. Select a tree for the network graph and include links (dotted).
2. Draw cut-set lines, each line cutting through only one tree branch and each line separating the graph into two parts.
3. Orient the branches of the graph by assumed direction arrows. The orientation of each cut-set line will be taken in the same

direction as the tree branch which it intersects. Each branch and cut-set line will be identified by appropriate lettering or numbering.

4. Prepare the $\bar{A}$ matrix row by row of $+1$, -1, and 0 elements, by relating branch voltages to independent node-pair voltages, or $\bar{V}_B = \bar{A}\,\bar{V}$. Again these independent node pair voltages of $\bar{V}$ are identical with tree branch voltages.

As an alternate to the row-by-row procedure, the A matrix may also be formed column by column, realizing that the source-current array $(\bar{I}_s)$ for the cut-sets is related to the branch source-current array $(\bar{I}_{BS})$ by the transpose of $\bar{A}$, or $\bar{I}_s = \bar{A}_t\bar{I}_{BS}$.

Example 6-4. Find the $\bar{A}$ matrix for the graph of Fig. 6-7 with the chosen tree indicated.

Directions are assumed for each branch and cut-set lines drawn in Fig. 6-7b. Each cut-set line is lettered and oriented according to the tree

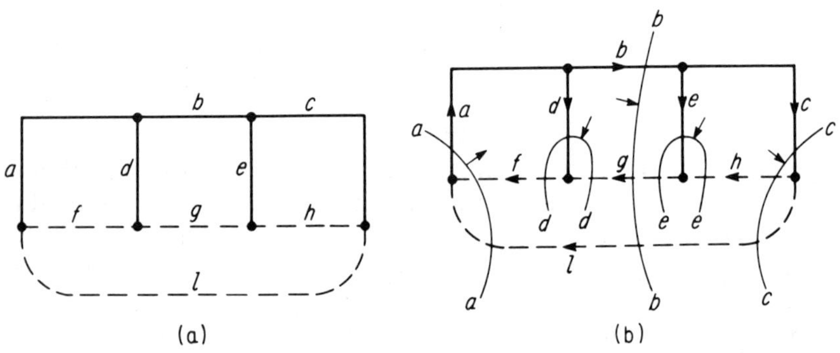

Fig. 6-7

branch which it intersects. For a better concept of the $\bar{A}$ matrix, the branch-node-pair voltage equations will be written out.

$$
\begin{aligned}
V_a &= V_a\\
V_b &= V_b\\
V_c &= V_c\\
V_d &= V_d\\
V_e &= V_e\\
V_f &= -V_d - V_a\\
V_g &= -V_e - V_b + V_d\\
V_h &= -V_c + V_e\\
V_l &= -V_c - V_b - V_a
\end{aligned}
\qquad
\underbrace{\begin{bmatrix} V_a\\ V_b\\ V_c\\ V_d\\ V_e\\ V_f\\ V_g\\ V_h\\ V_l \end{bmatrix}}_{\bar{V}_B}
=
\underbrace{\begin{bmatrix}
1 & 0 & 0 & 0 & 0\\
0 & 1 & 0 & 0 & 0\\
0 & 0 & 1 & 0 & 0\\
0 & 0 & 0 & 1 & 0\\
0 & 0 & 0 & 0 & 1\\
-1 & 0 & 0 & -1 & 0\\
0 & -1 & 0 & 1 & -1\\
0 & 0 & -1 & 0 & 1\\
-1 & -1 & -1 & 0 & 0
\end{bmatrix}}_{\bar{A}}
\underbrace{\begin{bmatrix} V_a\\ V_b\\ V_c\\ V_d\\ V_e \end{bmatrix}}_{\bar{V}}
$$

Actually the above steps may be bypassed once the graph is oriented. The $\overline{A}$ matrix is written out in one step by inspection of the graph. It is of some help to label rows and columns. The resulting $\overline{A}$ matrix is obviously

$$
\overline{A} = \begin{array}{c} \\ a \\ b \\ c \\ d \\ e \\ f \\ g \\ h \\ l \end{array}
\begin{array}{c} \begin{array}{ccccc} aa & bb & cc & dd & ee \end{array} \\
\begin{bmatrix} 1 & 0 & 0 & 0 & 0 \\ 0 & 1 & 0 & 0 & 0 \\ 0 & 0 & 1 & 0 & 0 \\ 0 & 0 & 0 & 1 & 0 \\ 0 & 0 & 0 & 0 & 1 \\ -1 & 0 & 0 & -1 & 0 \\ 0 & -1 & 0 & 1 & -1 \\ 0 & 0 & -1 & 0 & 1 \\ -1 & -1 & -1 & 0 & 0 \end{bmatrix} \end{array}
$$

The alternate formation (column-by-column) of $\overline{A}$ is readily accomplished from the knowledge of $\overline{I}_s = \overline{A}_t \overline{I}_{BS}$. If one were to sum the source currents entering each of the cut-set lines, the result would be

$$
\begin{aligned}
I_{aa} &= I_a - I_f - I_l \\
I_{bb} &= I_b - I_g - I_l \\
I_{cc} &= I_c - I_h - I_l \\
I_{dd} &= I_d - I_f + I_g \\
I_{ee} &= I_e - I_g + I_h
\end{aligned}
$$

$\overline{A}_t =$

	a	b	c	d	e	f	g	h	l
aa	1					-1			-1
bb		1					-1		-1
cc			1					-1	-1
dd				1		-1	1		
ee					1		-1	1	

The value of $\overline{A}_t$ checks with the previous result where the rows of $\overline{A}_t$ are merely the columns of $\overline{A}$. This means that $\overline{A}$ could be readily obtained directly by columns by noting the presence and direction of branches intersecting the cut-set lines.

6-6. General Application

The preceding coverage of transformation methods has emphasized two applications, namely the loop-current-to-branch-current transformation with connection matrix $\overline{C}$, and the node-pair-voltage-to-branch-voltage transformation with connection matrix $\overline{A}$. However, it should be understood that these transformation methods are by no means limited to the reference frames mentioned. Whenever any "old" system is to be transformed to a "new" system, with total real and reactive power being invariant, similar transformation equations apply. For just such an application the author refers the reader to Sec. 13-8 and Eqs. 13-37 through 13-39 for application to the economic dispatch problem.

Problems

6-1. Given the problem of Example 6-1, only with a different choice of loop currents as indicated in Prob. 6-1. Making use of the new connection matrix $\overline{C}$, solve for all branch currents using the branch-impedance transformation method.

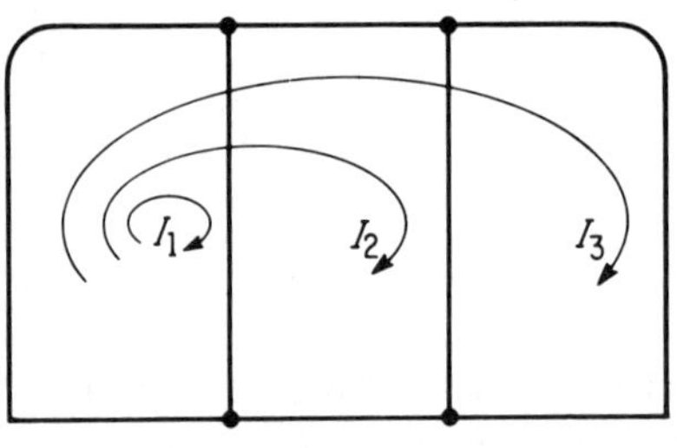

Prob. 6-1

6-2. In the circuit of Fig. 5-2, find the connection matrix $\overline{C}$ which relates branch currents to loop currents where loop currents are as shown in Fig. 5-3a.

6-3. Repeat Prob. 6-2 only where loop currents are as chosen in Fig. 5-3b.

6-4. Refer to Fig. 6-5. Choose a tree (other than that of Fig. 6-5c) which includes branches *a*, *b*, and *c*. Now find the connection matrix $\overline{A}$ for the branch-admittance transformation method.

6-5. Making use of the $\overline{A}$ matrix of Prob. 6-4, solve for all branch voltages using the branch-admittance transformation method.

6-6. Repeat Prob. 6-4, only with a tree which includes branches *d*, *b*, and *c*.

6-7. Find the $\overline{A}$ matrix for the graph of Fig. 6-6a where the chosen tree includes branches *a*, *b*, *f*, *g* and *l*. Label all cut-set lines and assumed directions clearly.

6-8. Repeat Prob. 6-7 with a tree which includes *f*, *g*, *h*, *d*, and *e*.

chapter 7

SYMMETRICAL FAULT STUDIES

7-1. Introduction

Faults in a three-phase system can be classified under the following headings:

1. Balanced three-phase faults
2. Single line-to-ground faults
3. Line-to-line faults
4. Double line-to-ground faults

About three faults in four are of the line-to-ground variety, and most of these are a result of insulator flashover during electrical storms. The balanced three-phase fault is the rarest in occurrence, accounting for only about 5 percent of the total, yet it is this fault which will be dealt with here. It is the least complex of all types of short-circuit studies insofar as the calculations are concerned. The last three unsymmetrical studies will require the knowledge and use of the tools of symmetrical components (treated in Chapter 8) or some other appropriate method of study.

Protective relays are employed to trip the circuit breakers under certain short-circuit conditions. The three-phase symmetrical fault could, in some locations, yield the smallest current of any type of fault. If one is to accommodate this rare type of fault, relays must be set accordingly. However, the interrupting capacity and momentary duty of the breakers will more likely be chosen to accommodate the largest of fault currents. It will be helpful to first gain a better understanding of some of the transient considerations of a generator undergoing a balanced short circuit.

7-2. Generator Transients During a Three-Phase Fault (Assuming Constant Inductance)

When a machine is generating a sinusoidal voltage, it is a known fact that the value of short-circuit current in the transient period will be partially dependent upon the instant in the cycle at which the short occurs. Perhaps the reader will recall in the study of a transformer's tran-

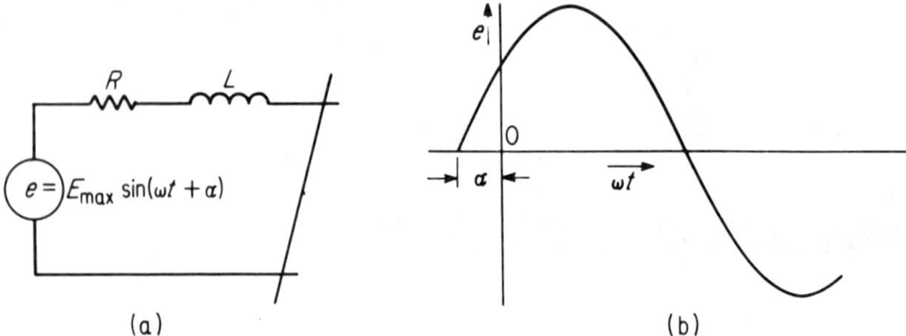

Fig. 7-1. Per phase representation of synchronous generator with balanced fault. Inductance is assumed constant.

sient magnetizing current that the inrush current was likewise a function of the time at which the primary switch was closed. If, in the case of the transformer, the switch was closed at the instant when the voltage was maximum, the initial current behaves much as the steady state current, right from the start. However, should the primary switch be closed at the instant when the voltage wave was passing through zero, the initial core flux would be twice the steady-state maximum value, which (with saturation effects) could demand a tremendous inrush current. Since it is impossible to predict at which instant a generator fault might occur, one must be prepared to deal with the worst situation. The mathematical proof for the above observation, regarding a shorted generator, is given as follows. Assume the balanced fault is placed upon the generator of Fig. 7-1 at time 0 with the generator initially on open circuit.

The instantaneous voltage equation for the circuit is

$$L\frac{di}{dt} + Ri = E_m \sin(\omega t + \alpha) \tag{7-1}$$

$$= E_m(\sin \omega t \cos \alpha + \sin \alpha \cos \omega t)$$

Taking the Laplace transform of the equation,

$$(Ls + R)I(s) = E_m\left[\cos \alpha\left(\frac{\omega}{s^2 + \omega^2}\right) + \sin \alpha\left(\frac{s}{s^2 + \omega^2}\right)\right]$$

$$I(s) = E_m\left[\frac{s(\sin \alpha) + \omega(\cos \alpha)}{(s^2 + \omega^2)(Ls + R)}\right]$$

$$I(s) = \frac{E_m}{L}\left[\frac{As + B}{s^2 + \omega^2} + \frac{C}{s + R/L}\right] \tag{7-2}$$

The partial-fraction form of the Laplace equation shown in Eq. 7-2 will be convenient for obtaining the time solution from the Laplace tables. The constants A, B, and C can be readily obtained and verification of these values will be left as a student exercise. They are

$$A = \frac{L}{Z}\sin(\alpha - \theta), \qquad B = \frac{\omega L}{Z}\cos(\alpha - \theta)$$

$$C = \frac{L}{Z}\sin(\alpha - \theta), \qquad Z = \sqrt{R^2 + \omega^2 L^2}, \qquad \text{and} \qquad \theta = \tan^{-1}\frac{\omega L}{R}$$

Substituting A, B and C into Eq. 7-2,

$$I(s) = \frac{E_m}{L}\left[\frac{s\,\frac{L}{Z}\sin(\alpha - \theta)}{s^2 + \omega^2} + \frac{\frac{\omega L}{Z}\cos(\alpha - \theta)}{s^2 + \omega^2} - \frac{\frac{L}{Z}\sin(\alpha - \theta)}{s + R/L}\right]$$

$$= \frac{E_m}{Z}\left[\sin(\alpha - \theta)\left(\frac{s}{s^2 + \omega^2}\right) + \cos(\alpha - \theta)\left(\frac{\omega}{s^2 + \omega^2}\right)\ \frac{\sin(\alpha - \theta)}{s + R/L}\right]$$

Obtaining the inverse transform for the time solution,

$$i(t) = \frac{E_m}{Z}\left[\sin(\alpha - \theta)\cos\omega t + \cos(\alpha - \theta)\sin\omega t - \sin(\alpha - \theta)e^{-Rt/L}\right]$$

Collecting sine and cosine terms,

$$i(t) = \frac{E_m}{Z}\left[\sin(\omega t + \alpha - \theta) - \sin(\alpha - \theta)e^{-Rt/L}\right] \qquad (7\text{-}3)$$

Now we have verified the statement that the transient current will depend to some degree upon the angle (α) of the voltage wave at the instant ($t = 0$) of the short circuit. For example, a unidirectional or d-c transient term will, in general, exist. Its magnitude at $t = 0$ may be as great as the magnitude of the steady-state current term. Suppose that at $t = 0$, the angle $(\alpha - \theta) = -90°$. The current waveshape would appear as seen in Fig. 7-2a. This would represent the worst possible transient situation where the current magnitude immediately following the short will approach twice the steady-state maximum value. If, however, $\alpha = \theta$ at $t = 0$, then the current waveshape follows the steady-state solution from the start, as depicted in Fig. 7-2b.

It must be realized that the current waveshapes of Fig. 7-2 are primarily of academic importance since we have assumed a constant L for the generator. In actual practice such will not be the case. However, even with our assumption we have succeeded in determining the general nature of the troublesome unidirectional current component, which must be considered in the selection of the circuit breaker. In practice it is

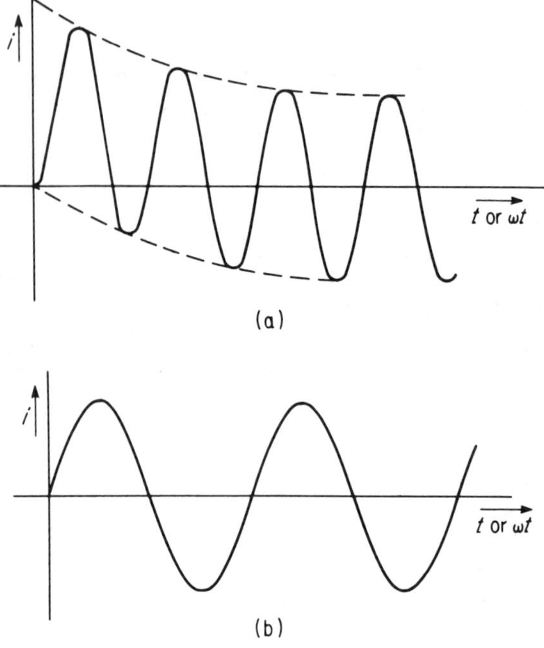

Fig. 7-2. (a) Short-circuit current waveshape for $\alpha - \theta = -90°$. (b) Short-circuit current waveshape for $\alpha = \theta$.

customary to first neglect the transient component in the calculation but consider the changing reactance following the fault. Then upon completion of the fault calculation the possible effect of unidirectional current will later be added by applying a multiplying factor to the switch gear duty. This factor will vary from 1.0 to 1.6, depending upon the type of breaker and its speed of operation. While it is true that the factor is an approximation, one might appreciate the fact that an exact consideration could prove most difficult in the actual case. This is especially true when it is realized that the angle α will be different in all three phases because of the 120° phase shift between generated voltages.

7-3. Symmetrical Faults on Synchronous Generators— Transient Reactances Considered and Unidirectional Currents Neglected

It was pointed out in Chapter 4 that the so-called synchronous reactance X_s of an alternator includes the combined effect of leakage reactance X_a and the reactance X_{AR} of armature reaction. Furthermore, for the salient-pole machine, X_s may often be replaced by the direct-axis synchronous reactance X_d. Justification for this, under short-circuit conditions, was given in Sec. 4-2.

The component X_{AR} of X_s might vary with saturation. However, under shorted conditions the effects of lagging armature mmf and field mmf are in such opposition as to make for an unsaturated situation. Under transient conditions new complications arise insofar as X_s (or X_d) is concerned. These complications are caused primarily by two factors. First the armature reaction flux (due to a rapidly changing armature current) will encounter a certain opposition to change from the built-in damper windings, should damper windings be present. It is recalled that damper windings in the pole faces of an alternator are often used (1) to aid in starting, and (2) to reduce the effects of hunting.

The second obstacle that the changing armature flux encounters as it attempts to make its way through the rotor poles is seen in the highly inductive field winding. Again the tendency is to increase the reluctance of the flux path, thereby decreasing the armature reactance at this stage.

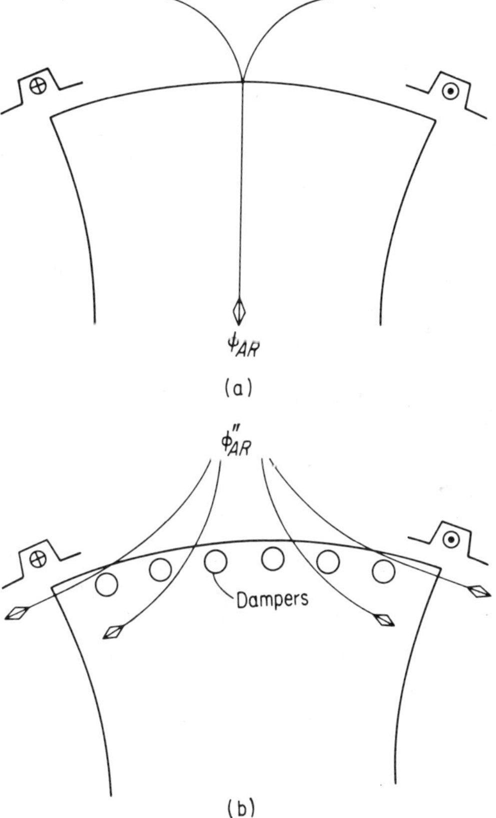

Fig. 7-3. (a) Steady-state path for $\Phi_{AR} \cdot X_d$ is high. (b) Subtransient effect of damper windings $X''_d < X'_d < X_d$.

Pictured in Fig. 7-3a is the normal steady-state path for the flux due to a lagging armature current. Short-circuit current will be essentially lagging in a system where $X \gg R$. Note a relatively low reluctance path and therefore the reactance of armature reaction $X_{d(AR)}$ will be high, meaning X_d for the steady state will be high, since $X_d = X_{d(AR)} + X_{d(a)}$, where $X_{d(a)}$ is the leakage reactance of the armature.

Now refer to Fig. 7-3b, where it is seen that a sudden increase in flux linkages is opposed by the shorted damper windings and, in fact, Φ''_{AR} is forced out of the pole to some extent, thereby increasing reluctance and yielding a reduced direct axis synchronous reactance X''_d.

Moving next to Fig. 7-4a, it is obvious that as time progresses and Φ'_{AR} moves on through the pole, the highly inductive field circuit will oppose the increase of flux inside of its turns. This yields a higher reactance

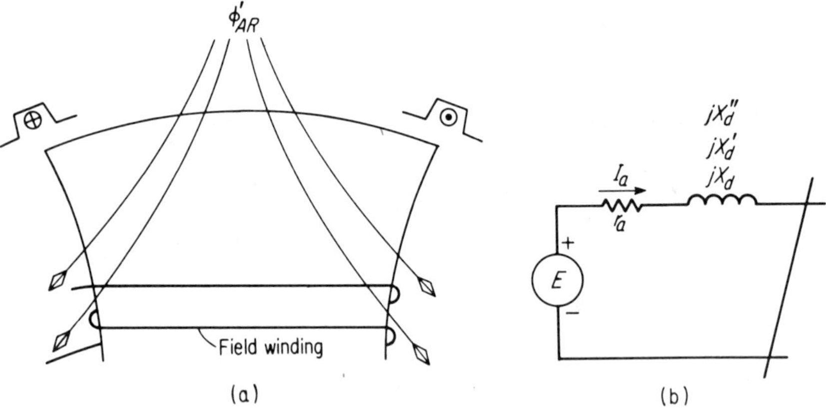

Fig. 7-4. Transient effect of field winding, $X'_d < X_d$.

X'_d than the first subtransient condition and yet lower than that of the steady-state condition or $X''_d < X'_d < X_d$. The relative magnitudes of these reactances will depend upon the individual machine designs. Various tests are possible for their determination. Figure 7-4b is a synchronous reactance schematic for approximating short-circuit conditions where the choice of X''_d, X'_d, or X_d will depend upon the time (after the fault) at which the study is to be made. Certainly the choice of X''_d would represent the first and worst condition.

The previous concept of the changing reactance is further justified when we view an oscillogram of the current in one phase of a synchronous generator undergoing a short circuit. Eliminate the unidirectional current by letting $\alpha = \theta$ for one phase. Figure 7-5 illustrates a typical current waveshape with three envelopes in evidence. In accordance with the foregoing theory of changing synchronous reactance, we can conclude that for

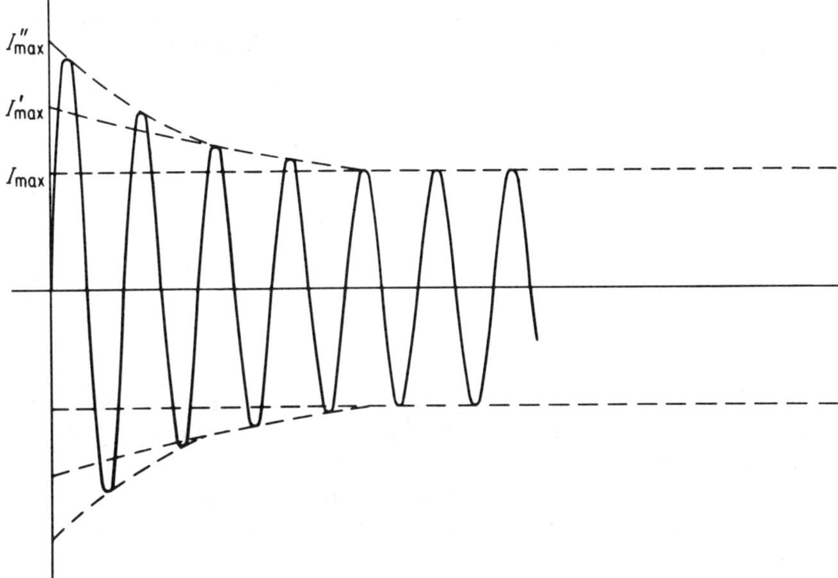

Fig. 7-5. Short-circuit current for one phase of a synchronous generator (with unidirectional current neglected).

the machine with damper windings, the subtransient reactance is

$$X''_d = E_{max}/I''_{max} \tag{7-4}$$

Likewise the effect of the field winding upon the transient reactance can be determined as

$$X'_d = E_{max}/I'_{max} \tag{7-5}$$

Finally, the steady-state reactance is

$$X_d = E_{max}/I_{max} \tag{7-6}$$

In each case above it was assumed that $X \gg r_a$ and therefore $X \doteq Z$.

Notice that the E_{max} of Eqs. 7-4, 7-5, and 7-6 are the same. This is the *no-load* line-to-neutral voltage of the machine, where the rms value is termed E_g. When a fault occurs on the terminals of an *unloaded* machine, the rms values of subtransient current (I''), transient current (I') and steady state current (I) are merely found as

$$I'' = \frac{E_g}{X''_d}, \quad I' = \frac{E_g}{X'_d}, \quad I = \frac{E_g}{X_d} \tag{7-7}$$

Example 7-1. A 65-mva, Y-connected, 15.5-kv synchronous generator is connected to a 15.5-kv to 120-kv, 65-mva, Δ-Y transformer. The subtransient reactance X''_d of the machine is 0.12 pu on a 65-mva base while the transformer reactance is 0.10 pu on the same base. The machine is unloaded when a symmetrical fault is suddenly placed at point f as shown in Fig. 7-6.

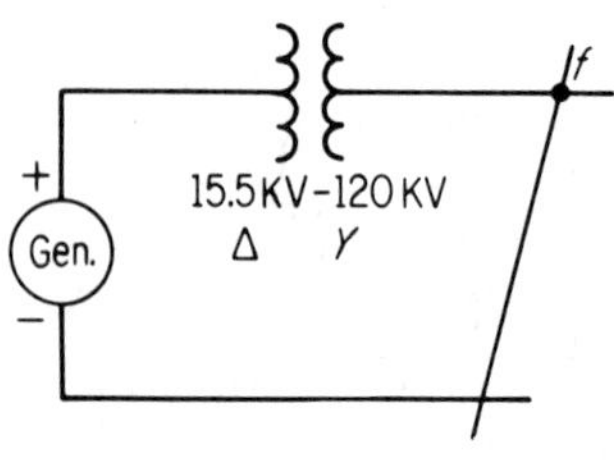

Fig. 7-6

(a) Find the subtransient symmetrical fault current in per unit amperes and actual amperes on both sides of the transformer. E_g of the generator = 1.0 pu.

(b) What is the maximum value possible of the d-c current component and what does the total instantaneous current approach as a maximum in pu?

Solution (a)

$$\text{pu } I_f'' = \frac{E_g}{X_d'' + X_{tr}}$$

$$= \frac{1.0}{j0.12 + j0.10}$$

$$= \frac{1.0}{j0.22} = \underline{\underline{4.54 \angle -90^\circ}} \text{ pu amp}$$

$$\text{Base amp (l-v side)} = \frac{\text{Base } VA}{\sqrt{3} \text{ Base } V} = \frac{65 \times 10^6}{\sqrt{3} \times 15.5 \times 10^3}$$

$$= 2{,}420 \text{ amp}$$

$$\text{Base amp (h-v side)} = \frac{65 \times 10^6}{\sqrt{3} \times 120 \times 10^3}$$

$$= 313 \text{ amp}$$

$$\text{Actual } I_f'' \text{ on 15.5-kv side} = \text{pu } I_f'' \times \text{base } I$$

$$= 4.54 \times 2420$$

$$= \underline{\underline{10{,}990}} \text{ amp}$$

$$\text{Actual } I_f'' \text{ on 120-kv side} = 4.54 \times 313$$

$$= \underline{\underline{1420}} \text{ amp}$$

Solution (b)

$$\text{pu } I_{d\text{-}c}\,(\text{max}) = \sqrt{2} \times \text{rms value of } I_f''$$
$$= \sqrt{2} \times 4.54 = \underline{\underline{6.43}} \text{ pu amp}$$

Max instantaneous $I\,(d\text{-}c + \text{symmetrical})$

$$\text{Max } I_{\text{tot}} = 2 \times \sqrt{2} \times I_f''$$
$$= \underline{\underline{12.86}} \text{ pu amp}$$

Equation 7-7 is satisfactory where the prefault load current either does not exist or where it is sufficiently small with respect to fault current as to be neglected. However, where prefault load currents are significant, either of two approaches might be used in the solution of symmetrical fault currents. These approaches yield identical results.

(a) Use of Thévenin's voltage (E_{thev}) and impedance (Z_{thev}) at the point of fault. Using this simple series equivalent, fault current is readily obtained. As an example, refer to the simple per phase diagram of Fig. 7-7a where a symmetrical fault is to be applied at f. Here the Thévenin's

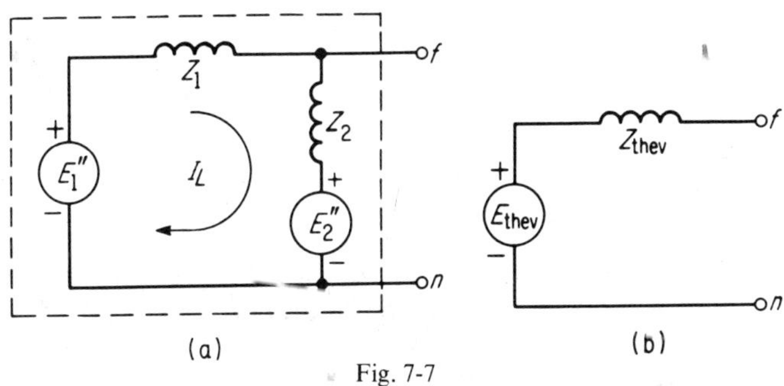

Fig. 7-7

voltage is the voltage from n to f before the fault occurs. This voltage is placed in series with the Thévenin's impedance as viewed from the fault point, or

$$Z_{\text{thev}} = \frac{Z_1 Z_2}{Z_1 + Z_2} \tag{7-8}$$

$$I_f'' = \frac{E_{\text{thev}}}{Z_{\text{thev}}} \tag{7-9}$$

Once I_f'' is found, the individual contributions of fault current may be determined in the branches of Fig. 7-7a.

$$I_{f1}'' = \frac{Z_2}{Z_1 + Z_2} I_f'' \tag{7-10}$$

$$I''_{f2} = \frac{Z_1}{Z_1 + Z_2} I''_f \tag{7-11}$$

While I''_f is total fault current, the values of I''_{f1} and I''_{f2} do not as yet include the prefault load currents by the Thévenin's method. Revising these branch currents to include the prefault load current (I_L) yields

$$I''_1 = I''_{f1} + I_L \tag{7-12}$$

$$I''_2 = I''_{f2} - I_L \tag{7-13}$$

(b) Use of machine internal voltages behind subtransient reactance. These voltages are equal to Thévenin's voltage only when no prefault load currents are flowing in the circuit. When prefault load current (I_L) does flow, three separate internal machine voltages may be visualized: the voltage behind synchronous reactance E, the voltage behind transient reactance E', and the voltage behind subtransient reactance E''.

Consider the loaded machine of Fig. 7-8a. The three possible internal voltages are depicted on the phasor diagram of Fig. 7-8b. The

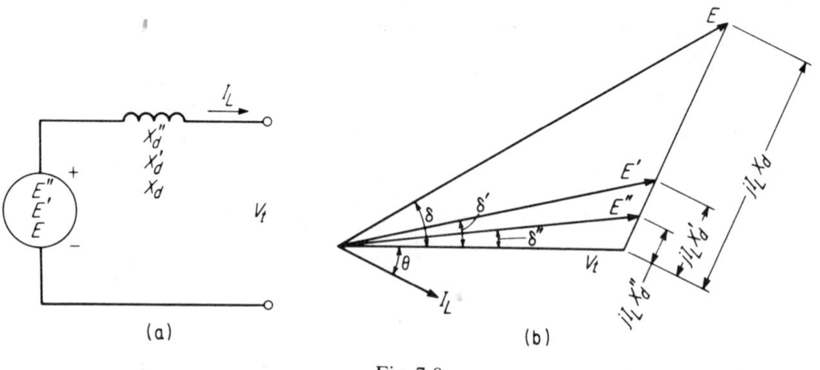

Fig. 7-8

proper choice of internal voltage depends upon the time elapsed since the fault. These internal voltages are

$$E'' = V_t + jI_L X''_d \tag{7-14}$$

$$E' = V_t + jI_L X'_d \tag{7-15}$$

$$E = V_t + jI_L X_d \tag{7-16}$$

For calculations immediately following the fault, we are primarily concerned with the subtransient voltage E''. Of course, for an *unloaded* machine, $E'' = E' = E$, which also equals the E_g of Eq. 7-7.

To demonstrate the fault calculation of method (b) for Fig. 7-7a, the subtransient voltages E''_1 and E''_2 corresponding to the given load current

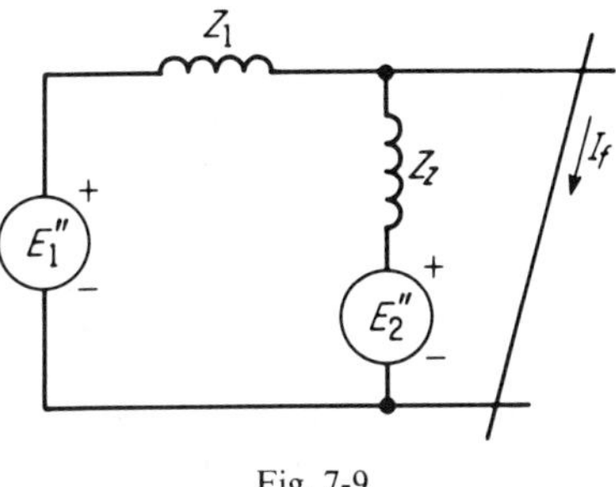

Fig. 7-9

will first be determined. The total currents (both fault and load components) for both branches may then be calculated directly from the circuit of Fig. 7-9 as

$$I_1'' = \frac{E_1''}{Z_1}, \quad I_2'' = \frac{E_2''}{Z_2} \tag{7-17}$$

and

$$I_f = I_1'' + I_2'' \tag{7-18}$$

The I_1'' and I_2'' of Eq. 7-17 should agree with the currents of Eqs. 7-12 and 7-13 found by the Thévenin's method (plus load component superposition).

Example 7-2. Given the machine of Example 7-1. A three-phase balanced load impedance (also 65-mva-base) of $0.8 + j0.6$ pu ohms is placed across the transformer 120-kv terminals. Later a three-phase fault is also placed at this point as shown in Fig. 7-10a. Figure 7-10b is the resulting impedance diagram.

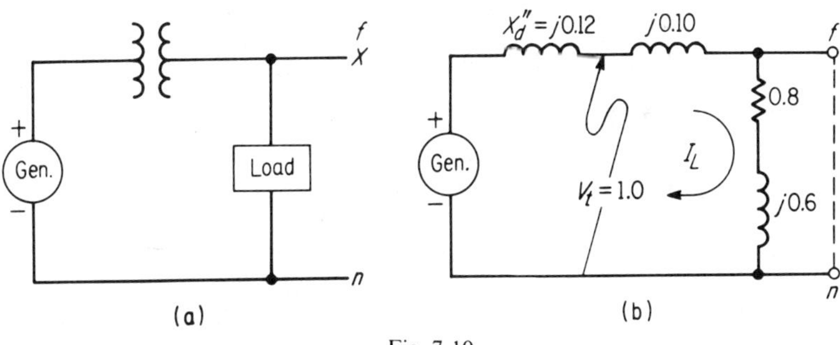

Fig. 7-10

Find the subtransient fault current and the subtransient generator current (load component included) using Thévenin's values at the point of fault [method (a)]. Before the fault, the generator terminal voltage was 1.0 pu volts.

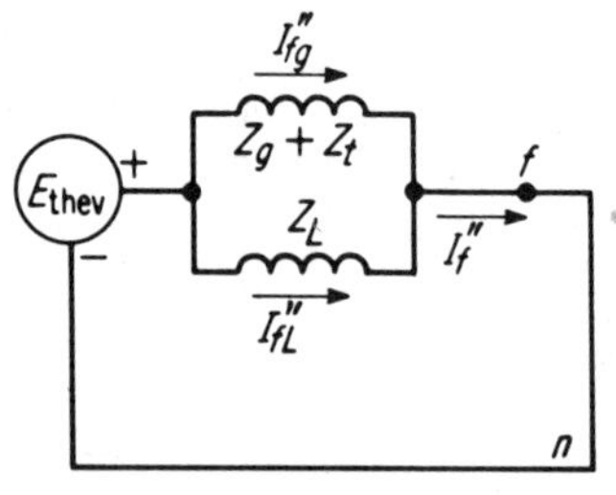

Fig. 7-11

Solution. The prefault load current referred to generator terminal voltage of 1.0 /0° is

$$I_L = \frac{1.0\ \underline{/0^\circ}}{0.8 + j(0.1 + 0.6)} = \frac{1.0\ \underline{/0^\circ}}{1.062\ \underline{/41.2^\circ}} = 0.94\ \underline{/-41.2^\circ}$$

$$E_{\text{thev}} = E_{nf} = V_t - I_L \underline{/Z_{\text{trans}}}$$

$$= 1.0\ \underline{/0^\circ} - 0.94\ \underline{/-41.2^\circ}\,(j0.10)$$

$$= 1.0\ \underline{/0^\circ} - 0.094\ \underline{/48.8^\circ}$$

$$= 1.0 - 0.0618 - j0.0707$$

$$= 0.9382 - j0.0707 = \underline{\underline{0.941\ /-4.32^\circ}}$$

$$Z_{\text{thev}} = \frac{(j0.12 + j0.10)(0.8 + j0.6)}{j0.12 + j0.10 + 0.8 + j0.6}$$

$$= \frac{-.132 + j.176}{0.8 + j0.82} = \frac{0.22\ \underline{/126.9^\circ}}{1.145\ \underline{/45.7^\circ}}$$

$$= \underline{\underline{0.192\ /81.2^\circ}}$$

$$I''_f = \frac{E_{\text{thev}}}{Z_{\text{thev}}} = \frac{0.941\ \underline{/-4.32^\circ}}{0.192\ \underline{/81.2^\circ}} = 4.90\ \underline{/-85.5^\circ}\ \text{pu}$$

The fault contribution

$$I''_{fg} = I''_f \frac{Z_L}{Z_L + Z_g + Z_{tr}}$$

$$= 4.90\ \underline{/-85.5^\circ} \left(\frac{0.8 + j0.6}{0.8 + j0.6 + j0.22} \right)$$

$$I''_{fg} = \frac{4.90\ \underline{/-85.5^\circ}(1.0\ \underline{/36.9^\circ})}{1.145\ \underline{/45.7^\circ}} = 4.28\ \underline{/-94.3^\circ}$$

Now, superimposing the load component upon the fault component yields

$$
\begin{aligned}
I_g'' &= I_{fg}'' + I_L = 4.28 \angle -94.3^\circ + 0.94 \angle -41.2^\circ \\
&= -0.321 - j4.27 + 0.707 - j0.619 \\
&= 0.386 - j4.89 = 4.90 \angle -85.5^\circ
\end{aligned}
$$

Note: The angle on I_g'' was referred to the terminal voltage before the fault.

Example 7-3. Repeat Example 7-2 using method (b) of this section

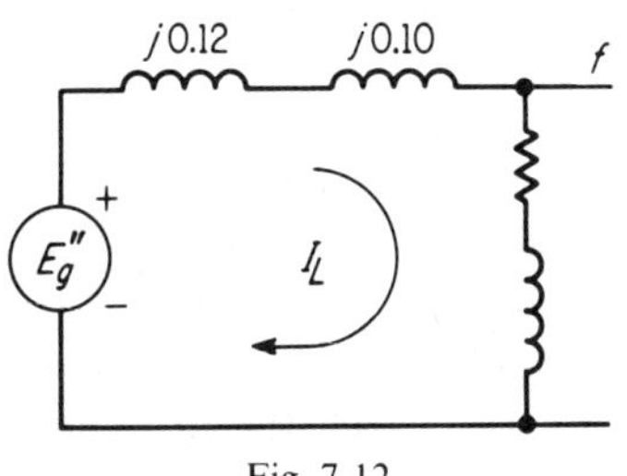

Fig. 7-12

(the internal voltage behind subtransient reactance). Refer to Fig. 7-12.

$$
\begin{aligned}
E_g'' &= V_t + I_L X_d'' \\
&= 1.0 + 0.94 \angle -41.2 \, (0.12 \angle 90^\circ) \\
&= 1.0 + 0.1126 \angle 48.8^\circ \\
&= 1.0 + 0.074 + j0.0846 \\
&= 1.074 + j0.0846
\end{aligned}
$$

$$
I_f'' = I_g'' = \frac{E_g''}{jX_d'' + jX_{tr}} = \frac{1.074 + j0.0846}{j0.22} = \underline{\underline{0.385 - j4.89}} \text{ pu}
$$

Results check with Example 7.2.

7-4. Interrupting the Fault

Having obtained the concept of three separate generator reactances corresponding to three different periods after the fault, some discretion must be used in the selection of X_d'', X_d' or X_d. Some general observations can be made in this regard although procedure may vary somewhat with various organizations. For example, in the determination of initial fault currents (within one half-cycle of the instant of fault), subtransient reactances should be used for both synchronous motors and generators. It is, of course, understood that the generated emf's of synchronous and induction motors will also contribute to the fault current. This initial or subtransient current is necessary in determining the so-called *momentary duty* of the circuit breaker. This momentary duty is not to be confused with the interrupting capacity of the breaker. Initial rms fault-current calculation for a particular point in the network is first necessary using subtransient reactances but neglecting the unidirectional current component discussed previously. Next a multiplying factor of 1.6 (for breakers above 5 kv) is applied to obtain the momentary duty of the breaker. This factor is reduced for voltages under 5 kv and, in any case, will be an approximate correction for the unidirectional current component. Circuits of 5,000 volts or less will require a factor of 1.5 (for momentary duty considerations) while air breakers of 600 volts or less require a factor of 1.25.

Calculation of the *interrupting* capacity of a circuit breaker will normally make use of subtransient reactance for synchronous generators and transient reactance for synchronous motors. As for the multiplying factor used on the rms-calculated current, this will depend upon the speed of the breaker, the ratio of X to R in the circuit, the distance between fault and generating station, etc. In general, the further the fault from the station, the smaller X/R becomes, the quicker the unidirectional component will decay, and therefore the smaller the multiplying factor. Of course, the faster the breaker, the higher the multiplying factor, other things being equal. For example, a two-cycle circuit breaker might require a factor of 1.4, whereas with an eight-cycle breaker a factor of 1.0 would be sufficient. This text will not attempt to furnish tables and detailed procedure in this regard.

7-5. Resume of a Symmetrical Fault Study

To summarize the procedure for three-phase symmetrical fault calculations, it may be said that a typical solution is carried out in the following manner:

1. Choose the base kva and represent the system on a per phase, per unit basis as outlined in Chapter 4, being careful to choose proper generator reactances (X''_d, X'_d, or X_d).
2. Make simplifying assumptions, where possible, of the type listed in Sec. 4-11. These might include such assumptions as: equal generator voltage magnitudes and phase angles, the disregarding of load currents (except for synchronous motors), the disregarding of transformer magnetizing currents, and deletion of shunt-line capacitance.
3. Proceed with the fault-current calculations using one of several methods.
   (a) Longhand calculation as described in problem Example 4-1.
   (b) D-c or a-c analyzer board methods, also described in Chapter 4.

(c) Digital-computer approach. Chapter 10 is devoted entirely to one of the widely used digital methods today for short circuit studies employing the "short-circuit matrix."

4. Information of interest generally includes the subtransient and/or transient currents in the network, the momentary duty of the breakers and the interrupting capacity of the breakers. The immediate results of the actual calculation would most likely yield the subtransient and/or transient currents. Multiplying factors would then be applied to allow for the possibility of an additional unidirectional transient current, as discussed in the preceding section.

At this point the reader should refer to Example 4-1 to review the calculating procedure. One could, if necessary, go a step further and include the effects of prefault load current, depending upon the accuracy desired.

Problems

7-1. Complete the steps required to determine the constants A, B, and C in Eq. 7-2.

7-2. Given a series R-L circuit with a sudden 60 Hz, sinusoidal voltage applied, where $R = 20$ ohms, $L = 0.3$ henry, and $E_{rms} = 120$ volts.

(a) The switch is closed at such a time as to permit maximum transient current. What is the instantaneous value of E upon closing the switch?

(b) What maximum value does the current approach in part (a)?

(c) What is the *actual* maximum value of current for part (b) and at what time does this occur after the switch is closed?

(d) Now let the switch be closed so as to yield minimum transient current. What instantaneous values of E and α correspond to this instant of closing the switch?

7-3. A synchronous generator delivers 1.0 pu current at 0.9 pf lagging (with respect to its own terminal voltage) and on its own base of 100 mva and 15.5 kv. The generator pu impedances are $X_d'' = 0.1$, $X_d' = 0.2$, $X_d = 1.4$. The generator terminal is at 1.0 pu volts. Consider the generator as feeding into a passive load impedance through a step-up transformer (15.5 kv to 120 kv). The transformer impedance is $j0.10$ pu ohms, also on a 100 mva base. A three-phase symmetrical fault is suddenly placed on the transformer secondary (120-kv side).

(a) What is the pu load impedance?

(b) Find the subtransient rms current in the fault and in the generator using Thévenin's method.

7-4. Repeat Prob. 7-3b using the voltages behind subtransient reactance.

7-5. Suppose the generator and transformer of Prob. 7-3 are unloaded with a terminal voltage of 1.0 pu on the secondary before the fault occurs.

(a) What is the value of subtransient rms fault current?
(b) What is the value of sustained rms fault current?
(c) What instantaneous value will the fault current approach as a maximum with full d-c offset?

7-6. What is the value of sustained fault current in the fault of Prob. 7-3?

7-7. Assume the generator (with transformer) of Prob. 7-3 is unloaded and is to be synchronized and paralleled to an infinite bus. Assume three of the four requirements for paralleling have been met, those of matching voltage magnitudes, frequency and phase sequence. However, assume the synchronizing procedure fails to match voltage phase angles. What out-of-phase condition would yield the highest transient current in the machine and transformer? What approximate value of subtransient, symmetrical rms current would you expect for this worst case of out-of-phase synchronizing?

7-8. The three-phase generator represented in Prob. 7-8 is rated at 60 mva and 6.9 kv with $X_d'' = 0.15$ pu ohms on its own base. The line between a and b is $j0.1$ pu on this same base. The motor is rated at 10 mva and 6.9 kv with

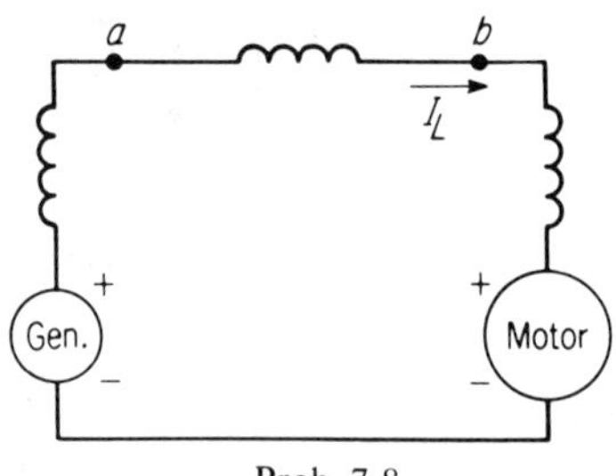

Prob. 7-8

$X_d'' = 0.2$ on the motor's base. The voltage at point b (the motor terminals) is 1.0 pu volts and the motor takes a load current I_L of 1.0 pu ampere (on the motor's base) at unity power factor. A symmetrical fault occurs at the generator terminals (point a). Determine the subtransient rms current at the fault, in the generator and in the motor. Use Thévenin's method.

7-9. Repeat Prob. 7-8 using the voltages E_g'' and E_m'' behind subtransient reactances.

chapter **8**

SYMMETRICAL COMPONENTS AND THE UNBALANCED FAULT

8-1. Introduction

Section 7-1 pointed out the various types of faults possible in the three-phase system. In Chapter 7 we dealt in particular with the first and simpler type of balanced fault which lends itself to a simple per phase, per unit approach. Reference might also be made to Example 4-1 for the balanced fault. However, things can become quite involved when we move to a fault of unbalanced nature. For example, if one were to attempt a three-phase mesh-current solution for the large unbalanced network, the labor involved would be quite prohibitive. Various approaches to the unbalanced (unsymmetrical) problem have been devised, but the most widely accepted calculation method involves the use of symmetrical components. This method was developed in 1918 by Dr. C. L. Fortescue. Whether one is to solve problems longhand, on the a-c or d-c analyzer board, or by digital computer, it is important to have a clear and working knowledge of this method. Use of the symmetrical-component method as applied to unbalanced faults allows once again a treatment of the problem on a simpler per phase basis. Symmetrical components are also useful in such work as the application of unbalanced voltages to rotating machinery and analysis of single-phase induction motors.

Another method of treating unbalanced faults is found in the use of Clarke components. There are applications which will permit a simpler solution with this method than with the symmetrical-component method.

To one already familiar with the subject of symmetrical components, it is quite possible that the heading of chapter eight will appear broad in scope. The author has purposely taken the broad approach at first in attempting the coverage of both theory and fault-study application in one chapter. The intention here is to spare the reader the various complexities and side issues related to this subject in the early stages of development. In this way it is felt that we can move more quickly and effectively to an early understanding of the heart of the problem. It will of course be necessary to pick up some of these additional related topics later which tend to complicate the problem (investigating sequence impedances, phase shift in transformers, etc.). Refer to Chapter 9.

8-2. The Concept of Symmetrical Components

The underlying observation of symmetrical component theory (as applied to three-phase systems) recognizes that any unbalanced three-phase system of phasors may be resolved into three balanced systems of phasors, namely:

1. The *positive-sequence system* consisting of a set of balanced three-phase components usually of the *same* phase sequence as the original unbalanced set. Positive-sequence phasors are equal in magnitude with the usual 120° phase-angle displacements. The (+) superscript will denote components of this system.
2. The *negative-sequence system* consisting of a balanced set of three-phase components of *opposite* phase sequence to the positive-sequence set. Phasors of the negative sequence are equal in

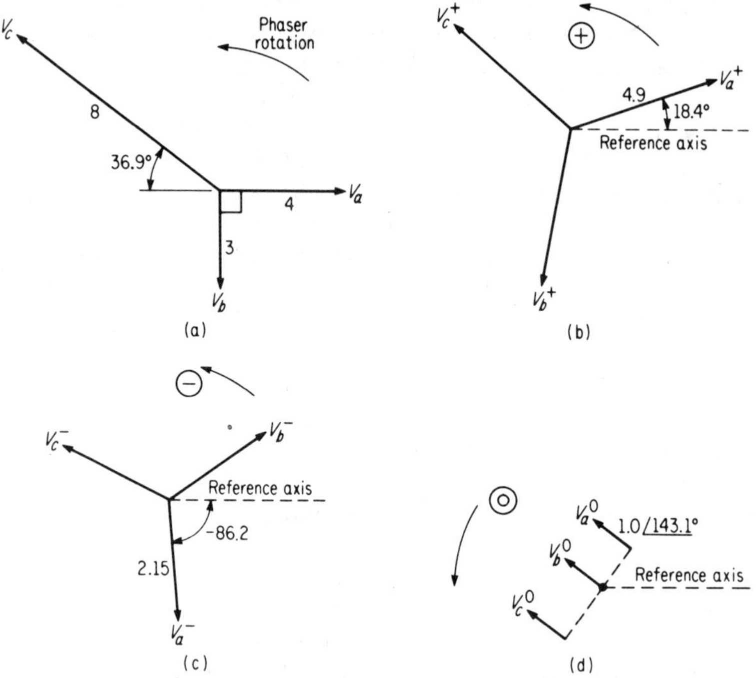

Fig. 8-1. (a) Original system of unbalanced phasors. (b) Positive-sequence components. (c) Negative-sequence components. (d) Zero-sequence components.

magnitude with 120° phase displacements. The (−) superscript will denote components of this system.

3. The *zero-sequence system* consisting of three single-phase components, all equal in magnitude but on the same phase angle.

The three phases of the original system will be designated a, b, and c and an original set of phasors might be said to have an abc phase sequence if phase b lags a and phase c lags b. In such a case it will also be true that the positive-sequence system will be of the abc sequence, while the negative-sequence system will be of the acb sequence. In one sense the method of symmetrical components is similar to the use of superposition principles upon a network, in that one system is replaced by the effect of three balanced systems, their effects being superimposed upon one another.

It might be well to illustrate an original unbalanced set of voltage phasors and the three corresponding balanced sets which, taken together, are equivalent to the original set. This illustration is made before we become involved with the mathematical detail in order to gain a concept of sequence components. Figure 8-1a illustrates an original unbalanced set, while Figs. 8-1b, c, and d represent the positive-, negative-, and zero-sequence sets respectively. It is not difficult, with a little imagination, to visualize that the sequence phasors add together to equal the original set according to the following equations:

$$V_a = V_a^+ + V_a^- + V_a^0 \tag{8-1}$$

$$V_b = V_b^+ + V_b^- + V_b^0 \tag{8-2}$$

$$V_c = V_c^+ + V_c^- + V_c^0 \tag{8-3}$$

Next we will define the operator a as a unit phasor which will rotate another phasor by 120°, or $a = 1\,\underline{/120°}$. This operator is used merely as a matter of convenience. Alternate ways of expressing this operator include $a = 1\,\underline{/120} = 1 \cdot e^{j2\pi/3} = -0.5 + j0.866$. It follows that $a^2 = 1\,\underline{/240}$ and $a^3 = 1\,\underline{/0°}$, $a^4 = a$, etc.

8-3. Determination of the Positive-, Negative-, and Zero-Sequence Components in Terms of the Original Phasors

It is yet to be determined as to how the sequence components of Figs. 8-1b, c, and d were obtained from the original set. It is known that an original phasor must equal the sum of all its parts, as indicated in Eqs. 8-1 through 8-3. Also, according to the three defining statements for positive-, negative-, and zero-sequence components we can say, for the balanced sets, that

$$V_b^+ = a^2 V_a^+, \quad V_c^+ = aV_a^+ \tag{8-4}$$

and

$$V_b^- = aV_a^-, \quad V_c^- = a^2 V_a^- \tag{8-5}$$

$$V_b^0 = V_a^0 = V_c^0 \tag{8-6}$$

Now by substitution of the Eqs. of 8-4 through 8-6, we can rewrite Eqs. 8-1 through 8-3 all in terms of phase *a* components:

$$V_a = V_a^+ + V_a^- + V_a^0 \tag{8-7}$$

$$V_b = a^2 V_a^+ + aV_a^- + V_a^0 \tag{8-8}$$

$$V_c = aV_a^+ + a^2 V_a^- + V_a^0 \tag{8-9}$$

Adding: $V_a + V_b + V_c = 0 + 0 + 3V_a^0$

or

$$V_a^0 = \frac{1}{3}(V_a + V_b + V_c) \tag{8-10}$$

Note that the positive- and negative-sequence terms in the addition went to zero, since the sum of phasors in a balanced three-phase system add to zero. Another way to state this is to say that $(1 + a + a^2)(V_a^+ + V_a^-) = 0$, since $(1 + a + a^2)$ always equals zero.

In order to find V_a^+ in terms of the original phasors, Eqs. 8-7 through 8-9 will be rewritten after having multiplied Eq. 8-8 by a and Eq. 8-9 by a^2:

$$V_a = V_a^+ + V_a^- + V_a^0$$

$$aV_b = a^3 V_a^+ + a^2 V_a^- + aV_a^0$$

$$a^2 V_c = a^3 V_a^+ + aV_a^- + a^2 V_a^0$$

Adding: $V_a + aV_b + a^2 V_c = 3V_a^+ + 0 + 0$

or

$$V_a^+ = \frac{1}{3}(V_a + aV_b + a^2 V_c) \tag{8-11}$$

To obtain V_a^- in terms of the original set, Eqs. 8-7 through 8-9 are again rewritten after having multiplied Eq. 8-8 by a^2 and Eq. 8-9 by a:

$$V_a = V_a^+ + V_a^- + V_a^0$$

$$a^2 V_b = aV_a^+ + a^3 V_a^- + a^2 V_a^0$$

$$aV_c = a^2 V_a^+ + a^3 V_a^- + aV_a^0$$

Adding: $V_a + a^2 V_b + aV_c = 0 + 3V_a^- + 0$

$$V_a^- = \frac{1}{3}(V_a + a^2 V_b + aV_c) \tag{8-12}$$

or

Equations 8-10, 8-11, and 8-12 are basic in our use of symmetrical components. There is a matrix that will yield the same results. This will also be given as an alternate approach. Writing Eqs. 8-7 through 8-9 in matrix form:

$$\begin{bmatrix} V_a \\ V_b \\ V_c \end{bmatrix} = \begin{bmatrix} 1 & 1 & 1 \\ 1 & a^2 & a \\ 1 & a & a^2 \end{bmatrix} \begin{bmatrix} V_a^0 \\ V_a^+ \\ V_a^- \end{bmatrix} \tag{8-13}$$

We will define the operator matrix as $[P]$ where

$$[P] = \begin{bmatrix} 1 & 1 & 1 \\ 1 & a^2 & a \\ 1 & a & a^2 \end{bmatrix} \tag{8-14}$$

The desired column matrix is the sequence component voltage matrix, where

$$\begin{bmatrix} V_a^0 \\ V_a^+ \\ V_a^- \end{bmatrix} = [P]^{-1} \begin{bmatrix} V_a \\ V_b \\ V_c \end{bmatrix} \tag{8-15}$$

Taking the inverse of $[P]$, using the conventional, three-step procedure as outlined in the Appendix,

1. First form the transpose of $[P]$ by interchanging rows and columns.

$$[P]_t = \begin{bmatrix} 1 & 1 & 1 \\ 1 & a^2 & a \\ 1 & a & a^2 \end{bmatrix}$$

2. Replace each element of $[P]_t$ by its cofactor.

$$\begin{bmatrix} +(a^4 - a^2) & -(a^2 - a) & +(a - a^2) \\ -(a^2 - a) & +(a^2 - 1) & -(a - 1) \\ +(a - a^2) & -(a - 1) & +(a^2 - 1) \end{bmatrix}$$

$$= \begin{bmatrix} j\sqrt{3} & j\sqrt{3} & j\sqrt{3} \\ j\sqrt{3} & j\sqrt{3}\underline{/210^\circ} & -\sqrt{3}\underline{/150^\circ} \\ j\sqrt{3} & -j\sqrt{3}\underline{/150^\circ} & \sqrt{3}\underline{/210^\circ} \end{bmatrix}$$

3. Each element of step 2 must be divided by the determinant Δ of $[P]$, where

$$\Delta = a^4 + a + a - a^2 - a^2 - a^2 = 3(a - a^2)$$
$$= 3j\sqrt{3}$$

Performing the division by Δ yields

$$[P]^{-1} = \frac{1}{3}\begin{bmatrix} 1 & 1 & 1 \\ 1 & a & a^2 \\ 1 & a^2 & a \end{bmatrix} \tag{8-16}$$

The desired result in matrix form is

$$\begin{bmatrix} V_a^0 \\ V_a^+ \\ V_a^- \end{bmatrix} = \frac{1}{3}\begin{bmatrix} 1 & 1 & 1 \\ 1 & a & a^2 \\ 1 & a^2 & a \end{bmatrix}\begin{bmatrix} V_a \\ V_b \\ V_c \end{bmatrix}$$

or written as separate questions, the results check with the previous method:

$$V_a^0 = \frac{1}{3}(V_a + V_b + V_c) \qquad [8\text{-}10]$$

$$V_a^+ = \frac{1}{3}(V_a + aV_b + a^2V_c) \qquad [8\text{-}11]$$

$$V_a^- = \frac{1}{3}(V_a + a^2V_b + aV_c) \qquad [8\text{-}12]$$

Example 8-1. Consider again the original set of voltage phasors pictured in Fig. 8-1a where $V_a = 4.0$, $V_b = 3\underline{/-90^\circ}$, and $V_c = 8\underline{/143.1^\circ}$. Find all of the voltage components for the positive, negative and zero-sequence systems.

$$V_a^0 = \frac{1}{3}(V_a + V_b + V_c)$$

$$= \frac{1}{3}[4.0 + (-j3) + 8(-0.8 + j0.6)]$$

$$= \frac{1}{3}[-2.4 + j1.8] = -0.8 + j0.6$$

$$= \underline{\underline{1 \angle 143.1^\circ}}$$

also,

$$V_b^0 = V_c^0 = V_a^0 = 1 \angle 143.1^\circ$$

$$V_a^+ = \frac{1}{3}[V_a + aV_b + a^2 V_c]$$

$$= \frac{1}{3}[4 \angle 0^\circ + (1 \angle 120^\circ)(3 \angle -90^\circ) + (1 \angle -120^\circ)(8 \angle 143.1^\circ)]$$

$$= \frac{1}{3}[4 + 3 \angle 30^\circ + 8 \angle 23.1^\circ] = \frac{1}{3}[13.96 + j4.64]$$

$$= 4.65 + j1.55 = \underline{\underline{4.90 \angle 18.4^\circ}}$$

$$V_b^+ = V_a^+ \angle -120^\circ = 4.90 \angle -101.6^\circ$$

$$V_c^+ = V_a^+ \angle +120^\circ = 4.90 \angle 138.4^\circ$$

$$V_a^- = \frac{1}{3}[V_a + a^2 V_b + aV_c]$$

$$= \frac{1}{3}[4.0 + (1 \angle -120^\circ)(3 \angle -90^\circ) + (1 \angle 120^\circ)(8 \angle 143.1^\circ)]$$

$$= \frac{1}{3}[4.0 + 3 \angle -210^\circ + 8 \angle 263.1^\circ]$$

$$= \frac{1}{3}[0.43 - j6.43] = 0.143 - j2.15$$

$$= \underline{\underline{2.15 \angle -86.2^\circ}}$$

$$V_b^- = V_a^- \angle +120^\circ = 2.15 \angle 33.8^\circ$$

$$V_c^- = V_a^- \angle -120^\circ = 2.15 \angle -206.2^\circ$$

The results can easily be checked either mathematically or graphically by realizing that each original phasor must be equal to the sum of its parts. For example, to check the results of phase *a* components,

$$V_a = V_a^+ + V_a^- + V_a^0$$

$$= (4.65 + j1.55) + (0.143 - j2.15) + (-0.8 + j0.6)$$

$$= 3.99(+) + j0$$

This checks with the original phasor which was given as 4.0.

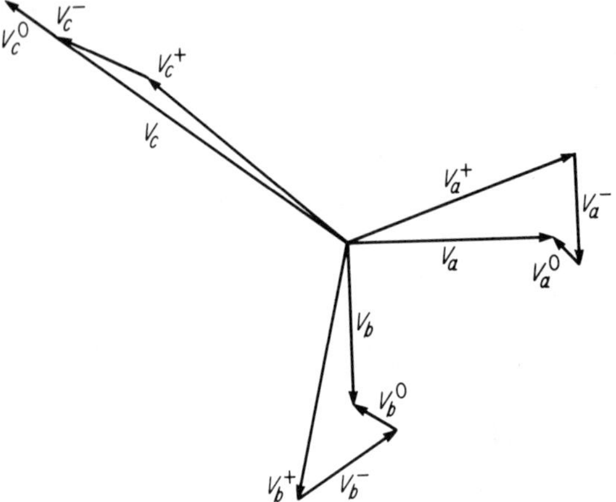

Fig. 8-2. Combining sequence components head-to-tail to form the original set of phasors. Refer also to Fig. 8-1.

A graphical verification of results is shown in Fig. 8-2, in which sequence component phasors are added head to tail in order to yield the original phasors.

The equations derived (Eqs. 8-10 through 8-12) relating sequence component voltages to the original unbalanced voltages will also apply to a system of unbalanced, three-phase currents. In other words,

$$I_a^0 = \frac{1}{3}\,[I_a + I_b + I_c] \tag{8-17}$$

$$I_a^+ = \frac{1}{3}\,[I_a + aI_b + a^2I_c] \tag{8-18}$$

$$I_a^- = \frac{1}{3}\,[I_a + a^2I_b + aI_c] \tag{8-19}$$

8-4. The General Nature of Zero-Sequence Currents

Notice Eq. 8-17 where $I_a^0 = \frac{1}{3}\,[I_a + I_b + I_c]$, and refer to Fig. 8-3. Since I_n also equals $I_a + I_b + I_c$, then

$$I_a^0 = \frac{1}{3}\,I_n, \qquad \text{or} \qquad I_n = 3I_a^0 \tag{8-20}$$

We conclude that whenever we have a three-wire, three-phase system *without* a neutral return, the zero-sequence component of line current is always zero. In particular, lines emanating from a three-wire Y-connection or a Δ connection cannot carry zero-sequence currents. However,

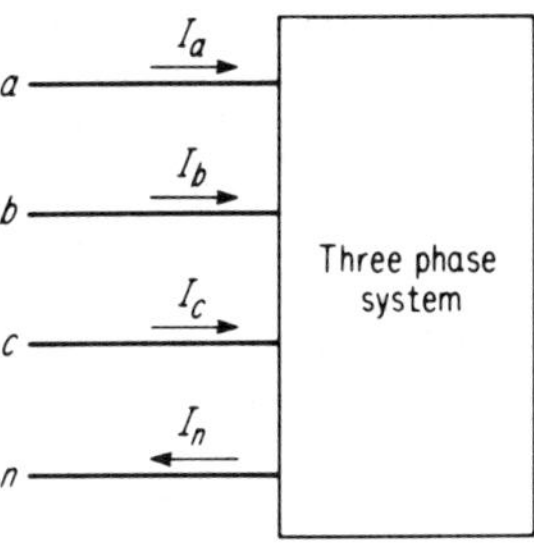

Fig. 8-3

zero-sequence currents may exist in a four-wire system. They may also be found to circulate within the delta of a transformer. This is possible, since it is known that zero-sequence currents of phases a, b, and c are equal in magnitude and in phase with one another.

Several examples of transformer configurations are studied in Fig. 8-4. In Fig. 8-4a, no zero-sequence current may flow from the lines of the secondary, since they have no return path. Yet the lines of the four-wire Y primary may carry zero-sequence currents as shown assuming the coils of the primary can find balancing ampere turns in the coils of the Δ secondary. These balancing ampere turns are available in Fig. 8-4a, since the in-phase zero sequence currents are free to circulate within the Δ.

Now moving to Fig. 8-4b, we find that, even though the primary has

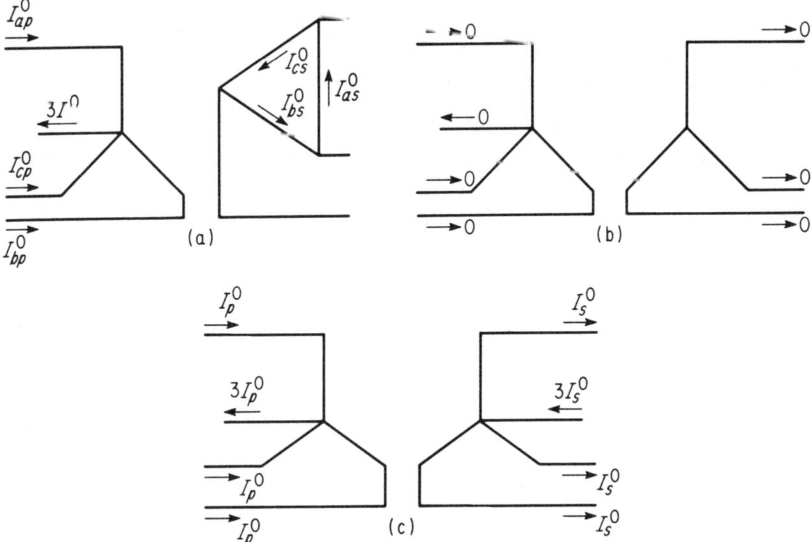

Fig. 8-4. Investigating various transformer configurations. Impossible situations for existence of zero-sequence currents are indicated.

a neutral return, zero-sequence currents are not possible due to the absence of balancing ampere turns in the three-wire Y secondary. Again, in phase zero-sequence currents cannot flow in the three-wire Y configuration.

Finally, the transformer of Fig. 8-4c may pass zero-sequence currents as shown due to the existence of a neutral return path for both primary and secondary windings.

The concept of *balancing ampere-turns* in a transformer is most important here, and in fact can be used to solve certain transformer unbalance problems in their entirety. This technique will be treated in more detail in Chapter 9.

It is fairly obvious from Eq. 8-20 that zero-sequence currents are involved primarily in the study of line-to-ground short circuits and unbalanced single-phase loading of four-wire Y systems. While some of the characteristics of zero-sequence currents resemble those of third-harmonic exciting currents, the two should not be confused. It is true that *a*, *b*, and *c* phase third-harmonic exciting currents are also in phase, but their flow does not require the condition of balancing ampere-turns in order that third harmonic core flux be suppressed.

Positive- and negative-sequence current flow is, of course, always possible whether the system is three- or four-wire. If a fourth wire (neutral return) does exist it will not carry positive sequence or negative-sequence currents since both of these systems are of a balanced three-phase nature and their line currents add to zero.

8-5. The Line-to-Ground Fault

In applying symmetrical components to various faults, one should keep in mind the basic principle of superposition. While this principle is

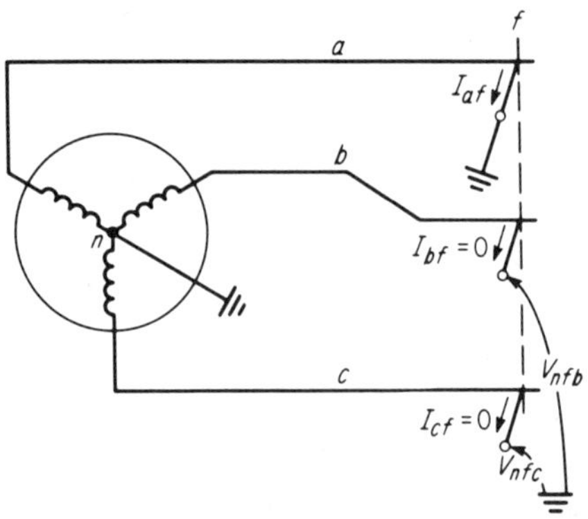

Fig. 8-5. Demonstrating the line-to-ground fault in a simple three-phase system.

perhaps not employed in simple form, yet it does provide a basis for our development.

Suppose that phase a of Fig. 8-5 is shorted to ground at the fault point f. By convention, fault current is taken to be positive when directed out of the fault point. In developing the line-to-ground calculating procedure, we write

$$I_{af}^0 = \frac{1}{3}\left(I_{af} + \cancelto{0}{I_{bf}} + \cancelto{0}{I_{cf}}\right) = \frac{1}{3} I_{af}$$

Likewise,

$$I_{af}^+ = \frac{1}{3}\left(I_{af} + a\cancelto{0}{I_{bf}} + a^2\cancelto{0}{I_{cf}}\right) = \frac{1}{3} I_{af}$$

and

$$I_{af}^- = \frac{1}{3}\left(I_{af} + a^2\cancelto{0}{I_{bf}} + a\cancelto{0}{I_{cf}}\right) = \frac{1}{3} I_{af}$$

Therefore

$$I_{af}^0 = I_{af}^+ = I_{af}^- = \frac{1}{3} I_{af} \tag{8-21}$$

The voltage above ground (V_{nfa}) at point f equals zero for phase a, or since the voltage equals the sum of all of its parts,

$$V_{nfa} = V_{nfa}^+ + V_{nfa}^- + V_{nfa}^0$$

or

$$0 = (E_a^+ - I_a^+ Z_a^+) + (\cancelto{0}{E_a^-} - I_a^- Z_a^-) + (\cancelto{0}{E_a^0} - I_a^0 Z_a^0) \tag{8-22}$$

where E_a^+, E_a^-, and E_a^0 are the sequence generated voltages of phase a. Since generators normally produce a balanced three-phase generated voltage, it is of a positive-sequence nature, whereas the negative-sequence and zero-sequence generated voltages are zero. For this reason, Eq. 8-22 has shown E_a^- and E_a^0 crossed out. Z_a^+, Z_a^-, and Z_a^0 are the impedances encountered by I_{af}^+, I_{af}^- and I_{af}^0 respectively.

Rewriting Eq. 8-22,

$$0 = E_a^+ - I_{af}^+ Z_a^+ - I_{af}^- Z_a^- - I_{af}^0 Z_a^0 \tag{8-23}$$

but since $I_a^+ = I_a^- = I_a^0$, we may rewrite Eq. 8-23 in terms of I_a^0

$$0 = E_a^+ - I_a^0(Z_a^+ + Z_a^- + Z_a^0)$$

or

$$I_a^0 = \frac{E_a^+}{Z_a^+ + Z_a^- + Z_a^0} \tag{8-24}$$

and

$$I_{af} = 3I_a^0 \tag{8-25}$$

Equations 8-24 and 8-25 are used in determining the fault current itself. However, we may also be interested in finding such information as the voltage above ground (or above the neutral bus) for phases *b* and *c* as well as various voltages and currents throughout a network. For this reason it is most convenient to realize that Eqs. 8-23 and 8-24 are completely satisfied by the line-to-ground series equivalent circuit of Fig. 8-6.

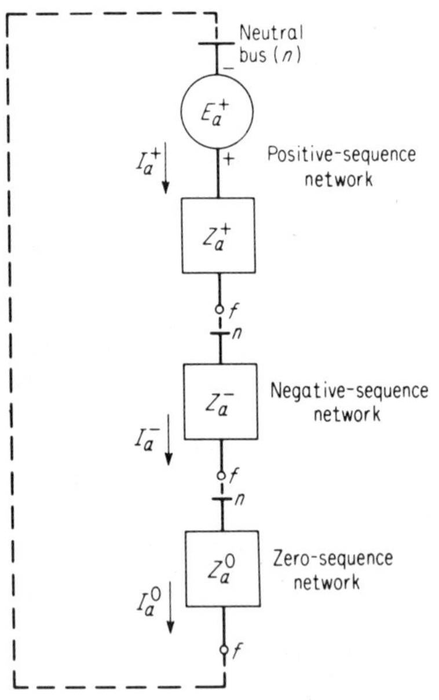

Fig. 8-6. Line-to-ground fault equivalent circuit for phase *a*.

Now as the problems grow more complex it is seen that one must merely place the three sequence networks in series to obtain a solution. In many practical applications, the positive and negative-sequence impedances are found to be equal. More detail will be offered in Chapter 9 concerning these impedances.

The voltage from neutral to fault for any phase and sequence can be determined by subtracting the impedance drop of that phase and sequence from the generated voltage of that same sequence. For example, refer to Fig. 8-7.

$$V_{nfa}^+ = E_a^+ - I_a^+ Z_a^+ \tag{8-26}$$

$$V_{nfa}^{-} = 0 - I_a^{-} Z_a^{-} \tag{8-27}$$

$$V_{nfa}^{0} = 0 - I_a^{0} Z_a^{0} \tag{8-28}$$

Since $V_{nfa} = 0$, Eqs. 8-26 through 8-28 should add to zero. V_{nfb} can now

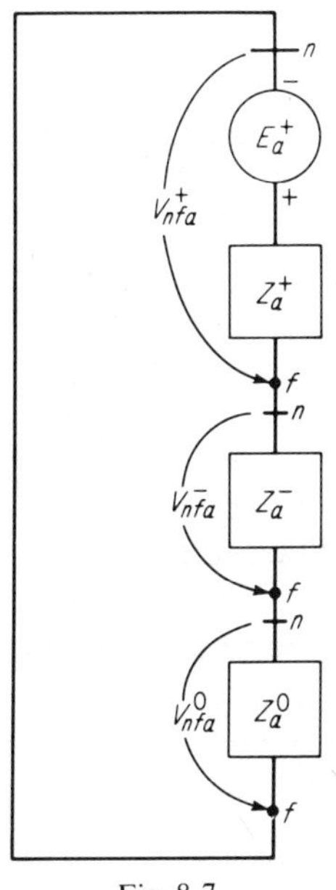

Fig. 8-7

be found as

$$V_{nfb} = V_{nfb}^{+} + V_{nfb}^{-} + V_{nfb}^{0} \tag{8-29}$$

Now relating the phase *b* components to the known phase *a* voltages,

$$V_{nfb} = V_{nfa}^{+} \underline{/-120^\circ} + V_{nfa}^{-} \underline{/+120^\circ} + V_{nfa}^{0} \tag{8-30}$$

Likewise,

$$V_{nfc} = V_{nfc}^{+} + V_{nfc}^{-} + V_{nfc}^{0}$$

or

$$V_{nfc} = V_{nfa}^{+} \underline{/+120^\circ} + V_{nfa}^{-} \underline{/-120^\circ} + V_{nfa}^{0} \tag{8-31}$$

Example 8-2. Given the circuit of Fig. 8-8 with per unit, balanced phase impedances as follows:

$$\text{pu } Z_{\text{gen}}^{+} = \text{pu } Z_{\text{gen}}^{-} = j0.10 \qquad \text{pu } Z_{\text{tr}}^{+} = \text{pu } Z_{\text{tr}}^{-} = \text{pu } Z_{\text{tr}}^{0} = j0.05$$

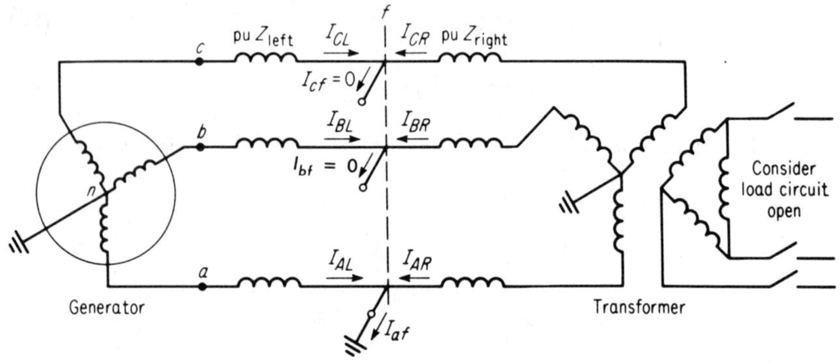

Fig. 8-8. Circuit for Example 8-2a.

$$\text{pu } Z^0_{\text{gen}} = j0.05 \qquad \text{pu } Z^+_R = \text{pu } Z^-_R = j0.2$$

$$\text{pu } Z^+_L = \text{pu } Z^-_L = j0.2 \qquad \text{pu } Z^0_R = j0.4$$

$$\text{pu } Z^0_L = j0.4 \qquad \text{pu } E_{\text{gen}} = 1.0\,\underline{/0^\circ}$$

For a line-to-ground fault on phase *a* as shown, find all line currents feeding into the fault point and voltages at the fault when (a) the generator is grounded as shown, (b) the generator is ungrounded.

Solution (*a*). Figures 8-9a, b, and c represent the three sequence networks for this problem. It is seen that positive- and negative-sequence

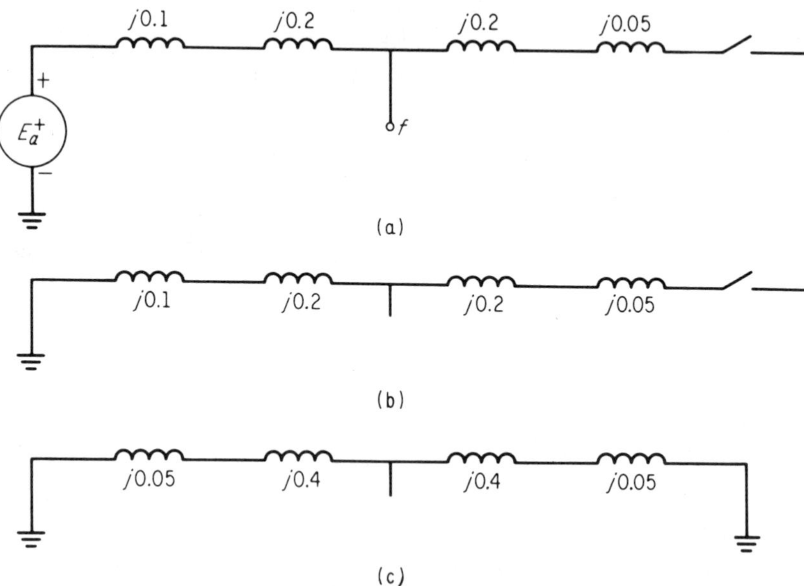

Fig. 8-9. (a) Positive-sequence network. (b) Negative-sequence network. (c) Zero-sequence network.

networks see an open circuit to the right of the fault. However a clear path to zero-sequence currents is considered to exist back to the neutral bus. This is due to the freedom of zero-sequence currents to circulate in the Δ path as explained in Sec. 8-4.

The sequence networks have been combined in Fig. 8-10 for the first step in the solution of I_a^0. A somewhat pictorial display of the left and

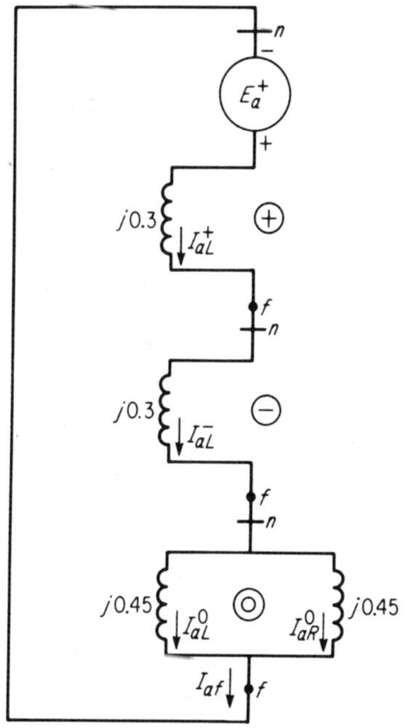

Fig. 8-10. Combining the sequence networks for solution of I_{af}^0.

right branches is maintained as it is easier to visualize the individual contributions of the left and right branches which are feeding the fault.

$$I_{af}^0 = \frac{E_a^+}{Z_a^+ + Z_a^- + Z_a^0}$$

$$= \frac{1.0\,\underline{/0^\circ}}{j0.3 + j0.3 + j0.225} = \frac{1.0}{j0.825}$$

$$= 1.212\,\underline{/-90^\circ}$$

$$I_{af} = 3I_{af}^0 = 3.636\,\underline{/-90^\circ}$$

$$I_{aL}^0 = I_{af}^0/2 = 0.606\,\underline{/-90^\circ}$$

$$I_{aL}^+ = I_{af}^0 = 1.212\,\underline{/-90^\circ}$$

$$I_{aL}^- = I_{af}^0 = 1.212\,\underline{/-90^\circ}$$

The total a-phase current contributed by the left branch is then

$$\begin{aligned} I_{aL} &= I_{aL}^{+} + I_{aL}^{-} + I_{aL}^{0} \\ &= 1.212\,\underline{/-90^\circ} + 1.212\,\underline{/-90^\circ} + 0.606\,\underline{/-90^\circ} \\ &= \underline{\underline{3.03\,/-90^\circ}} \end{aligned}$$

The a-phase fault contribution coming from the right branch is

$$\begin{aligned} I_{aR} &= I_{aR}^{+} + I_{aR}^{-} + I_{aR}^{0} \\ &= 0 + 0 + \underline{\underline{0.606\,/-90^\circ}} \end{aligned}$$

Determining next the line current for phase b,

$$\begin{aligned} I_{bL} &= I_{bL}^{+} + I_{bL}^{-} + I_{bL}^{0} \\ &= I_{aL}^{+}\,\underline{/-120^\circ} + I_{aL}^{-}\,\underline{/+120^\circ} + I_{aL}^{0} \\ &= (1.212\,\underline{/-90^\circ})(1\,\underline{/-120^\circ}) + (1.212\,\underline{/-90^\circ}) \\ &\qquad \times (1\,\underline{/+120^\circ}) + 0.606\,\underline{/-90^\circ} \\ &= 0 + j0.606 = \underline{\underline{0.606\,/90^\circ}} \end{aligned}$$

Similarly in phase c,

$$\begin{aligned} I_{cL} &= I_{cL}^{+} + I_{cL}^{-} + I_{cL}^{0} \\ &= I_{aL}^{+}\,\underline{/120^\circ} + I_{aL}^{-}\,\underline{/-120^\circ} + I_{aL}^{0} \\ &= (1.212\,\underline{/-90^\circ})(1\,\underline{/+120^\circ}) + (1.212\,\underline{/-90^\circ}) \\ &\qquad \times (1\,\underline{/-120^\circ}) + 0.606\,\underline{/-90^\circ} \\ &= 0 - j0.606 = \underline{\underline{0.606\,/-90^\circ}} \end{aligned}$$

In determining the voltages above ground at the fault location, it is apparent that $V_{nfa} = 0$. However we do need the phase a component voltages to determine V_{nfb} and V_{nfc}:

$$\begin{aligned} V_{nfa}^{+} &= E_a^{+} - I_a^{+}Z_a^{+} = 1.0\,\underline{/0^\circ} - (1.212\,\underline{/-90^\circ})(0.3\,\underline{/+90^\circ}) \\ &= 1.0 - 0.3636 = 0.636 \\ V_{nfa}^{-} &= 0 - I_a^{-}Z_a^{-} = -(1.212\,\underline{/-90^\circ})(0.3\,\underline{/90^\circ}) \\ &= -0.3636 \\ V_{nfa}^{0} &= 0 - I_a^{0}Z_a^{0} = -(1.212\,\underline{/-90^\circ})(0.225)\,\underline{/90^\circ}) \\ &= -0.273 \end{aligned}$$

Checking,

$$\begin{aligned} V_{nfa} &= V_{nfa}^{+} + V_{nfa}^{-} + V_{nfa}^{0} = 0.636 - 0.3636 - 0.273 \\ &= 0 \qquad \underline{\text{O.K.}} \end{aligned}$$

Now

$$\begin{aligned} V_{nfb} &= V^+_{nfb} + V^-_{nfb} + V^0_{nfb} \\ &= V^+_{nfa}\,\underline{/-120^\circ} + V^-_{nfa}\,\underline{/+120^\circ} + V^0_{nfa} \\ &= 0.636\,\underline{/-120^\circ} - 0.3636\,\underline{/120^\circ} - 0.273 \\ &= -0.409 - j0.865 = \underline{\underline{0.957\,/244.7^\circ}} \end{aligned}$$

$$\begin{aligned} V_{nfc} &= V^+_{nfc} + V^-_{nfc} + V^0_{nfc} \\ &= V^+_{nfa}\,\underline{/120^\circ} + V^-_{nfa}\,\underline{/-120^\circ} + V^0_{nfa} \\ &= -0.409 + j0.865 \\ &= \underline{\underline{0.957\,/115.3^\circ}} \end{aligned}$$

Having completed the solution to part (a), it is well to study the original circuit with numerical results included as per Fig. 8-11. A factor of $(-j)$ has been omitted from each of the currents for simplification. This

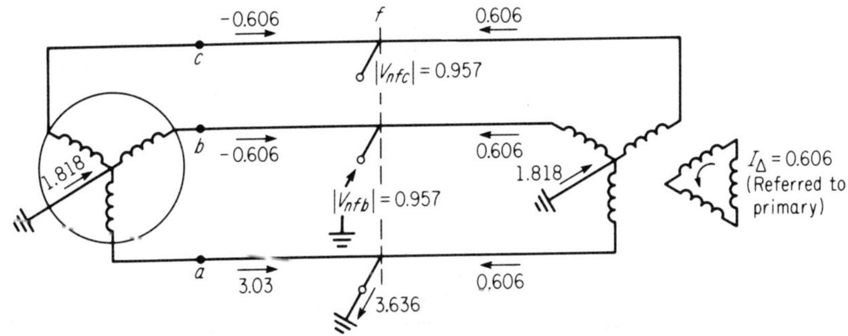

Fig. 8-11. Original circuit with numerical results of Example 8-2a.

could have been done at the beginning of the problem by omitting the factor (j) in the impedances and solving as though the circuit were purely resistive. This can be done whenever all impedances are considered purely inductive as was explained in Chapter 4. Notice that the circulating current of the Δ secondary has provided the necessary balancing ampere turns for the transformer.

Solution (*b*). With generator ungrounded the line to ground equivalent has changed since the per phase zero-sequence impedance for the left branch has gone to infinity, the generator having no return path for zero-sequence currents. Figure 8-10 is altered as shown in Fig. 8-12, where

$$I^0_{af} = \frac{1.0}{j0.3 + j0.3 + j0.45} = 0.952\,\underline{/-90^\circ}$$

$$I_{af} = 3I^0_{af} = 2.856\,\underline{/-90^\circ}$$

$$\begin{aligned} I_{aL} &= I^+_{aL} + I^-_{aL} + I^0_{aL} \\ &= 0.952\,\underline{/-90^\circ} + 0.952\,\underline{/-90^\circ} + 0 \\ &= \underline{\underline{1.904\,/-90^\circ}} \end{aligned}$$

$$\begin{aligned} I_{aR} &= I^+_{aR} + I^-_{aR} + I^0_{aR} \\ &= 0 + 0 + \underline{\underline{0.952\,/-90^\circ}} \end{aligned}$$

$$\begin{aligned} I_{bL} &= I^+_{bL} + I^-_{bL} + I^0_{bL} \\ &= I^+_{aL}\,\underline{/-120^\circ} + I^-_{aL}\,\underline{/+120^\circ} + I^0_{aL} \\ &= (0.952\,\underline{/-90^\circ})(1\,\underline{/-120^\circ}) + (0.952\,\underline{/-90^\circ})(1\,\underline{/+120^\circ}) + 0 \\ &= \underline{\underline{0.952\,/90^\circ}} \end{aligned}$$

$$I_{bR} = -I_{bL} = \underline{\underline{-0.952\,/90^\circ}}$$

The procedure for finding the voltages above ground at the fault lo-

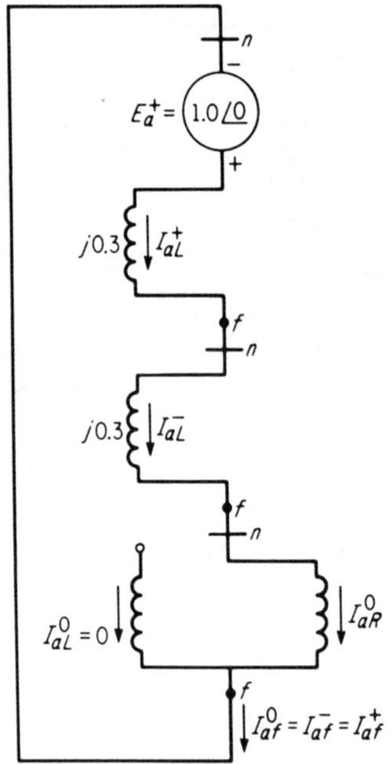

Fig. 8-12. Line-to-ground equivalent for Example 8-2b.

cation are identical to the procedures of part (a). The student can verify the fact that

$$V_{nfb} = \underline{\underline{1.08\,/-126.5^\circ}} \qquad \text{and} \qquad V_{nfc} = \underline{\underline{1.08\,/+126.5^\circ}}$$

8-6. The Line-to-Line Fault

Now let phase a be the unfaulted phase. Figure 8-13 shows a three-phase system with a line-to-line short circuit between phases b and c. We

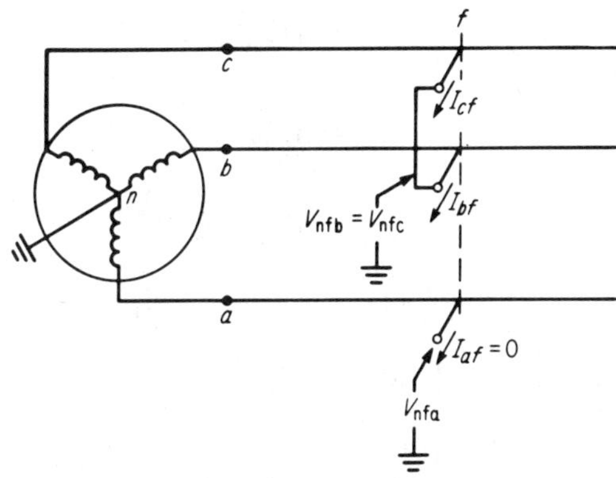

Fig. 8-13. Example of a line-to-line fault.

know that $I_{af} = 0$ and $I_{bf} = -I_{cf}$. Then

$$I_{af}^0 = \frac{1}{3}[\overset{0}{\cancel{I_{af}}} + I_{bf} + I_{cf}] \quad \text{or} \quad I_{af}^0 = 0 \tag{8-32}$$

Zero-sequence currents and voltages are absent here. Solving for I_{af}^+, where $I_{bf} = -I_{cf}$:

$$\begin{aligned} I_{af}^+ &= \frac{1}{3}[\overset{0}{\cancel{I_{af}}} + aI_{bf} + a^2I_{cf}] \\ &= \frac{1}{3}[I_{bf}\angle 120^\circ - I_{bf}\angle -120^\circ] \\ &= \frac{I_{bf}\angle 90^\circ}{\sqrt{3}} \end{aligned} \tag{8-33}$$

$$\begin{aligned} I_{af}^- &= \frac{1}{3}[\overset{0}{\cancel{I_{af}}} + a^2I_{bf} + aI_{cf}] \\ &= \frac{1}{3}[I_{bf}\angle -120^\circ - I_{bf}\angle +120^\circ] \\ &= \frac{I_{bf}\angle -90^\circ}{\sqrt{3}} \end{aligned} \tag{8-34}$$

Comparing Eqs. 8-33 and 8-34 yields

$$I_{af}^- = -I_{af}^+ \tag{8-35}$$

It is apparent from Fig. 8-13 that

$$V_{nfb} = V_{nfc} \tag{8-36}$$

Now solving for the sequence voltages at the fault location,

$$\begin{aligned} V_{nfa}^{+} &= \frac{1}{3}\,[V_{nfa} + aV_{nfb} + a^2 V_{nfc}] \\ &= \frac{1}{3}\,[V_{nfa} + aV_{nfb} + a^2 V_{nfb}] \\ &= \frac{1}{3}\,[V_{nfa} - V_{nfb}] \end{aligned} \tag{8-37}$$

Also

$$\begin{aligned} V_{nfa}^{-} &= \frac{1}{3}\,[V_{nfa} + a^2 V_{nfb} + aV_{nfc}] \\ &= \frac{1}{3}\,[V_{nfa} + a^2 V_{nfb} + aV_{nfb}] \\ &= \frac{1}{3}\,[V_{nfa} - V_{nfb}] \end{aligned} \tag{8-38}$$

Comparing Eqs. 8-37 and 8-38,

$$V_{nfa}^{+} = V_{nfa}^{-} \tag{8-39}$$

Equations 8-32, 8-35 and 8-39 suggest a simple equivalent circuit in the solution of the line-to-line fault. The circuit satisfies these equations in every respect. Refer to Fig. 8-14 in which the neutral buses of positive and negative networks are joined together as shown.

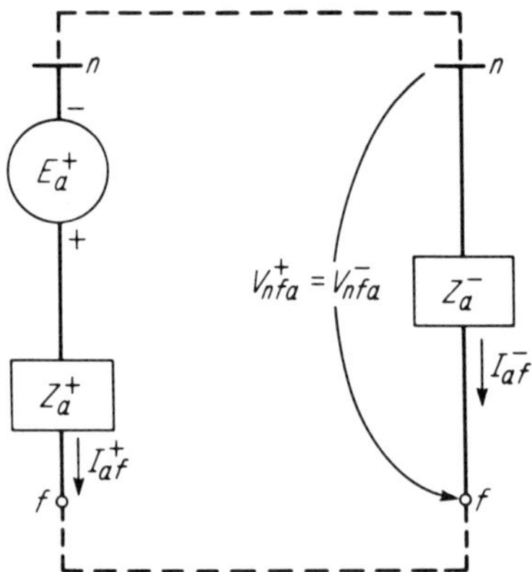

Fig. 8-14. Line-to-line fault equivalent circuit for phase *a*.

The solution procedure is outlined as follows:

1. Obtain I_{af}^{+} and I_{af}^{-}. From the equivalent circuit,

$$I_{af}^{+} = \frac{E_a^{+}}{Z^{+} + Z^{-}} \tag{8-40}$$

and

$$I_{af}^{-} = -I_{af}^{+} \qquad [8\text{-}35]$$

2. Find the fault currents $I_{bf} = -I_{cf}$,

$$\begin{aligned} I_{bf} &= I_{bf}^{+} + I_{bf}^{-} + I_{bf}^{0} \\ &= I_{af}^{+}\angle -120^\circ + I_{af}^{-}\angle +120^\circ \end{aligned}$$

substituting,

$$I_{bf} = I_{af}^{+}\angle -120^\circ - I_{af}^{+}\angle 120^\circ$$

$$I_{bf} = \sqrt{3}I_{af}^{+}\angle -90^\circ = -I_{cf} \tag{8-41}$$

3. Find the voltages above ground at the fault,

$$V_{nfa} = V_{nfa}^{+} + V_{nfa}^{-} + V_{nfa}^{0}$$

$$V_{nfa} = 2V_{nfa}^{+} = 2V_{nfa}^{-} \tag{8-42}$$

where

$$V_{nfa}^{+} = E_a^{+} - I_{af}^{+}Z^{+}$$

or

$$V_{nfa}^{+} = -I_{af}^{-}Z^{-}$$

[*Note:* In the special case where $Z^{+} = Z^{-}$, half of E_a^{+} will drop across Z_a^{-} and $V_{nfa} = 2V_{nfa}^{-} = 2\,\dfrac{E_a^{+}}{2} = E_a^{+}$.]

4. Finding the voltage above ground for phases b and c,

$$\begin{aligned} V_{nfc} = V_{nfb} &= V_{nfb}^{+} + V_{nfb}^{-} + V_{nfb}^{0} \\ &= a^2V_{nfa}^{+} + aV_{nfa}^{-} \\ &= V_{nfa}^{+}(a^2 + a) \end{aligned}$$

$$V_{nfc} = V_{nfb} = -V_{nfa}^{+} \tag{8-43}$$

5. Line-to-line voltages at the point of fault may be needed as well. If so,

$$\begin{aligned} V_{ab} &= V_{fna} + V_{nfb} \\ &= -V_{nfa} + V_{nfb} \end{aligned}$$

substituting results of Eqs. 8-42 and 8-43,

$$V_{ab} = -2V^{+}_{nfa} - V^{+}_{nfa}$$

$$V_{ab} = -3V^{+}_{nfa} \tag{8-44}$$

$$V_{bc} = V_{fnb} + V_{nfc} = -V_{nfb} + V_{nfc}$$

$$= +V^{+}_{nfa} - V^{+}_{nfa}$$

$$V_{bc} = 0 \tag{8-45}$$

Equation 8-45 is also obvious since lines b and c are shorted together.

$$V_{ca} = V_{fnc} + V_{nfa} = -V_{nfc} + V_{nfa}$$

$$= -(-V^{+}_{nfa}) + 2V^{+}_{nfa}$$

$$V_{ca} = 3V^{+}_{nfa} = -V_{ab} \tag{8-46}$$

Example 8-3. Given the one-line diagram of Fig. 8-15 with a line-to-line short circuit at point f on phases b and c. All per phase imped-

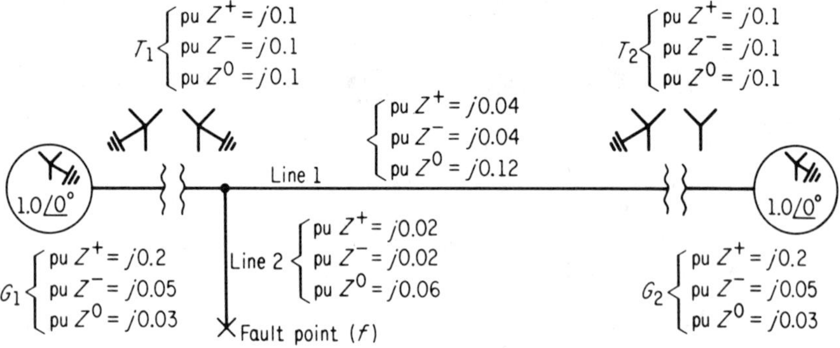

Fig. 8-15. One-line diagram for Example 8-3. All values of per unit impedances are referred to common base.

ances have been referred to a common-base kva as discussed in Chapter 4. Find the per unit values of I_b, V_{na}, V_{nb}, and V_{ab}, all at the fault location. Also find the per unit current contributions from generator G_1.

Since all impedances in the problem are purely inductive, the j factor will be common to each impedance term. To simplify the reduction procedure, the j term will be dropped as if the circuit were purely resistive.

The phase a equivalent network for solving the line-to-line fault is represented in Fig. 8-16, where the positive and negative-sequence networks have been joined together. The network of Fig. 8-16 has been progressively reduced in Figs. 8-17a through c. The reductions are of the

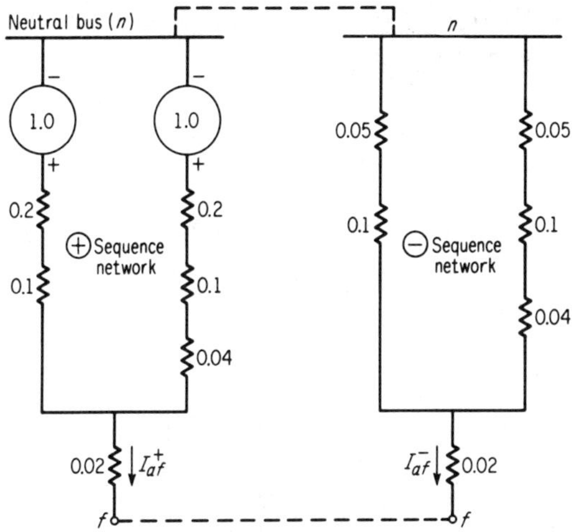

Fig. 8-16. Sequence network equivalent for line-to-line fault of Example 8-3.

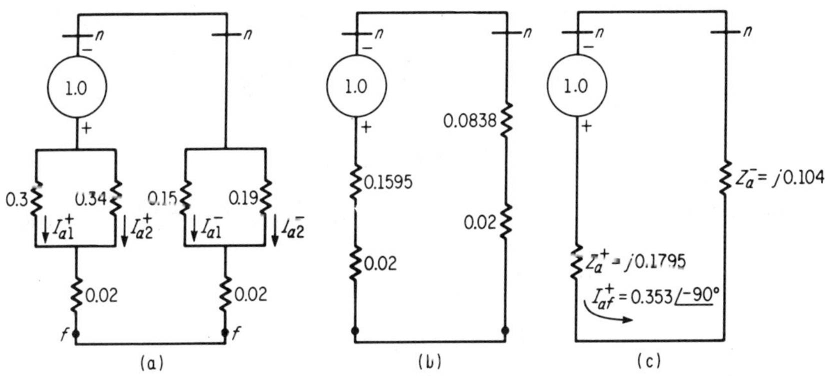

Fig. 8-17. Steps in the reduction of network equivalent.

simplest forms, involving only series and parallel impedances combinations. For this reason, only the results are shown on the drawings.

$$I_{af}^{+} = \frac{E_a^{+}}{Z^{+} + Z^{-}} = \frac{1.0}{j0.1795 + j0.104} = \frac{1.0}{j0.2835}$$

$$= \underline{\underline{3.53 \angle -90°}}$$

Of course $I_{af} = 0$ but I_{bf} can be found either from the knowledge that $I_{bf} = I_{bf}^{+} + I_{bf}^{-}$ or more directly from Eq. 8-41 already derived, where

$$I_{bf} = \sqrt{3}\, I_{af}^{+} \angle -90^\circ = \sqrt{3}(3.53 \angle -90^\circ)(1 \angle -90^\circ)$$
$$= \underline{\underline{6.12 \angle 180^\circ}} \text{ pu amp}$$

and

$$I_{cf} = -I_{bf} = \underline{\underline{6.12 \angle 0^\circ}} \text{ pu amp}$$

The positive sequence phase a voltage above neutral is

$$V_{nfa}^{+} = I_{af}^{+} Z^{-} = (3.53 \angle -90^\circ)(0.104 \angle +90^\circ)$$
$$= 0.367 \angle 0^\circ \text{ pu volts}$$

According to Eq. 8-42,

$$V_{nfa} = 2V_{nfa}^{+} = \underline{\underline{0.734 \angle 0^\circ}} \text{ pu volts}$$

From Eq. 8-43

$$V_{nfc} = V_{nfb} = -V_{nfa}^{+} = -\underline{\underline{0.367}} \text{ pu volts}$$

The line-to-line voltages at the point of fault may be obtained directly from Eqs. 8-44, 8-45, and 8-46.

$$V_{ab} = -3V_{nfa}^{+} = -3(0.367) = -\underline{\underline{1.101}} \text{ pu volts}$$
$$V_{bc} = 0$$
$$V_{ca} = -V_{ab} = +\underline{\underline{1.101}} \text{ pu volts}$$

The per unit current contributions from $G_1(I_{a1}, I_{b1}$ and $I_{c1})$ can be found by working back into Fig. 8-17a.

Using the current-divider principle,

$$I_{a1}^{+} = I_{af}^{+}\frac{0.34}{0.3 + 0.34} = (0.353 \angle -90^\circ)\,\frac{0.34}{0.64}$$
$$= 0.1875 \angle -90^\circ$$

Likewise,

$$I_{a1}^{-} = I_{af}^{-}\frac{0.19}{0.15 + 0.19} = (0.353 \angle +90^\circ)\,\frac{0.19}{0.34}$$
$$= 0.197 \angle +90^\circ$$
$$I_{a1} = I_{a1}^{+} + I_{a1}^{-} = 0.1875 \angle -90^\circ + 0.197 \angle +90^\circ$$
$$= \underline{\underline{0.0095 \angle 90^\circ}}$$
$$I_{b1} = I_{b1}^{+} + I_{b1}^{-} = I_{a1}^{+} \angle -120^\circ + I_{a1}^{-} \angle +120^\circ$$
$$= (0.1875 \angle -90^\circ)(1 \angle -120^\circ) + (0.197 \angle +90^\circ)(1 \angle +120^\circ)$$
$$= -0.3325 - j0.0047 = \underline{\underline{0.333 \angle 180.8^\circ}}$$

$$\begin{aligned} I_{c1} &= I_{c1}^{+} + I_{c1}^{-} = I_{a1}^{+} \angle 120^\circ + I_{a1}^{-} \angle -120 \\ &= (0.1875 \angle -90^\circ)(1 \angle 120^\circ) + (0.197 \angle 90^\circ)(1 \angle -120^\circ) \\ &= 0.3325 - j0.0047 = \underline{\underline{0.333 \angle -0.8^\circ}} \end{aligned}$$

As a further check on results, $I_{a1} + I_{b1} + I_{c1}$ should add to zero. This is easily verified.

8-7. The Double Line-to-Ground Fault

The simple three-phase system of Fig. 8-18 shows phase a as the unfaulted phase, whereas phases b and c are both shorted to ground at the location f.

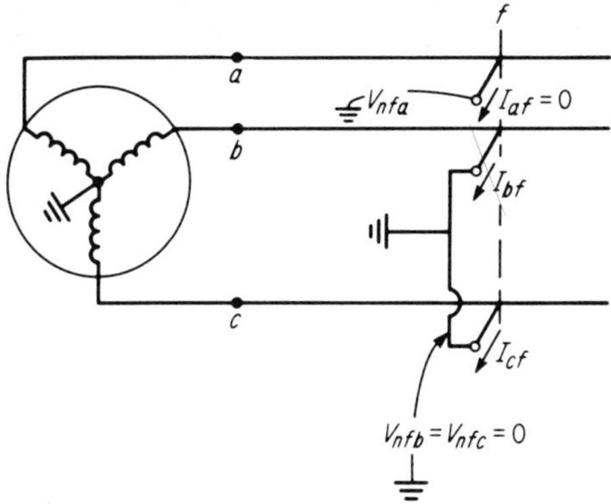

Fig. 8-18. Circuit with double line-to-ground fault.

The voltage above ground at the fault location is obviously zero for phases b and c, or

$$V_{nfb} = V_{nfc} = 0 \tag{8-47}$$

Also

$$I_{af} = I_{af}^{+} + I_{af}^{-} + I_{af}^{0} = 0 \tag{8-48}$$

The sequence component voltages are

$$\begin{aligned} V_{nfa}^{0} &= \frac{1}{3}[V_{nfa} + \overset{0}{\cancel{V_{nfb}}} + \overset{0}{\cancel{V_{nfc}}}] \\ &= \frac{1}{3} V_{nfa} \\ V_{nfa}^{+} &= \frac{1}{3}[V_{nfa} + a\overset{0}{\cancel{V_{nfb}}} + a^2\overset{0}{\cancel{V_{nfc}}}] \\ &= \frac{1}{3} V_{nfa} \end{aligned}$$

$$V_{nfa}^{-} = \frac{1}{3}[V_{nfa} + a^2 \overset{0}{\cancel{V_{nfb}}} + a \overset{0}{\cancel{V_{nfc}}}]$$

$$= \frac{1}{3} V_{nfa}$$

It follows that

$$V_{nfa}^{0} = V_{nfa}^{+} = V_{nfa}^{-} \tag{8-49}$$

From Eqs. 8-48 and 8-49 the double line-to-ground sequence network of Fig. 8-19 is obtained, a network which satisfies the equations in every respect. Again, currents coming out of the fault points are considered positive.

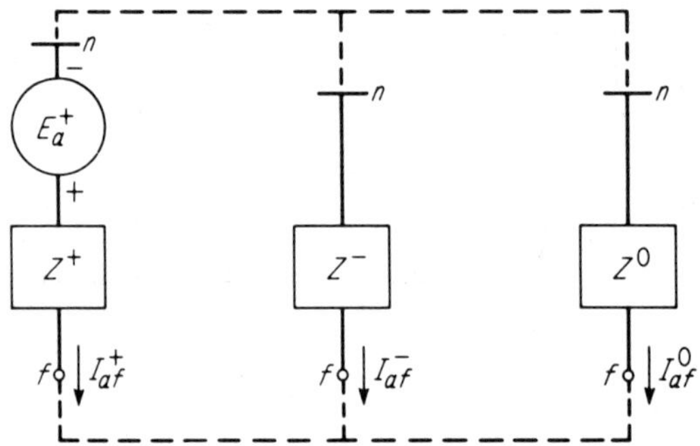

Fig. 8-19. Sequence network for double line-to-ground fault.

The calculation procedure for the double line-to-ground fault is outlined below. It is some help to maintain a physical concept of the actual situation as it might appear at the point of fault. For this reason Fig. 8-20 is given.

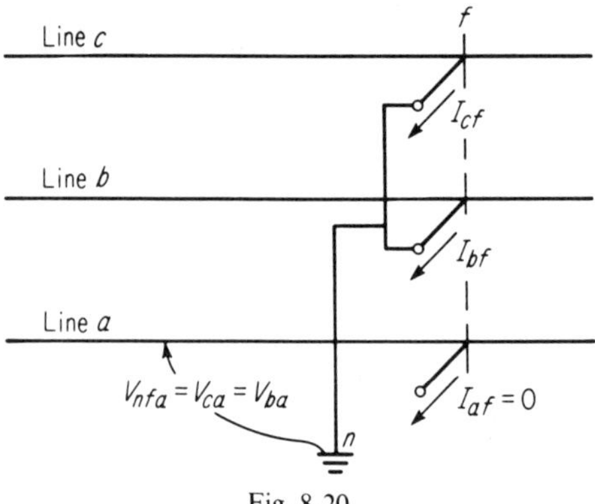

Fig. 8-20

1. First, the two fault currents (I_{bf} and I_{cf}) will be found. From Fig. 8-19,

$$I_{af}^{+} = \frac{E_a^{+}}{Z^{+} + \dfrac{Z^{-}Z^{0}}{Z^{-} + Z^{0}}} \tag{8-50}$$

$$V_{nfa}^{+} = E_a^{+} - I_{af}^{+}Z^{+} \tag{8-51}$$

$$I_{af}^{-} = \frac{-V_{nfa}^{+}}{Z^{-}}, \qquad \text{and} \qquad I_{af}^{0} = \frac{-V_{nfa}^{+}}{Z_0} \tag{8-52}$$

I_{af}^{-} and I_{af}^{0} could also have been determined using the current-divider principle.

$$\begin{aligned} I_{bf} &= I_{bf}^{+} + I_{bf}^{-} + I_{bf}^{0} \\ &= I_{af}^{+}\underline{/-120^\circ} + I_{af}^{-}\underline{/+120^\circ} + I_{af}^{0} \end{aligned} \tag{8-53}$$

$$\begin{aligned} I_{cf} &= I_{cf}^{+} + I_{cf}^{-} + I_{cf}^{0} \\ &= I_{af}^{+}\underline{/+120^\circ} + I_{af}^{-}\underline{/-120^\circ} + I_{af}^{0} \end{aligned} \tag{8-54}$$

2. Again refer to Fig. 8-20. From this we see that

$$V_{nfa} = V_{ca} = -V_{ab} \tag{8-55}$$

where

$$V_{nfa} = V_{nfa}^{+} + V_{nfa}^{-} + V_{nfa}^{0}$$

or

$$V_{nfa} = 3V_{nfa}^{+} \tag{8-56}$$

likewise,

$$V_{ca} = 3V_{nfa}^{+}, \qquad \text{and} \qquad V_{ab} = -3V_{nfa}^{+}$$

Example 8-4. The same circuit (Fig. 8-15) will be used as was given in Example 8-3. However a double line-to-ground fault now exists at f, on phases b and c. Note that even though the zero-sequence impedances on T_2 and G_2 are shown finite on the figure, the absence of a neutral path between G_2 and T_2 will render Z^0 infinite at this point.

Find the phase b and phase c fault currents as well as the line and phase voltages at the fault location (f).

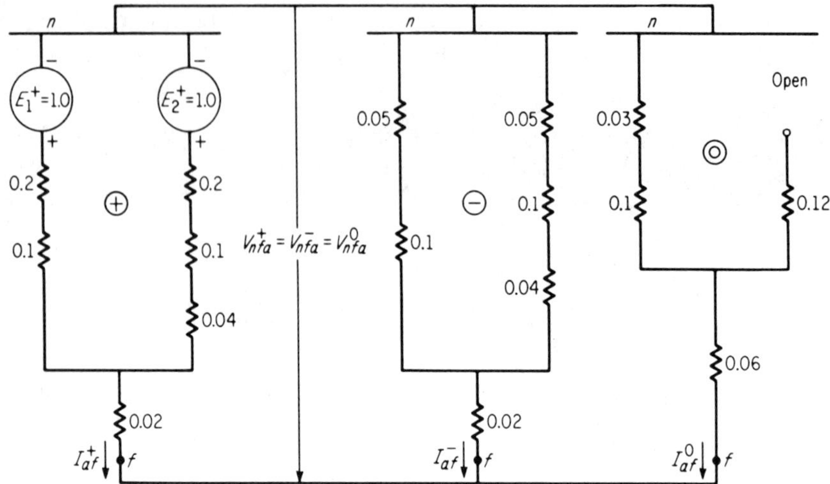

Fig. 8-21. Sequence network for Example 8-4.

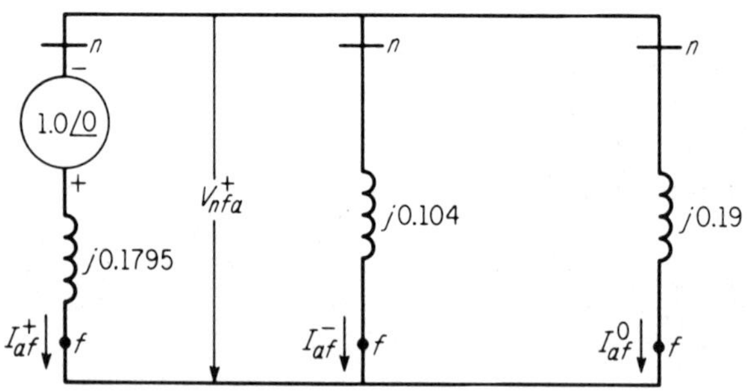

Fig. 8-22. Sequence network (reduced) for Example 8-4.

Figure 8-21 represents the sequence networks for phase a, connected together in the proper manner as discussed in Sec. 8-7. The reduced sequence network is shown in Fig. 8-22. First, the sequence component currents will be determined.

$$I_{af}^+ = \frac{E_a^+}{Z^+ + \dfrac{Z^-Z^0}{Z^- + Z^0}} = \frac{1.0\,\underline{/0^\circ}}{j0.1795 + \dfrac{(j0.104)(j0.19)}{j0.104 + j0.19}}$$

$$= \frac{1.0}{j0.247} = \underline{\underline{-j4.05}}$$

$$V_{nfa}^+ = E_a^+ - I_{af}^+ Z^+ = 1.0 - (-j4.05)(j0.1795)$$

$$= 1.0 - 0.727 = 0.273$$

$$I_{af}^- = \frac{V_{nfa}^+}{Z^-} = \frac{-0.273}{j0.104} = +j2.62$$

$$I_{af}^{0} = \frac{-V_{nfa}^{+}}{Z^{0}} = \frac{-0.273}{j0.19} = +j1.44$$

$$\begin{aligned} I_{bf} &= I_{af}^{+} \angle -120^\circ + I_{af}^{-} \angle +120^\circ + I_{af}^{0} \\ &= (-j4.05)(1 \angle -120^\circ) + (j2.62)(1 \angle 120^\circ) + j1.44 \\ &= 4.05(-0.866 + j0.5) + 2.62(-0.866 - j0.5) + j1.44 \\ &= -5.78 + j2.16 = \underline{\underline{6.16 \angle 159.5^\circ}} \text{ pu amp} \end{aligned}$$

$$\begin{aligned} I_{cf} &= I_{af}^{+} \angle +120^\circ + I_{af}^{-} \angle -120^\circ + I_{af}^{0} \\ &= (-j4.05)(1 \angle 120^\circ) + (j2.62)(1 \angle -120^\circ) + j1.44 \\ &= (4.05)(0.866 + j0.5) + (2.62)(0.866 - j0.5) + j1.44 \\ &= 5.78 + j2.16 = \underline{\underline{6.16 \angle 20.5^\circ}} \text{ pu amp} \end{aligned}$$

The total fault current flowing into the neutral is

$$I_n = I_{bf} + I_{cf} = -5.78 + j2.16 + 5.78 + j2.16 = j4.32$$

As a check, $I_n = 3I_{af}^{0} = 3 \times j1.44 = j4.32$ O.K.

$$\begin{aligned} V_{nfa} &= 3V_{nfa}^{+} = 3(0.273) = \underline{\underline{0.819}} \text{ pu volt} \\ V_{ab} &= -V_{nfa} = -\underline{\underline{0.819}} \text{ pu volt} \\ V_{ca} &= V_{nfa} = \underline{\underline{0.819}} \text{ pu volt} \end{aligned}$$

8-8. Summary

Chapter 8 has developed the basic method of symmetrical components and indicated its wide application to unbalanced short-circuit studies. Highlights of the chapter will be summarized briefly as follows:

(a) The concept of resolving an original unbalanced three-phase system into three balanced sets was set forth in Sec. 8-2. In each case an original phasor must equal the sum of all its parts, or $I_a = I_a^{+} + I_a^{-} + I_a^{0}$.

(b) Given the original unbalanced set of phasors, the positive-, negative-, and zero-sequence sets may be determined. For example,

$$V_a^{0} = \frac{1}{3}(V_a + V_b + V_c) \qquad [8\text{-}10]$$

$$V_a^{+} = \frac{1}{3}(V_a + aV_b + a^2V_c) \qquad [8\text{-}11]$$

$$V_a^{-} = \frac{1}{3}(V_a + a^2V_b + aV_c) \qquad [8\text{-}12]$$

Knowing the a-phase components, the b and c phase components are then given by equations 8-4 through 8-6.

(c) Zero-sequence currents may circulate within the delta of a transformer or flow into (or away from) a Y connection with neutral return. However, in either case, the test of available ampere-turns must be met for the transformer. Chapter 9 will treat the latter subject in more detail.

(d) While the principle of superposition vaguely underlies the symmetrical component approach, each type of unbalance problem is in reality approached directly through the use of a sequence-network equivalent. The author has often referred in the classroom to these equivalents as "gimmick" networks, since in every case the network may not appear obvious. Yet at the same time these most useful equivalents will meet all of the conditions of the fault in operation. This chapter has proven this to be the case. These "gimmick" networks are always related to phase *a*. The author refers to the three networks of Figs. 8-6, 8-14, and 8-19. Phase *a* is the faulted phase for the line-to-ground fault, but is the unfaulted phase for both the line-to-line and double line to ground faults. Phase *a* voltage and current components are readily found from these sequence networks and the information transferred to phases *b* and *c* through known angular relationships of the balanced sets. Procedures and example problems have been offered for each type of fault.

Problems

8-1. Find the symmetrical components for the three currents $I_a = 100 \underline{/0^\circ}$, $I_b = 150 \underline{/-110^\circ}$, $I_c = 120 \underline{/+130^\circ}$.

8-2. Find the symmetrical components for the three currents $I_a = 0$, $I_b = 10 \underline{/0^\circ}$, $I_c = 10 \underline{/180^\circ}$. Show phasor diagrams for all nine components.

8-3. A three-phase induction motor will overheat with excessive voltage unbalance at its terminals. A few percent of negative sequence voltage with its resulting counter revolving field can be critical. In a three-wire, three-phase motor the rms voltage readings (magnitudes only) indicate $|E_{ab}| = 120$, $|E_{bc}| = 114$, $|E_{ca}| = 122$. Assume *abc* sequence.

(a) Find, using the law of cosines, the phase angles on the given line-to-line voltages. Use E_{ab} as reference.

(b) Solve part (a) graphically.

(c) Determine the symmetrical components for both line to line and line to neutral voltages.

(d) Express the negative sequence voltage magnitude in percent of base voltage, where base voltage is 120 volts. This is one significant measure of motor unbalance.

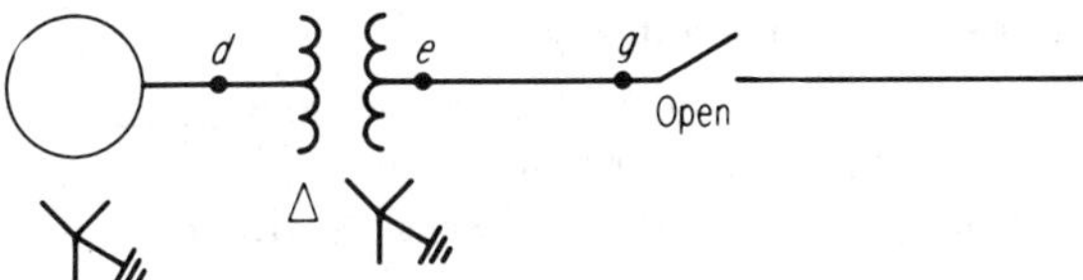

Prob. 8-4

8-4. The 60-Hz, 50-mva, 15-kv, Y-connected (solid ground) turbogenerator of Prob. 8-4 is unloaded and running at rated voltage with no system ties beyond point *g*. Reactance values are $X''_d = X^- = 0.11$ per unit ohms and $X^0 = 0.04$ pu, all on a 50-mva base. The transformer shown is connected 15-kv to 120-kv, Δ-Y with neutral solidly grounded. The transformer impedance is $Z^+ = Z^- = Z^0 = j0.10$ pu ohms, also on a 50-mva base. The line impedance from *e* to *g* (also on a 50-mva base) is: $Z^+ = Z^- = j0.03$ pu; $Z^0 = j0.09$ pu. Taking point *d* as a fault point, find the subtransient fault current for (a) a three-phase fault, (b) a line-to-ground fault. Also express this as a ratio of line-to-ground fault current to three-phase fault current of part (a). (c) Find the subtransient fault current for a line-to-line fault. Also express this as a percent of the three-phase fault current. (d) Find the subtransient fault current for a double line-to-ground fault.

8-5 Solve all parts of Prob. 8-4 with the new fault point of Prob. 8-4 at point *e*.

8-6. Solve Prob. 8-4 with the new fault point at point *g*.

8-7. Rework Prob. 8-4 with an impedance of $j0.5$ ohms (actual) in the generator neutral. Note especially the reduction of the magnitude of line-to-ground faults.

8-8. Rework Prob. 8-5 with an impedance of $j0.5$ ohms (actual) in the generator neutral.

8-9. Determine all three line currents (pu) in the generator of Prob. 8-4 with a line to ground fault at point *e*. Several methods are possible in this solution where transformer phase shift exists. While this problem is not beyond the scope of Chapter 8, more detailed treatment of such problems is offered in Secs. 9-2 and 9-8.

8-10. On the right-hand side of line *eg* of Prob. 8-4, close the switch and add a transformer and generator, duplicates of those described in Prob. 8-4. Refer to Prob. 8-10. Neglect prefault currents and consider no load voltages as 1.0 pu. With point *d* as the fault point, find the subtransient fault current for (a) a 3-phase fault, (b) a line-to-ground fault, (c) a line-to-line fault.

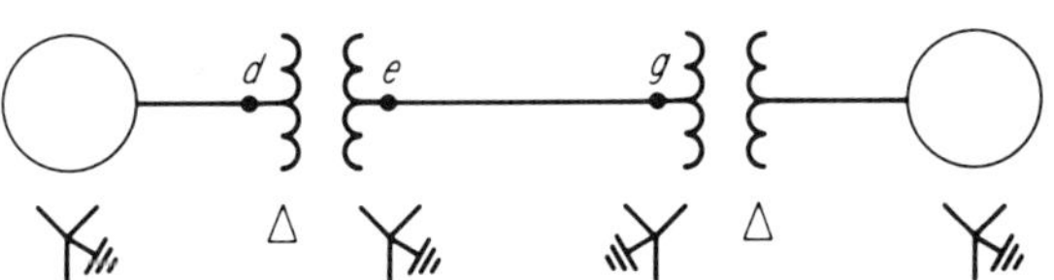

Prob. 8-10

8-11. Repeat Prob. 8-10 for a fault at *g*.

8-12. Given the circuit of Prob. 8-12 with the fault point at *f*. All tabulated reactance data to be used for fault calculations is given on a 100-mva base and referred to nominal system voltages. Neglect prefault currents.

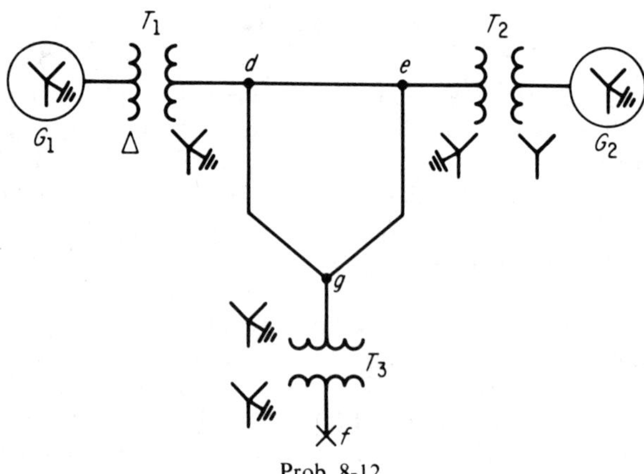

Prob. 8-12

Given: A three-phase symmetrical fault at *f*.
Find: The fault current and the current in line *dg*. (Example 4-1 makes good reference here.)

	$\% X^+$	$\% X^-$	$\% X^0$
Generator No. 1	40	40	10
Generator No. 2	40	40	10
Line *de*	50	50	200
Line *dg*	30	30	100
Line *eg*	30	30	100
Transformer T_1	20	20	20
Transformer T_2	20	20	20
Transformer T_3	15	15	15

8-13. Determine the line-to-neutral voltages for all phases at point *d* for Prob. 8-12.

8-14. Repeat Prob. 8-12 with a line-to-ground fault at point *f*. Also determine all three line-to-neutral voltages at point *d*.

8-15. Repeat Prob. 12 with a line-to-line fault at *f*.

chapter **9**

FURTHER CONSIDERATIONS FOR UNSYMMETRICAL FAULTS AND LOADS

9-1. Introduction

Chapter 8 purposely sidestepped some aspects of the unbalanced problem, such as the determination of impedance to positive-, negative-, and zero-sequence currents, phase shift in transformers, etc. This was done in order to move more directly to an understanding of the basic fault-calculating procedure. Chapter 9 will attempt to pick up a number of these separate considerations, all of which are closely related to the system of unbalanced fault or load. Note in particular Sec. 9-2, which employs a most basic principle, yet with its use one may bypass a more tedious procedure for certain types of unbalance problems.

9-2. Technique of Balancing Ampere-Turns for Problem Solving

From principles of electrical machinery, we know that, insofar as load or fault currents are concerned, the primary ampere-turns of a two-winding transformer must be balanced with the secondary ampere-turns, or $(NI)_p \doteq (NI)_s$. This is a fair approximation, the only difference being in the relatively small value of ampere-turns needed to excite the core with flux and provide core losses. In Sec. 8-4 we made use of the balancing ampere-turn requirement in determining whether a zero-sequence path was present for various transformer configurations. There are, however, applications for which the use of this simple concept will yield a quick and simple solution. Suppose for example, one would attempt to place a single-phase load on the transformer of Fig. 9-1. This would not be a workable situation because if I_{as} were to flow then I_{ap} would be required in the same amount. But I_{ap} cannot leave the junction into coils b and/or c because coils b and c of the primary are unable to find balancing ampere-turns in the open b and c phase secondary coils. It may in fact be possible to throw a direct short across the phase a sec-

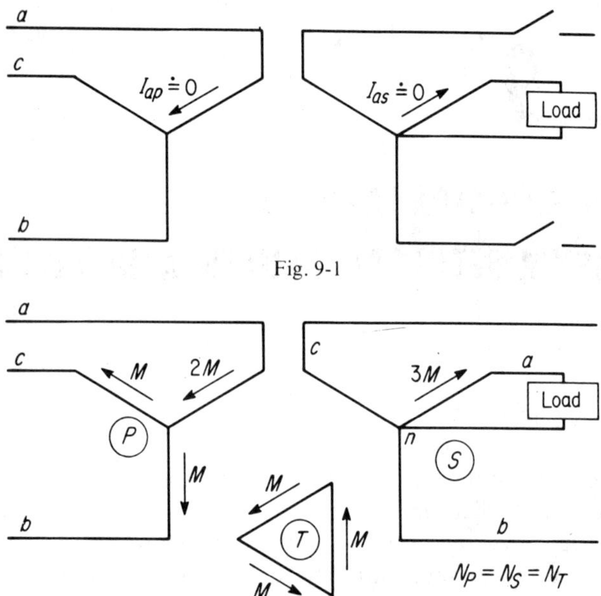

Fig. 9-1

Fig. 9-2. Analyzing current distribution of transformer with single-phase loading.

ondary without appreciable current flow. The electrical neutral will shift so as to cause V_{nas} to approach 0. We are also prepared to analyze the condition of Fig. 9-1 on the basis of zero-sequence current flow. We would say that the single-phase load is of a zero-sequence nature since $I_{as} = I_n = 3I^\circ$. But $I^\circ = 0$ since it sees an infinite impedance path in the primary with no neutral return. An exception to this appears in subsection (b) on page 181.

Figure 9-2 shows the addition of a delta tertiary. It is recalled from transformer theory that a Δ tertiary may serve one or more purposes such as the suppression of third-harmonic voltages, providing auxiliary loading, etc. We will show how the addition of the Δ tertiary will also make possible single-phase, unbalanced, secondary loading, even though a primary neutral return is absent. First we will analyze the distribution of currents in every coil of the transformer. Let M be the circulating current of the delta. Two legs of the delta can find no balancing ampere-turns in the secondary b and c coils, so if they exist at all they must find their counterparts in the primary (also marked M). Kirchhoff's current law at the junction of the primary requires that coil a of the primary carry $2M$ amperes. The sum of the phase a amperes of primary plus tertiary is $M + 2M = 3M$ which must be counterbalanced by $3M$ in phase a of the secondary. Now if the secondary current were known to equal 60 amperes then $3M = 60\,\underline{/0^\circ}$, $2M = 40\,\underline{/0^\circ}$, and $M = 20\,\underline{/0^\circ}$. The problem has been solved with a minimum of effort. While symmetrical components will also solve this problem it is more tedious in this case where I_{load} is known.

Example 9-1. Check the results of the preceding paragraph (and Fig. 9-2) using symmetrical components. I_{load} is again given as 60 amp.

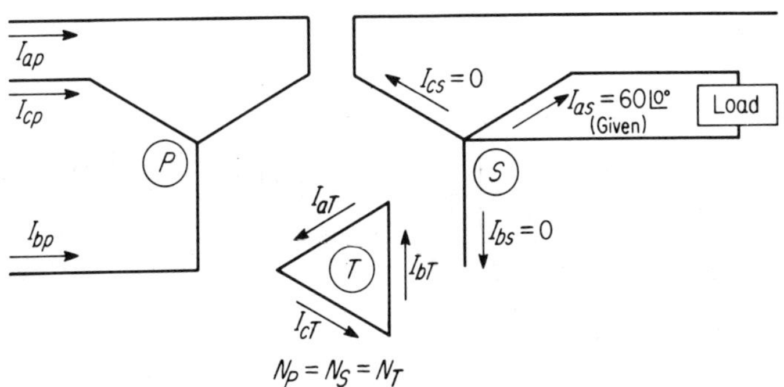

Fig. 9-3. Circuit for Example 9-1.

The figure is redrawn with the more conventional three-phase assumed current directions as shown in Fig. 9-3.

$$I_{as}^0 = \frac{1}{3}(I_{as} + \cancelto{0}{I_{bs}} + \cancelto{0}{I_{cs}}) = \frac{1}{3}(60\,\underline{/0^\circ}) = 20\,\underline{/0^\circ}$$

$$I_{as}^+ = \frac{1}{3}(I_{as} + a\cancelto{0}{I_{bs}} + a^2\cancelto{0}{I_{cs}}) = \frac{1}{3}(60\,\underline{/0^\circ}) = 20\,\underline{/0^\circ}$$

$$I_{as}^- = \frac{1}{3}(I_{as} + a^2\cancelto{0}{I_{bs}} + a\cancelto{0}{I_{cs}}) = 20\,\underline{/0^\circ}$$

Since no zero sequence is permitted in the primary winding, all zero-sequence ampere-turns of the secondary must be balanced by the tertiary. The delta, in fact, carries only the equal zero sequence currents, or $I_{at} = I_{bt} = I_{ct} = I_{as}^0 = 20$ amp.

The positive- and negative-sequence secondary currents are balanced by primary ampere turns or $I_{ap}^+ = 20\,\underline{/0}$, $I_{bp}^+ = 20\,\underline{/-120}$, $I_{cp}^+ = 20\,\underline{/+120^\circ}$, $I_{ap}^- = 20\,\underline{/0^\circ}$, $I_{bp}^- = 20\,\underline{/120^\circ}$, $I_{cp}^- = 20\,\underline{/-120^\circ}$.

$$I_{ap}^0 = I_{bp}^0 = I_{cp}^0 = 0$$

$$I_{ap} = I_{ap}^+ + I_{ap}^- + I_{ap}^0 = 20 + 20 + 0 = \underline{\underline{40\,\underline{/0^\circ}}}$$

$$I_{bp} = I_{bp}^+ + I_{bp}^- + I_{bp}^0 = 20\,\underline{/-120^\circ} + 20\,\underline{/+120^\circ} = \underline{\underline{-20\,\underline{/0^\circ}}}$$

$$I_{cp} = I_{cp}^+ + I_{cp}^- + I_{cp}^0 = 20\,\underline{/120^\circ} + 20\,\underline{/-120^\circ} = \underline{\underline{-20\,\underline{/0^\circ}}}$$

Results agree with the previous method.

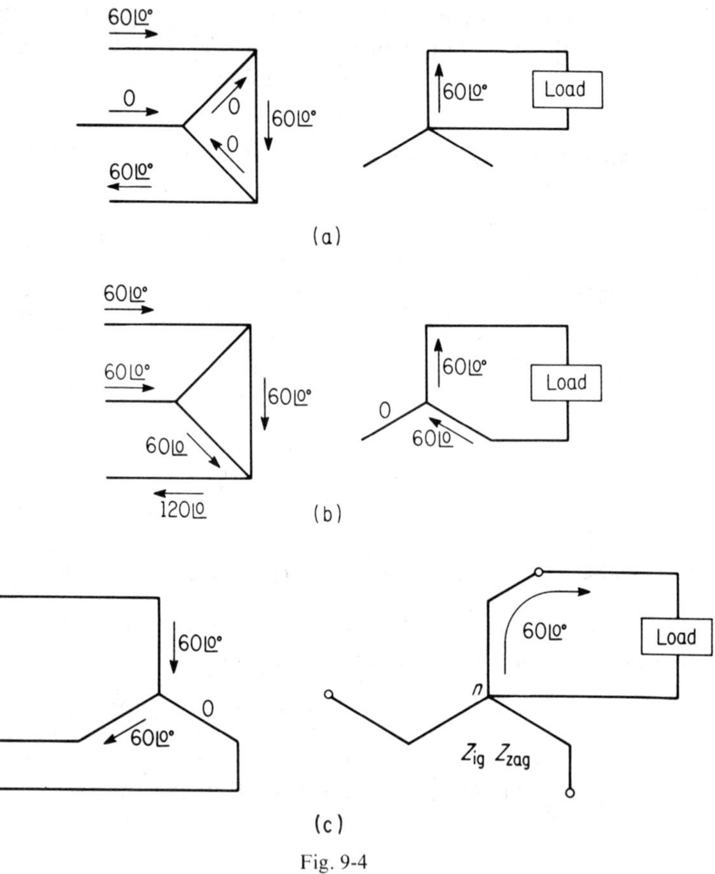

Fig. 9-4

Several other examples of unbalanced loading are given in Figs. 9-4a–c. In each case, the load current is assumed to be 60 amperes. From here it requires merely an "eyeball exercise" to obtain results using the balancing-ampere-turns method. Results of the current distribution are placed on the drawing without further comment. All windings have equal turns.

9-3. Sequence Impedances—General

Considerable confusion often prevails when one refers to sequence impedances. This is due to the fact that there are two entirely separate sets of impedances to which one could make reference. First of all, it is possible to resolve three original impedances into positive-, negative-, and zero-sequence impedance components. The components would take the same general form as Eqs. 8-10 through 8-12 where the voltages are replaced by impedances. This, however, is *not* the type of sequence impedance referred to in this text. A more proper term for the sequence impedances of this text would be *impedance to positive-, negative-, or zero-sequence currents.* These impedances are applied to three phase systems with balanced phase impedances. For this reason it is assumed from this point on that the impedance to positive sequence current is the same for each phase. Therefore, the sequence impedance phase subscript will be dropped as was done in Chapter 8. In other words,

$$Z_a^+ = Z_b^+ = Z_c^+ = Z^+ \tag{9-1}$$

and

$$Z_a^- = Z_b^- = Z_c^- = Z^- \tag{9-2}$$

$$Z_a^0 = Z_b^0 = Z_c^0 = Z^0 \tag{9-3}$$

The concept of phase impedance to the positive-sequence current is not difficult to visualize as it merely implies that Z_a^+ is the ratio of some voltage in the positive-sequence network to some current in that network, or

$$Z_a^+ = V_a^+/I_a^+ \tag{9-4}$$

9-4. Sequence Impedances for Transmission Lines

It is not the intent here of repeating a detailed treatise on transmission-line impedance. Rather, a more general look at impedance to sequence currents will be taken. Refer to Appendix A for a treatment of line parameters.

First we observe that the per phase line impedances to positive- and negative-sequence currents are the same, or $Z_{\text{line}}^+ = Z_{\text{line}}^-$. It is recalled from Secs. 3-3 and 4-5 that the short line can be represented by one series impedance and, in fact, can often be further simplified to one series line reactance, when $X_{\text{line}} \gg R_{\text{line}}$. Again, we assume that line impedances are equal (balanced) in all phases, a fairly good assumption when proper transposition of conductors has been made. The actual calculation of R_{line} and X_{line}, based on the conductor material and dimensions, will not be included here. The primary objective of this text from the beginning is to deal with the network solution, where circuit parameters have been given.

Per phase line impedance (Z^0) to zero sequence current is not, in general, equal to the line impedance to positive- and negative-sequence currents. Recalling that all zero-sequence currents are in phase, the neutral return path must be included as a part of the impedance. Refer to Fig. 9-5 in which zero-sequence currents are seen to travel through three lines in parallel, then returning together through the neutral path. Two general observations will be made with regard to Fig. 9-5.

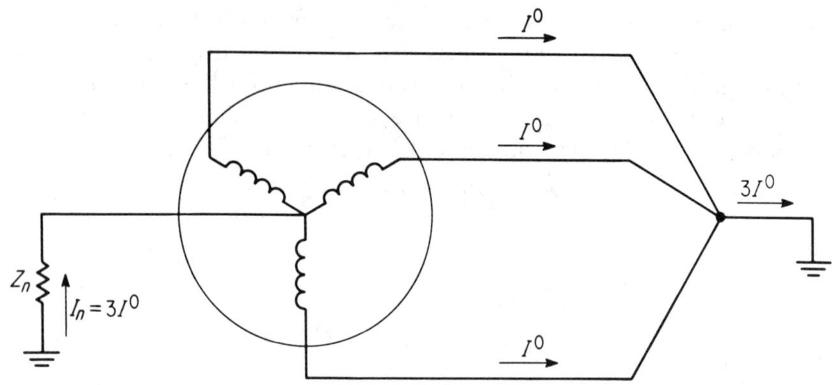

Fig. 9-5. Flow of zero-sequence currents.

1. The zero-sequence voltage drop across the neutral impedance Z_n is proportional to three times I^0. Therefore neutral impedance to zero sequence current of the generator is considered as $3Z_n$.

2. The Z^0 of the line itself is not a simple matter to calculate, even for the short line. All three lines are mutually coupled together, and this mutual coupling will, of itself, offer greater reactance to I^0 than is offered to I^+ or I^-. The reason for this is found in the fact that zero-sequence currents are all in phase and therefore corresponding fluxes are in phase. This will obviously cause more mutual coupling (and higher mutual inductance) than for the case of either positive- or negative-balanced currents where individual phase currents (and corresponding fluxes) are out of phase by 120°. It should also be realized that this mutual inductance between phases is always included as part of the total phase inductance of the line, along with the self-inductance per phase. This effect of increased magnetic coupling may cause Z^0_{line} to be several times as large as Z^+ or Z^-.

The final observation regarding sequence line impedance also concerns the general theory of *mutual impedance* between two three-phase lines running along in parallel. They may or may not have common busses at the line ends. Refer to Secs. 3-3 and 4-15 for an explanation of how we include such mutual impedance in the network. When impedance parameters are furnished for the network solution, it is noted that the per phase mutual impedance to zero-sequence may be significant while the mutual impedance to positive- and negative-sequence currents is often taken to be zero. Figure 9-6 illustrates the two sets (K and K') of three-

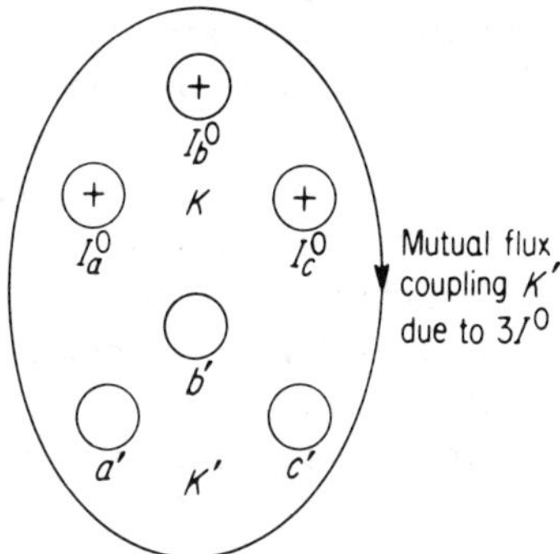

Fig. 9-6. Two three-phase lines running together.

phase lines running together. The difference between mutual impedance to zero-sequence and mutual impedance to positive- and negative-sequence becomes quite clear when it is realized that balanced currents (both positive- and negative-sequence) flowing in the K lines sum to zero, therefore the total flux linkages of the K set tend to cancel one another. At the same time, the zero-sequence currents of the K system are in phase and their fluxes which couple K' are additive. For convenience it has been assumed that there is one simple mutual flux path in Fig. 9-6, which is an oversimplification.

9-5. Sequence Impedances of Generators

Positive-sequence reactance is generally taken as the direct-axis synchronous reactance X_d, since for short circuit studies, the power factor of the short-circuit current is nearly zero and lagging. This was disucssed in Chapter 7. It is assumed when X_d is used that one is concerned with the determination of sustained (or steady-state) short-circuit currents. If subtransient or transient periods are under study, X_d'' or X_d' will be used, respectively.

Impedances to positive- and negative-sequence currents are not the same in synchronous generators. This is not difficult to imagine considering that the revolving field (due to positive-sequence armature currents) rotates *with* the rotor while the negative-sequence armature currents (of *acb* sequence) set up a revolving field rotating at the same speed but in opposite direction to the rotor. It is not to be expected that these two counterrotating fluxes will encounter the same opposition. Consider, for example, the case of the negative-sequence field which passes the poles, damper windings, and field windings at twice synchronous speed, thereby encountering higher opposition and lower effective reactance. In reality this negative-sequence reactance will vary almost sinusoidally with time between maximum and minimum values as it encounters an ever-changing rotor configuration.

In fact, the situation encountered by the negative-sequence flux is much like that which is met by the flux of a suddenly changing armature current immediately after undergoing a short circuit. Recall from Sec. 7-3 that opposition to sudden flux change is offered by field windings and damper windings, if present.

Figure 9-7a demonstrates the physical situation which exists for the

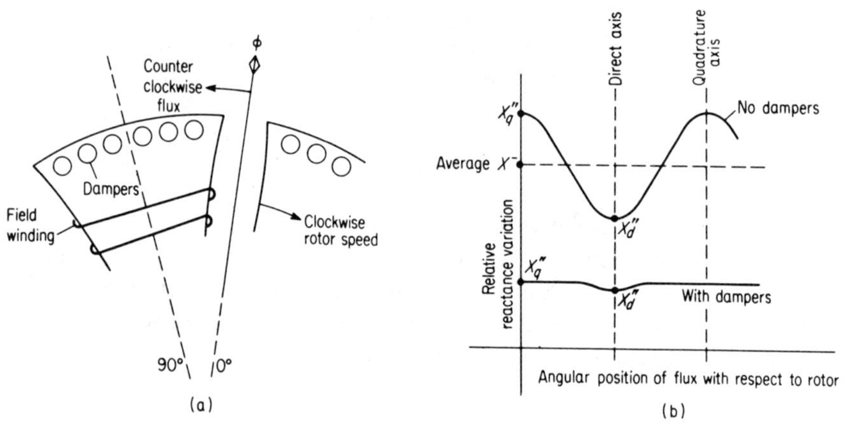

Fig. 9-7

flux due to negative-sequence armature currents. Note that as time progresses the flux will move alternately through the quadrature, then the direct axis.

The top curve of Fig. 9-7b shows the variation of subtransient reactance for various positions of the rotor (*without* dampers) relative to the position of negative-sequence flux. The pronounced effect of the field-winding inductance is clearly in evidence. Notice that the reactance fluctuates between the values of subtransient quadrature-axis reactance X_q'' and subtransient direct-axis reactance X_d''. Despite this fluctuation, an average negative-sequence reactance is used where

$$X^- = \frac{X_d'' + X_q''}{2} \tag{9-5}$$

Next refer to the bottom curve of Fig. 9-7b. If damper windings are present, the negative-sequence flux meets constant opposition from the dampers. As a result, this flux is forced even further from the rotor path, increasing effective reluctance and lowering effective reactance. This is much the same treatment given to the flux of subtransient short-circuit currents, treated in Sec. 7-3. Again, the average negative-sequence reactance for machines *with damper windings* is given by Eq. 9-5 using reactances from the bottom curve of Fig. 9-7b. Refer to Table 9-1 for typical machine reactances.

Table 9-1*

TYPICAL CONSTANTS OF THREE-PHASE SYNCHRONOUS MACHINES

(Reactances in per unit. Values below the line give the normal range of values, while those above give an average value.)

	1	2	3	4	5	6
	X_d (Unsat)	X_q Rated Current	X_d' Rated Voltage	X_d'' Rated Voltage	X^- Rated Current	X^0 Rated Current
2-pole turbine generators	1.20 0.95–1.45	1.16 0.92–1.42	0.15 0.12–0.21	0.09 0.07–0.14	$= X_d''$	0.03 0.01–0.08
4-pole turbine generators	1.20 1.00–1.45	1.16 0.92–1.42	0.23 0.20–0.28	0.14 0.12–0.17	$= X_d''$	0.08 0.015–0.14
Salient-pole generators and motors (with dampers)	1.25 0.60–1.50	0.70 0.40–0.80	0.30 0.20–0.50	0.20 0.13–0.32	0.20 0.13–0.32	0.18 0.03–0.23
Salient-pole generators (without dampers)	1.25 0.60–1.50	0.70 0.40–0.80	0.30 0.20–0.50	0.30 0.20–0.50	0.48 0.35–0.65	0.19 0.03–0.24

*Table reproduced by permission of Westinghouse Electric Corporation, from "Electrical Transmission and Distributor Reference Book," 4th ed.

It is worthwhile to note from Fig. 9-7b that, with or without dampers, $X_q'' > X_d''$. At the same time, we recall that for the steady-state case of direct and quadrature synchronous reactance, $X_d > X_q$ for the salient-pole machine.

The reasoning is as follows. Refer to Figs. 7-3 and 7-4. X_d exceeds X_q because of a lower reluctance path to the steady state armature flux. However, X_q' exceeds X_d' because of the high field inductance present when the changing, transient flux is centered on the direct axis. In other words, for negative-sequence, transient, or subtransient considerations, the effect of field inductance and/or damper-winding inductance will predominate, even for the salient-pole machine. A transient reactance variation curve would take the general shape of Fig. 9-7 (top curve).

Several methods are available for the actual determination of X^- by test. The method given here is simple and direct. Drive the generator at rated speed with a line-to-line short between phases b and c, as shown

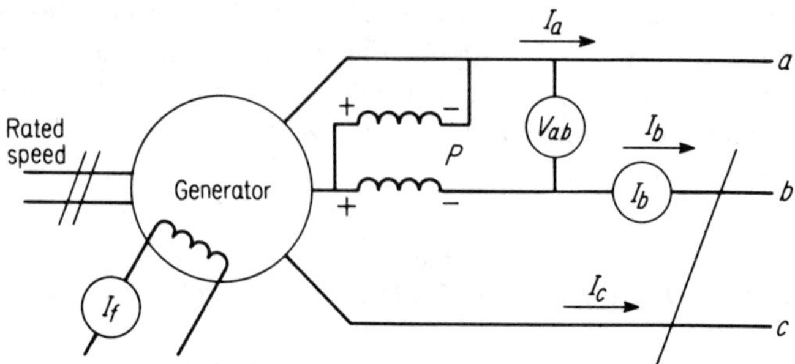

Fig. 9-8. Circuit for determination of X^-.

in Fig. 9-8. Adjust I_f to yield approximately rated I_b. Record $|V_{ab}|$, P and $|I_b|$ as shown. We know that

$$Z^- = \frac{-V_{na}^-}{I_a^-} \tag{9-6}$$

This was demonstrated in Fig. 8-14 for the line-to-line fault. But from Eq. 8-44,

$$V_{ab} = -3V_{na}^+ = -3V_{na}^-$$

or

$$V_{na}^- = \frac{-V_{ab}}{3} \tag{9-7}$$

From Eq. 8-41,

$$I_b = \sqrt{3}I_a^+ \angle -90 = \sqrt{3}I_a^- \angle +90^\circ$$

or

$$I_a^- = \frac{I_b}{j\sqrt{3}} \tag{9-8}$$

Substituting Eqs. 9-7 and 9-8 into Eq. 9-6,

$$Z^- = \frac{V_{ab}/3}{I_b/j\sqrt{3}} = \frac{jV_{ab}}{\sqrt{3}I_b} \tag{9-9}$$

or

$$Z^- = \frac{|V_{ab}|}{\sqrt{3}|I_b|}(\sin\theta + j\cos\theta) \tag{9-10}$$

where θ is the angle between the voltage and current, or

$$\theta = \cos^{-1} \frac{P}{|V_{ab}|\,|I_b|} \tag{9-11}$$

$$R^- = \frac{|V_{ab}|}{\sqrt{3}\,|I_b|} \sin\theta ; \; X^- = \frac{|V_{ab}|}{\sqrt{3} I_b} \cos\theta \tag{9-12}$$

Notice that it is not safe in this test to assume that X^- equals Z^-, as is often done for positive-sequence reactances. A typical value of X^- may be on the order of about 20 percent of X^+, which could cause R^- to be a significant part of Z^-.

Generator impedance to zero-sequence is smaller yet than the impedance to negative-sequence. For example, it would not be unusual for the value of X^0 to be only 5 percent of X^+. The explanation for this lies in the fact that zero-sequence armature currents of the three phases are all in phase but physically displaced from each other by 120° electrically. Theoretically the sum total of the three sinusoidally distributed mmf's is zero, which would theoretically result in zero reactance. However, some slot and end-connection leakage reactance is present. Also the actual mmf distribution will not be perfectly sinusoidal, the waveshape depending to some degree upon the pitch and distribution factors. Zero-sequence resistance will compare favorably with the positive-sequence resistance. As mentioned previously, this resistance is often neglected in the fault calculation.

In order to test for Z^0, one need merely to connect the three phases of the armature in series and apply a single-phase voltage, recording V, P, and I as in Fig. 9-9. The rotor may either be turning at synchronous speed

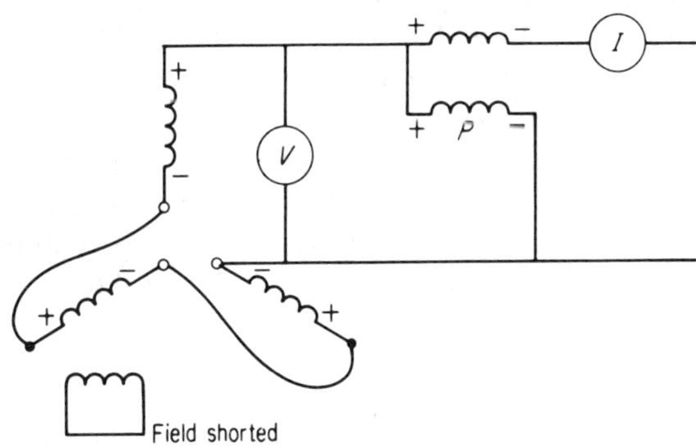

Fig. 9-9. Circuit for determining Z^0.

or average readings can be taken with rotor blocked in both the direct axis and quadrature axis positions. The field circuit is short circuited. Then

$$Z^0 = \frac{|V|}{3|I|}(\cos\theta + j\sin\theta) \tag{9-13}$$

where

$$\theta = \cos^{-1} P/VI \tag{9-14}$$

Again, it is understood that a neutral path must be provided for the synchronous generator or the impedance to zero-sequence current becomes infinite.

9-6. Sequence Impedances of Transformers

The impedance of a transformer to both positive- and negative-sequence currents is the same—as was the case for the transmission line. This fact is apparent when one realizes that the only difference in the nature of currents in these two systems lies in their phase sequence. A differing phase sequence would obviously not change the per phase impedance any more than would the interchange of two lines leading to the transformer.

One general observation is often made regarding the zero-sequence impedance of conventional two-winding transformers. It is often assumed that if the transformer permits zero-sequence current flow at all, the per phase impedance to zero-sequence is merely the ordinary series equivalent impedance (Z_{tr}) of the transformer, and

$$Z^0 = Z^+ = Z^- = Z_{tr} \tag{9-15}$$

If, on the other hand, zero-sequence current is not permitted, then Z^0 is *infinite*. Equation 9-15 is somewhat of an oversimplification in some instances, as will be pointed out. To review some of the conditions necessary for permitting the flow of zero-sequence current, the reader should refer back to Sec. 8-4 and Example 8-2.

Figure 9-10 will serve as a reference for an individual treatment of some of the more common transformer configurations and their impedances to zero-sequence. (*Note:* Wherever a switch is encountered, it will be considered closed only if the side to which the switch corresponds has a grounded neutral.)

(a) *Y-Y transformation* with either a bank of three single-phase transformers or one three-phase shell type. Here, the per phase impedance to zero-sequence equals the ordinary series transformer impedance (Z_{tr}). However, we see by the presence of S_p and S_s that a zero-sequence current cannot pass through unless both primary and secondary neutrals are grounded. A return path for these in phase currents is not sufficient merely on one side of the transformer because the requirement of balancing ampere-turns must be met.

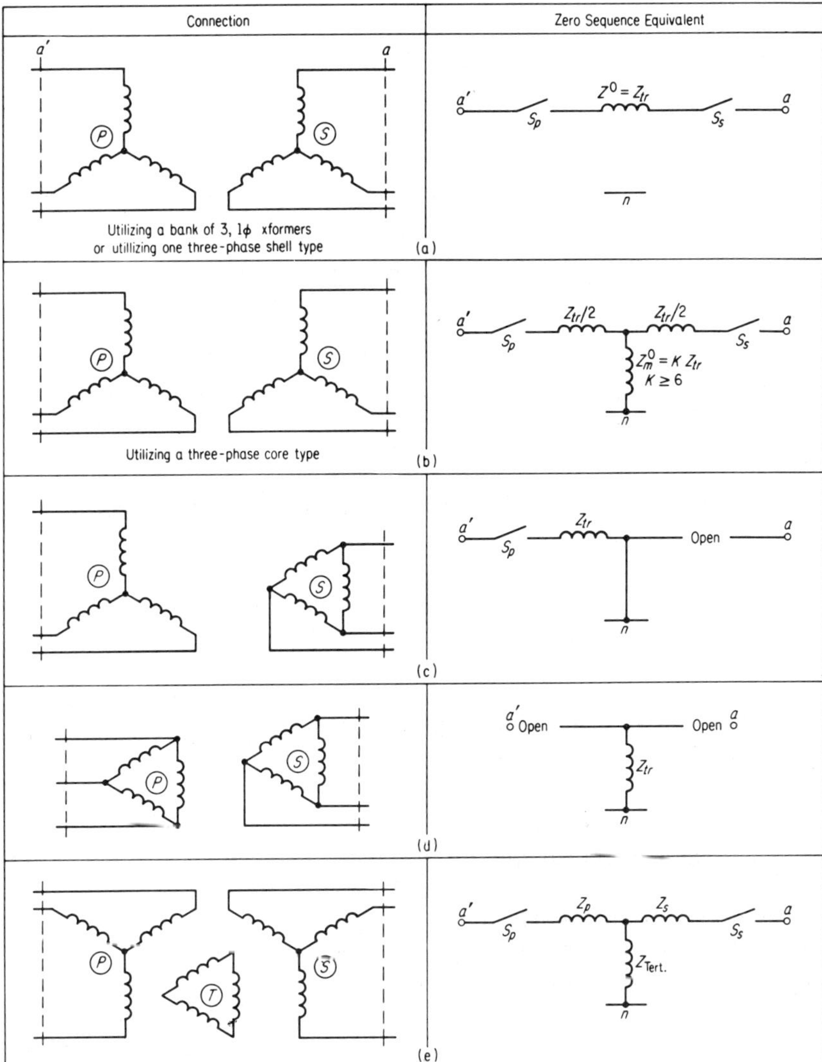

Fig. 9-10. Common transformer configurations and their corresponding zero-sequence equivalents. (*Note:* a switch is considered closed only if the side to which the switch corresponds has a grounded neutral.)

(b) *Y-Y transformation* with three-phase core type construction. In order to explain the difference between the equivalent of Figs. 9-10a and 9-10b we must look at the path of zero-sequence flux in the core. Refer to Figs. 9-11a and 9-11b. When transformation is made with either the three-phase shell type of Fig. 9-11a or a bank of three single-phase units, a fairly independent low-reluctance iron path is offered in each phase for

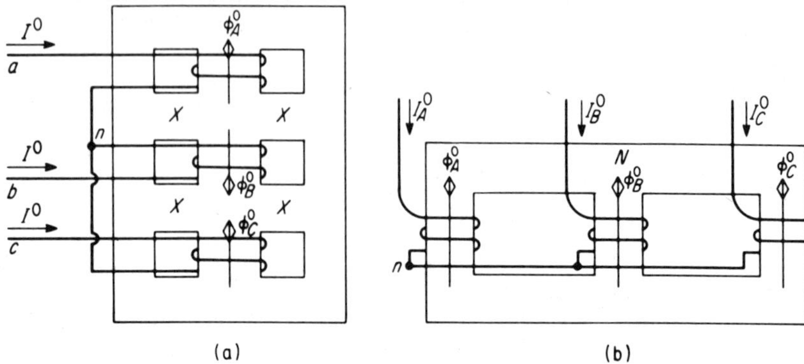

Fig. 9-11. (a) Tracing the path for zero-sequence flux in the three-phase shell type transformer. (b) Tracing the path for zero-sequence flux in the three-phase core type transformer.

the in-phase zero-sequence flux. This low reluctance means high magnetizing shunt impedance, which is neglected in Figs. 9-10a and 9-11a. However, ϕ_A^0, ϕ_B^0, and ϕ_C^0 of the core-type unit (Fig. 9-11b) all join together at the N-junction and $\phi_{\text{total}}^0 = 3\phi_A^0$. Yet the only return path for this flux is through the tank, air and/or oil, etc., of the transformer. This obviously is a path of high reluctance, meaning zero-sequence magnetizing current is high and shunt-magnetizing impedance may be too low to disregard. For example, if the value of per unit $Z_{tr} = 0.05$, a typical value of shunt-magnetizing impedance (Z_m) might be 0.3, meaning $Z_m \doteq 6Z_{tr}$. Refer to Fig. 9-10b. Because this shunt path has a similar effect to that of a delta tertiary, the Y-Y core type is often regarded as though it contained a fictitious delta of relatively high impedance. The equivalent then corresponds somewhat to the configuration of the Y-Y-Δ of Fig. 9-10e. The Y-Y core type will likewise serve to suppress third harmonic voltages (line-to-neutral) even in the absence of a neutral or delta tertiary. True, some will continue to neglect the shunt impedance of Fig. 9-10b for the sake of simplifying the problem of unbalance loading or line-to-ground faults. Actually the series impedance to zero-sequence is also somewhat less than that for the positive- and negative-sequence.

Another interesting sidelight to the shell configuration of Fig. 9-11a concerns the intentional reversal of the winding in leg B. The reason for this is generally treated in detail in the machine theory. In brief, considerable core-iron saving is made possible. It can easily be shown that the summation of flux for balanced, three-phase systems is considerably less in certain core regions (marked with an X) when the middle phase is reversed.

(c) *Y-Δ transformation.* This permits zero-sequence current to flow in from the left when the neutral is grounded. In such a case the per phase zero-sequence current is considered to complete its path back to the neutral bus through the relatively low series equivalent impedance of the transformer. Of course the impedance to zero-sequence of the secondary line side is infinite as is the case in any such three-wire system with no return. A considerable study was made of the Y-Δ arrangement in Example 8-2.

(d) *A Δ-Δ transformation* is open-circuited to incoming zero-sequence currents on both primary and secondary side, again since their is no neutral path by which these currents can return.

(e) *The Y-Y with Δ tertiary* permits zero-sequence current to flow in from either Y line (if neutrals are grounded). As an example, assume a line to neutral short circuit were placed on the Y secondary. Any zero-sequence current entering the primary would then divide inversely at the zero-sequence impedances of secondary and tertiary.

The foregoing principles concerning zero-sequence impedances for transformers are most important in setting up the complete network for the solution of line-to-ground and double line-to-ground faults. For a more complete listing of transformer configurations (and the corresponding sequence impedances) refer to the Westinghouse publication, *Transmission and Distribution Reference Book*.

9-7. Measurement of Sequence Voltages and Currents

It is sometimes necessary to be selective in either the metering or relaying of sequence currents and voltages. For example, a relay may be required to detect only the zero-sequence current flowing in a three-phase line. Zero sequence is therefore said to be segregated or filtered from the line currents. While the reader should be aware of the existence of such segregating schemes, this text will not undertake a detailed treatment of them. In general, the particular method must meet the requirements of Eqs. 8-10 through 8-12 and 8-17 through 8-19. This section will present

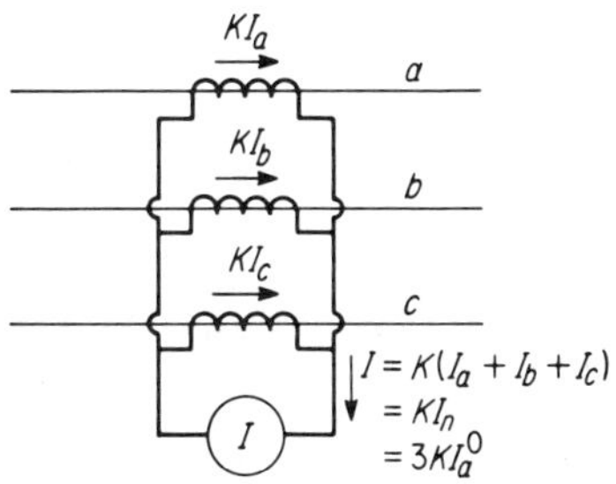

Fig. 9-12. Measuring zero-sequence current.

only two such schemes for detecting zero-sequence current and zero-sequence voltage. For a more detailed discussion of this subject, refer to J. G. Tarboux *Introduction to Electric Power Systems*, International Textbook Company, and Westinghouse *Transmission and Distribution Reference Book*.

Measurement of zero-sequence current can be achieved with the aid of three current transformers connected in parallel as in Fig. 9-12. This circuit satisfies Eq. 8-17, which states that I_a^0 is proportional to the sum of the three line currents. This is true, since $I = K(I_a + I_b + I_a) = KI_n = 3KI_a^0$.

Zero-sequence voltage is measured with the aid of the three potential transformers connected as in Fig. 9-13. The voltmeter will give an indication proportional to V_{na}^0, since $V = K_1(V_{na} + V_{nb} + V_{nc}) = 3K_1V_{na}^0$.

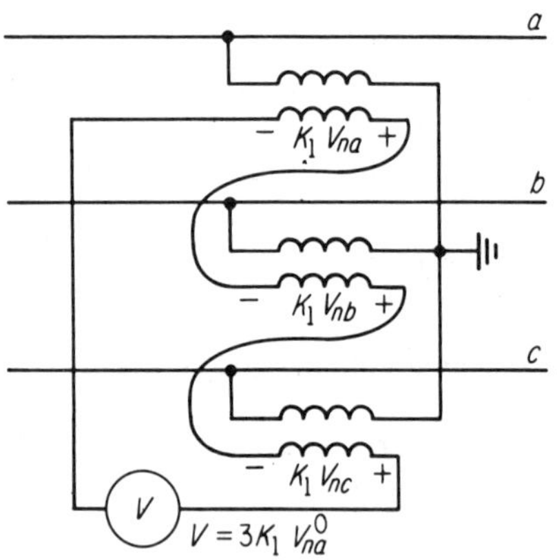

Fig. 9-13. Measuring zero-sequence voltage.

9-8. Transformer Polarity and Phase-Shift Considerations

The possibility of a slight voltage phase shift due to the per phase series equivalent impedances of a transformer is not in question here, since it is of little consequence in the short circuit problem. All short-circuit example problems of Chapters 7 and 8 assumed zero transformer phase shift all of the way from the point of fault back to the generator. This assumption is permissible for Y-Y or Δ-Δ transformations. However, the Δ-Y or Y-Δ transformer will not yield zero phase shift of voltages for the balanced three-phase system (including the positive- and negative-sequence systems). Basic machine theory is careful to point out the fact that Y-Δ transformers of 30° phase shift should not be paralleled with Y-Y or Δ-Δ transformers of zero phase shift.

Even though phase-shift may occur in a large short-circuit problem, the study is first made without regard to phase shift. This will yield a correct solution at the point of the fault, from which point one may work back into the network, accounting for phase shift where necessary. While the Δ-Y and Y-Δ phase shift problem complicates the procedure, it is not as difficult as it might appear at first glance.

Standard markings for a *single-phase transformation* will require that H_1 be positive with respect to H_2 at the same instant that X_1 is positive with respect to X_2, as in Fig. 9-14a. In power-transformer problems we

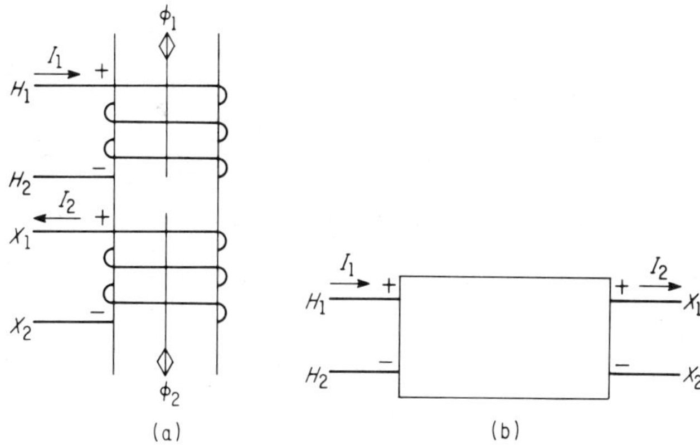

Fig. 9-14

often think of current as entering the positive terminal of the primary and leaving the positive terminal of the secondary. Power then flows from left to right through the circuit of Fig. 9-14b. The currents I_1 and I_2 set up ϕ_1 and ϕ_2 in opposite directions. The only difference between ϕ_1 and ϕ_2 is the constant, pulsating, magnetizing flux needed to satisfy Faraday's law $\left(e = N \dfrac{d\phi}{dt}\right)$.

The terms "additive" and "subtractive" polarity merely have reference to the relative physical locations of the X and H terminals. Figure

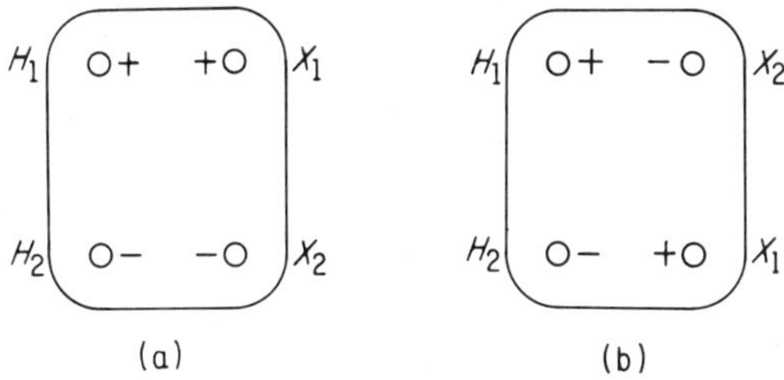

Fig. 9-15. (a) Subtractive polarity. (b) Additive polarity.

9-15a represents terminals of subtractive polarity, while Fig. 9-15b is additive. If one were to connect across the top terminals of the drawing and then excite one winding, the voltage measured across the bottom terminals will be either the sum or difference of the H and X winding voltages. Figure 9-15a would measure the difference of voltages from H_2 to X_2 and

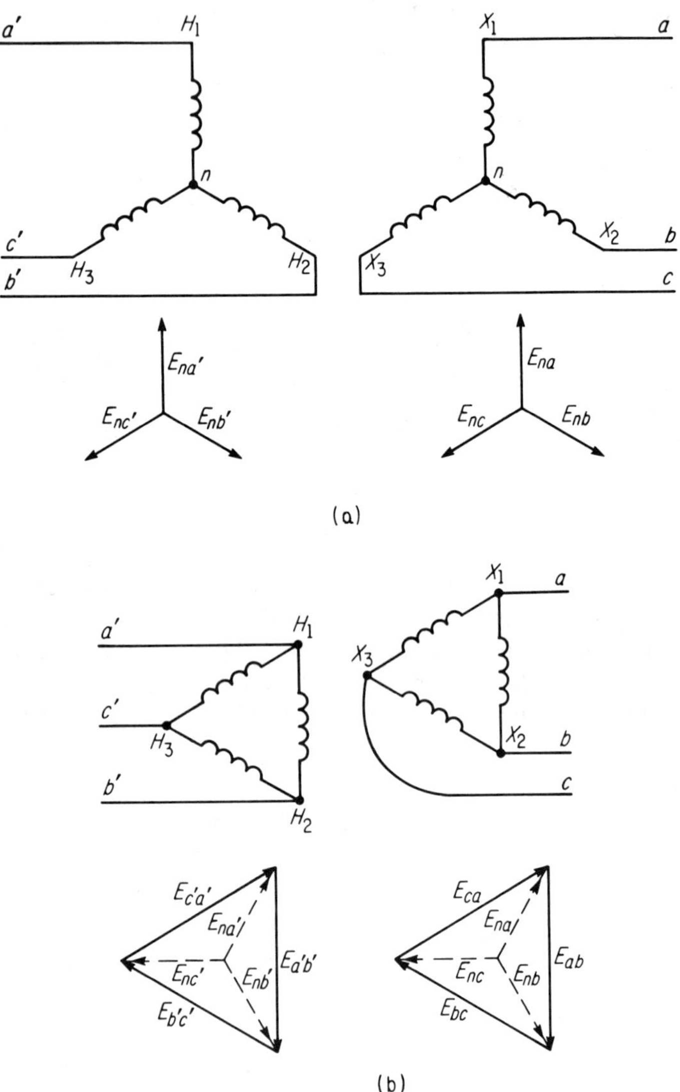

Fig. 9-16. Standard markings and voltage phasor diagrams for Y-Y and Δ-Δ transformations with 0° phase shift.

is therefore subtractive. By the same token, the voltage from H_2 to X_1 of Fig. 9-15b is additive.

Standard conventions have also been adopted for the markings of *three-phase transformer* terminals. The high-voltage line terminals are marked H_1, H_2, and H_3 and the low-voltage terminals are X_1, X_2 and X_3. ASA standards specify that, for the Y-Y and Δ-Δ connections, the voltages from neutral to H_1, H_2, and H_3 will be in phase with voltages from neutral to X_1, X_2, and X_3 respectively. There is, of course, the possibility of connecting Y-Y and Δ-Δ transformers for either 0° or 180° phase shift. Typical Y-Y and Δ-Δ connections are pictured in Figs. 9-16a and 9-16b with phasor diagrams below the schematics. The positioning of the primary and secondary coils on the drawing are significant in that a vertical primary coil couples a vertical secondary coil, etc.

Next we turn our attention to the more troublesome Y-Δ and Δ-Y connections. ASA Standards require that voltages from neutral to H_1, H_2, and H_3 lead the secondary voltages from neutral to X_1, X_2, and X_3 (respectively) by 30°, for both Y-Δ and Δ-Y configurations. This does not, however, imply the primary a', b', and c' notation must correspond to secondary a, b, and c notation in the same way. This latter consideration is a matter of personal choice as will be explained later. Figure 9-17 is meant to demonstrate how the 30° phase-shift requirement is met in a

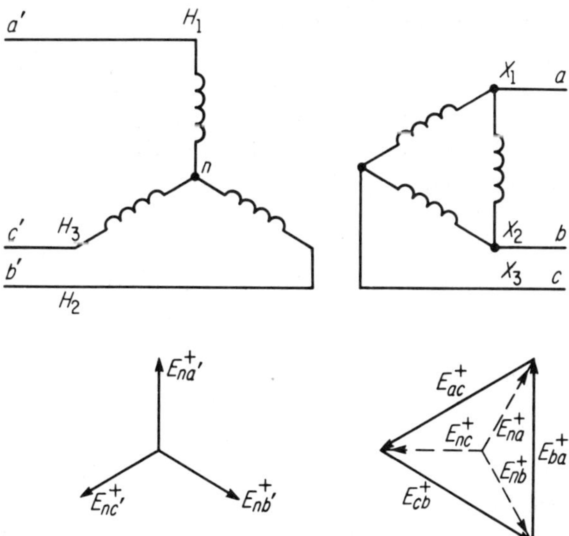

Fig. 9-17. Positive sequence phasor diagrams for Y-Δ transformer with standard H and X markings. $E^+_{na'}$ leads E^+_{na} by 30°.

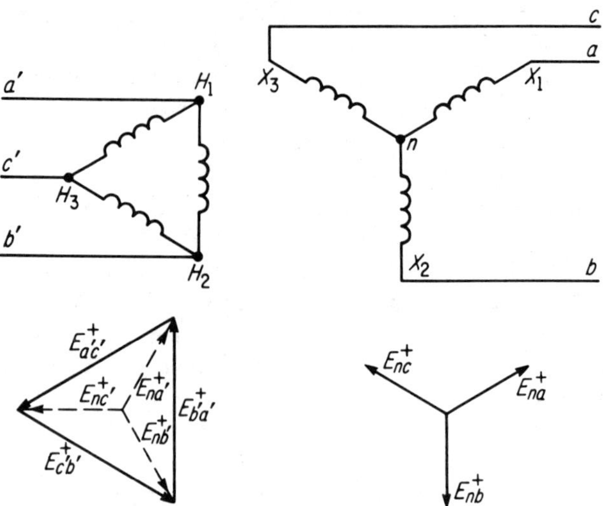

Fig. 9-18. Positive-sequence phasor diagrams for Δ-Y transformer with standard H and X markings. $E^+_{na'}$ leads E^+_{na} by 30°.

Y-Δ transformation, while Fig. 9-18 shows the phase shift of a Δ-Y. Notice the secondary Y-connected coils of the Δ-Y transformer have been reversed (compared with the Y- of Fig. 9-17) to accomplish a 30° lag in secondary voltages with respect to the primary voltages.

Our main concern with phase shift is in connection with unbalanced fault currents. If the line currents for the Y side of a Y-Δ transformer are known, there are several ways to find the line currents on the Δ side. One obvious way would be to find the corresponding ampere-turns in each leg of the Δ and then procede with the Kirchhoff's current law to obtain the line currents at the nodes of the Δ. Another approach would be to employ symmetrical components, being careful to shift both positive- and negative-sequence voltages and currents by the appropriate angles (zero-sequence currents will not move out into the line on the Δ side). The positive-sequence system of voltages experiences the 30° phase shift already described in either Fig. 9-17 or Fig. 9-18, whichever is the case. The phasor relationship between primary and secondary voltages for the *negative-sequence* system is pictured in Fig. 9-19. This figure applies to the Y-Δ of Fig. 9-17. $E^-_{na'}$ will not necessarily have the same phase angle as E^+_{na}. Notice that the diagram of Fig. 9-19 shows the high-voltage side lagging the secondary by 30°, whereas the positive-sequence voltages of the high-voltage side were leading the secondary voltages by 30°.

We have labeled Figs. 9-17 and 9-18 so that a', b', and c' correspond to H_1, H_2, and H_3 and a, b, and c correspond to X_1, X_2, and X_3.

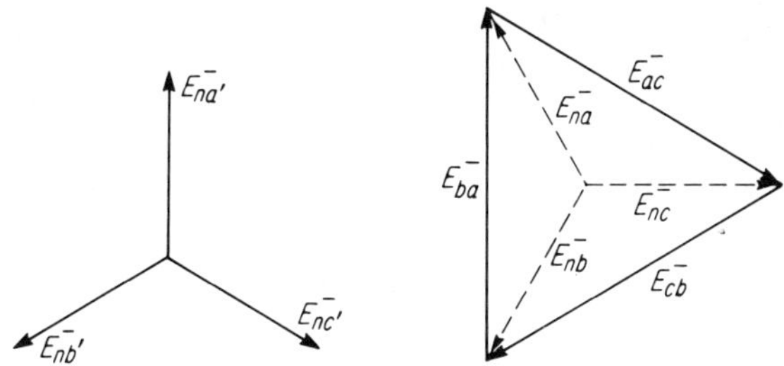

Fig. 9-19. Negative-sequence phasor diagram for the Y-Δ of Fig. 9-17.

This means that $E_{na'}^+$ leads E_{na}^+ by 30°, just as E_{nH_1} leads E_{nx_1} by 30°. However, as previously mentioned, this choice of line labeling (a, b, c, a', b', c', etc.) is an arbitrary one. Two alternate methods for labeling lines will be presented. While the method of Figs. 9-17 and 9-18 may be simpler to visualize, an alternate choice is preferred by some for the sake of convenience. Refer to Fig. 9-20 in which the standard H and X notation is

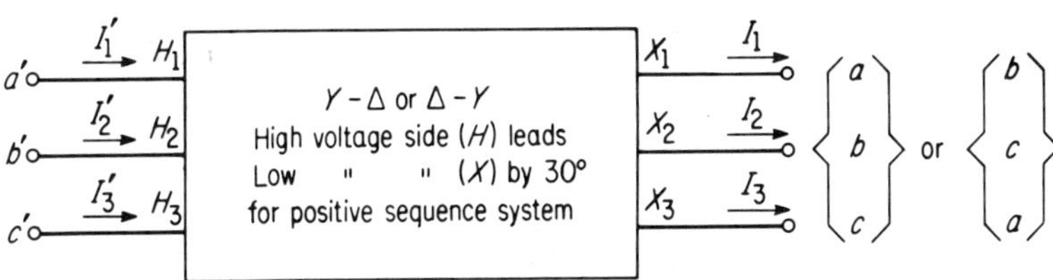

Fig. 9-20. Alternate methods for labeling low-voltage side of Y-Δ or Δ-Y transformers.

used. To the right of the drawing is a choice of secondary line markings (a, b, c, or b, c, a). Either Y-Δ or Δ-Y could be represented in the box, since in either case the positive-sequence phase voltages on side X will lag corresponding phase voltages on the H side by 30°. It should be noted that when a voltage displacement is encountered in moving through the transformer from H to X, than an equal current displacement is encountered for the same H and X terminals. This is true if we have assumed our primary and secondary current directions in the manner suggested by both Figs. 9-14 and 9-20. In arriving at primary and secondary phasor relationships for the *a-b-c choice* we will use Figs. 9-17 and 9-19 as a guide. These relationships are tabulated for the positive- and negative-sequence voltages and currents of Fig. 9-20 as follows (*abc* choice):

$$E_{na'}^+ = E_{na}^+\underline{/30^\circ}; \quad E_{nb'}^+ = E_{nb}^+\underline{/30^\circ}; \quad E_{nc'}^+ = E_{nc}^+\underline{/30^\circ}$$
$$I_{a'}^+ = I_a^+\underline{/30^\circ}; \quad I_{b'}^+ = I_b^+\underline{/30^\circ}; \quad I_{c'}^+ = I_c^+\underline{/30^\circ} \qquad (9\text{-}16)$$

$$E^-_{na'} = E^-_{na}\angle{-30^\circ}; \quad E^-_{nb'} = E^-_{nb}\angle{-30^\circ}; \quad E^-_{nc'} = E^-_{nc}\angle{-30^\circ}$$
$$I^-_{a'} = I^-_{a}\angle{-30^\circ}; \quad I^-_{b'} = I^-_{b}\angle{-30^\circ}; \quad I^-_{c'} = I^-_{c}\angle{-30^\circ} \tag{9-17}$$

The *b-c-a* notation for the secondary lines of Fig. 9-20 may be chosen by some with the aim of simplifying calculations. This is explained by taking another look at the phasor diagrams of Fig. 9-17 where the *abc* choice was made. Notice that the phasor E^+_{nc} is leading that of $E^+_{na'}$ by 90°. It should then be easier to relate $E_{na'}$ of Fig. 9-17 to E_{nc} by a simple factor of $-j$ than to relate to E_{na} with a factor of $1\angle{30} = 0.866 + j0.5$. Similarly note that $E^+_{nb'}$ is related to E^+_{na} by $-j$ and $E^+_{nc'}$ to E_{nb} by $-j$. So we merely relabel the *abc* secondary lines of Fig. 9-20 as *bca*, where *a* replaces *c*, *c* replaces *b*, and *b* replaces *a*. With the lines relabeled, the new positive-sequence phasor diagrams are as depicted in Fig. 9-21a. The negative-sequence phasor diagrams are found by checking Fig. 9-19, and again replacing *c* with *a*, *b* with *c*, and *a* with *b*.

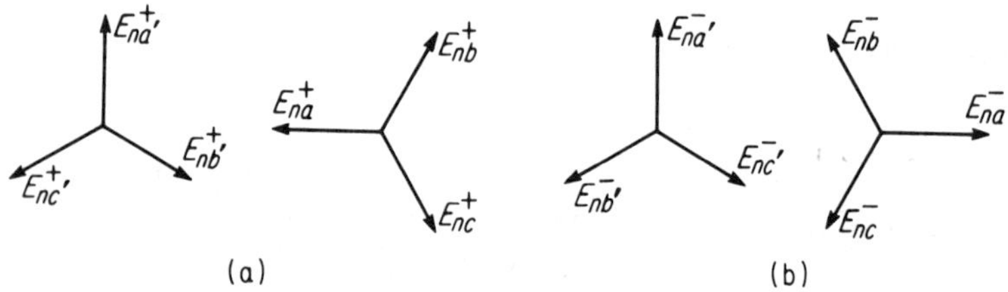

Fig. 9-21. Positive and negative sequence phasor diagrams using the choice of *bca* in Fig. 9-20.

The new negative-sequence phasor diagram is shown in Fig. 9-21b. From Fig. 9-21 we can tabulate the simplified relationships which exist between primary and secondary sequence components (*bca* choice).

$$E^+_{na'} = -jE^+_{na}; \quad E^+_{nb'} = -jE^+_{nb}; \quad E^+_{nc'} = -jE^+_{nc}$$
$$I^+_{a'} = -jI^+_{a}; \quad I^+_{b'} = -jI^+_{b}; \quad I^+_{c'} = -jI^+_{c} \tag{9-18}$$

$$E^-_{na'} = +jE^-_{na}; \quad E^-_{nb'} = +jE^-_{nb}; \quad E^-_{nc'} = +jE^-_{nc}$$
$$I^-_{a'} = +jI^-_{a}; \quad I^-_{b'} = +jI^-_{b}; \quad I^-_{c'} = +jI^-_{c} \tag{9-19}$$

Example 9-2. Shown in Fig. 9-22 is a Y-Δ transformer (standard 30° phase shift between *H* and *X* terminals). A line-to-line fault exists somewhere in the system and to the right of the Δ secondary. A symmetrical component solution has determined the per unit secondary line currents as $I_c = 4.0$, $I_b = -4.0$, and $I_a = 0$.

(a) Find the per unit currents in the primary lines, using symmetrical components and the phase shift information of Eqs. 9-18 and 9-19. Notice that the *bca* method of labeling secondary lines has been chosen.

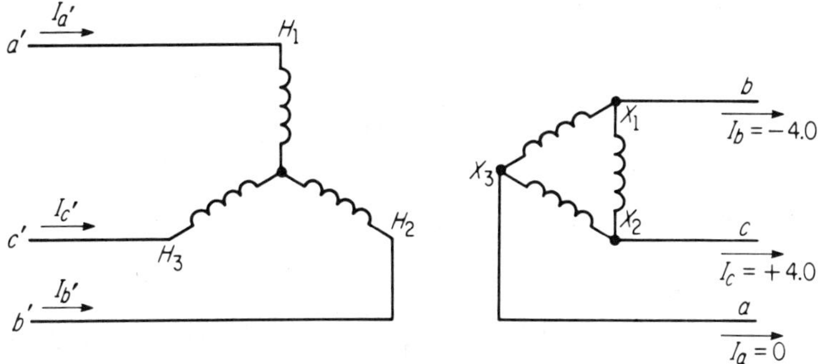

Fig. 9-22. Circuit for Example 9-2.

(b) Check the results of (a) with the method of balancing ampere-turns.

Solution (a)

$$I_a^+ = \frac{1}{3}[I_a + aI_b + a^2 I_c]$$

$$= \frac{1}{3}[0 + (1\underline{/120^\circ})(-4.0) + (1\underline{/-120^\circ})(4.0)]$$

$$I_a^+ = \frac{1}{3}[-j4.0(\sqrt{3})] = 4/\sqrt{3}\underline{/-90^\circ}$$

$$I_b^+ = I_a^+\underline{/-120^\circ} = 4/\sqrt{3}\underline{/-210^\circ}$$

$$I_c^+ = I_a^+\underline{/+120^\circ} = 4/\sqrt{3}\underline{/30^\circ}$$

$$I_a^- = \frac{1}{3}[I_a + a^2 I_b + aI_c]$$

$$- \frac{1}{3}[0 + (1\underline{/-120^\circ})(-4.0) + (1\underline{/120^\circ})(4.0)]$$

$$= 4/\sqrt{3}\underline{/90^\circ}$$

$$I_b^- = I_a^-\underline{/+120^\circ} = 4/\sqrt{3}\underline{/210^\circ}$$

$$I_c^- = I^-\underline{/-120^\circ} = 4/\sqrt{3}\underline{/-30^\circ}$$

$$I_a^0 = I_b^0 = I_c^0 = 0$$

From Eq. 9-18,

$$I_{a'}^+ = -jI_a^+ = -j(4/\sqrt{3}\underline{/-90^\circ}) = -4/\sqrt{3}$$

$$I_{b'}^+ = -jI_b^+ = -j(4/\sqrt{3}\underline{/-210^\circ}) = 4/\sqrt{3}\underline{/60^\circ}$$

$$I_{c'}^+ = -jI_c^+ = -j(4/\sqrt{3}\underline{/30^\circ} = 4/\sqrt{3}\underline{/-60^\circ}$$

From Eq. 9-19,

$$I_{a'}^- = +jI_a^- = j(4/\sqrt{3}\,\underline{/90^\circ}) = \underline{-4/\sqrt{3}}$$
$$I_{b'}^- = jI_b^- = j(4/\sqrt{3}\,\underline{/210^\circ}) = \underline{4/\sqrt{3}\,/-60^\circ}$$
$$I_{c'}^- = jI_c^- = j(4/\sqrt{3}\,\underline{/-30^\circ}) = \underline{4/\sqrt{3}\,/60^\circ}$$

Solving for the primary line currents,

$$I_{a'} = I_{a'}^+ + I_{a'}^- = \underline{-4/\sqrt{3}} - \underline{4/\sqrt{3}} = \underline{\underline{-8/\sqrt{3}}}$$
$$I_{b'} = I_{b'}^+ + I_{b'}^- = 4/\sqrt{3}\,\underline{/60^\circ} + 4/\sqrt{3}\,\underline{/-60^\circ} = \underline{\underline{4/\sqrt{3}}}$$
$$I_{c'} = I_{c'}^+ + I_{c'}^- = 4/\sqrt{3}\,\underline{/-60^\circ} + 4/\sqrt{3}\,\underline{/+60^\circ} = \underline{\underline{4/\sqrt{3}}}$$

Solution (b). To confirm the above answers, the balancing-ampere-turns method is used. We reason as follows: Let M and N of Fig. 9-23 be

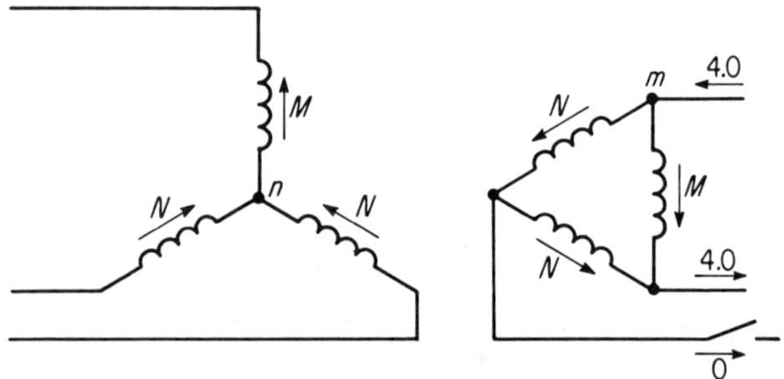

Fig. 9-23. Circuit for Solution (b).

the unknown Δ coil currents which must find their counterparts (balancing ampere-turns) in the Y coils as shown. The two unknowns are found by employing Kirchhoff's current law at junctions n and m.

(1) At n: $M = 2N$

(2) At m: $M + N = 4.0$

Substituting (1) into (2),

$$3N = 4.0$$
$$N = \underline{\underline{4/3}} \text{ pu (sec. line base); pu } N(\text{coil base}) = 4/\sqrt{3}$$
$$(1)\ M = 2(4/3) = \underline{\underline{8/3}} \text{ pu (sec. line base); pu } M(\text{coil base}) = 8/\sqrt{3}$$

There are a few instances where power utilities have departed from the ASA or NEMA standards with respect to transformer phasing. For example, The Detroit Edison Company has standardized on a 30° lead on the low-voltage winding with respect to the high-voltage winding—for YΔ and ΔY transformers. The company standard for the Y-Y and Δ-Δ transformations specifies a 180° phase shift between high- and low-voltage terminals. While such a choice is arbitrary for a utility, this departure from standards must be considered in the ordering of transformers from the manufacturer. A NEMA standard ΔY or YΔ (*LV* lags *HV* by 30°) transformer can be made to conform to a 30° lead on the *LV* in one of two ways:

1. Reversing of either the *X* or *H* coils. This could require an internal reconnection.
2. Instead of applying an *abc* voltage sequence to H_1, H_2, H_3, the *cba* sequence is applied to H_1, H_2, and H_3 respectively. The X_1, X_2, and X_3 terminals will now have the *cba* sequence, and a 30° lead with respect to the HV winding.

Problems

9-1. Consider all coils of the YΔ transformer of Prob. 9-1 have equal turns. Will the transformer easily support the unbalance load and maintain voltage? If so, what current flows in each of the 6 coils? Use the balancing-ampere-turns method.

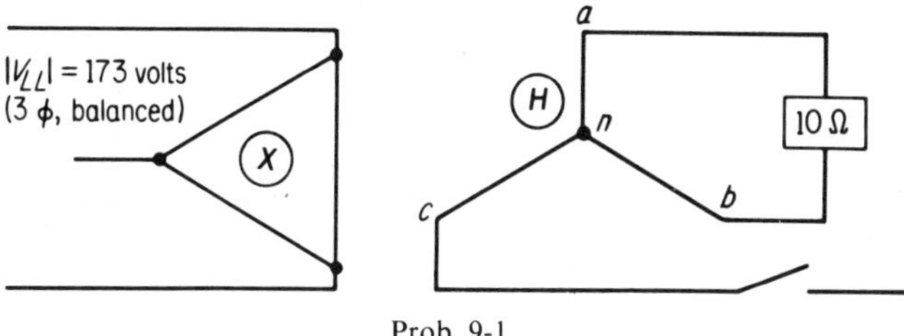

Prob. 9-1

9-2. Find the primary currents of Prob. 9-1 using symmetrical components and the principles of transformer phase shift discussed in Sec. 9-8. Assume an *abc* phase sequence and the NEMA standard phase shift.

9-3. Assume all coils of equal turns. What current flows in each coil of the three-winding transformer of Prob. 9-3?

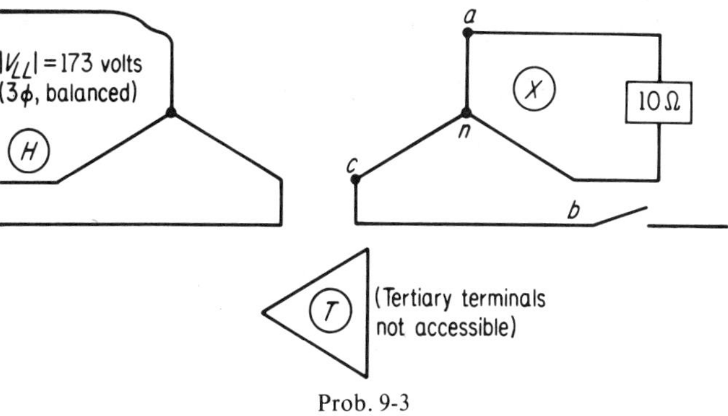

Prob. 9-3

9-4. Repeat Prob. 9-3 with the 10Ω resistor placed from *a* to *n*.

9-5. (a) Will the zigzag transformer connection of Prob. 9-5 support the unbalanced load of 10 ohms? If so, what current flows in each of the 9 coils? Assume all coils have equal turns. Use the balancing-ampere-turns technique.

(b) Answer part (a) where the end-coils of Prob. 9-5 (coils *d-a*, *e-b*, and *f-c*) have half the turns of the other 6 coils.

(c) Again, assume the end coils of the secondary circuit have half the turns of the other 6 coils. Now place the 10-ohm resistor across the terminals *a* to *n*. Will this arrangement support such a load and if so, give all coil currents.

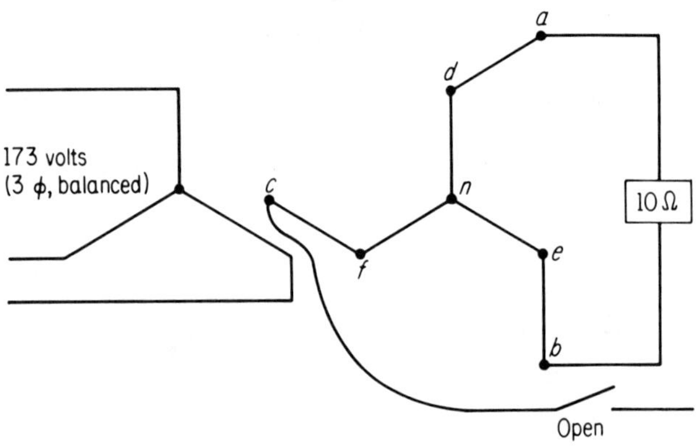

Prob. 9-5

9-6. Given the generator and transformer described in Prob. 8-4. A line-to-line fault is applied at point *g* of Prob. 8-4.

(a) Determine the pu subtransient fault current.

(b) Find the pu current in all lines of the generator, using the balancing-ampere-turns approach and the results of part (a).

(c) Repeat part (b) using symmetrical components and the phase shifting methods of Sec. 9-8.

(d) Determine the line-to-neutral voltage on all phases of the transformer secondary.

(e) Determine the line-to-neutral voltages on all phases of the transformer primary.

chapter **10**

THE BUS IMPEDANCE MATRIX METHOD FOR FAULT STUDIES

10-1. Introduction

In the calculation of power systems problems, the short-circuit studies in particular have often been simplified considerably with the use of the so-called short-circuit matrix method, also referred to by some as the *bus impedance matrix method.* The whole of Chapter 10 is devoted to this method as it is the author's feeling that the many advantages which it offers make it an almost indispensable tool today in the area of fault studies. The method has come to the forefront largely as a result of the emphasis upon the use of the digital computer for system studies in preference to the analyzer board. References have been made to the short-circuit equivalent in Secs. 2-4, 2-5, and 4-13. It is strongly suggested that these sections be reviewed before continuing on with Chapter 10. While the previous sections mentioned serve to introduce and compare the equivalent with other equivalents, the present chapter will stress an understanding and application of this impedance equivalent and its corresponding impedance matrix.

The per phase short-circuit matrix equivalent of a large power system is shown in Fig. 10-1. It is recalled from Chapter 2 that an n-bus or n-node passive network (plus reference node 0) may be represented by the combination of self- or driving-point impedances (Z_{11}, Z_{22}, Z_{33}, etc.) and transfer or mutual impedances (Z_{12}, Z_{13}, Z_{23}, etc.). In a discussion of the short-circuit matrix equivalent with Robert Cline of the Detroit Edison Company, it was his suggestion that a shorter name be given to this circuit. Since the star and mesh equivalents are both named according to their shape or appearance, his suggestion was that the circuit be tagged the "rake" equivalent in preference to either the short-circuit matrix equivalent or the bus impedance matrix equivalent. Actually this rake equivalent is a direct result of the short-circuit matrix to be derived in Sec. 10-3.

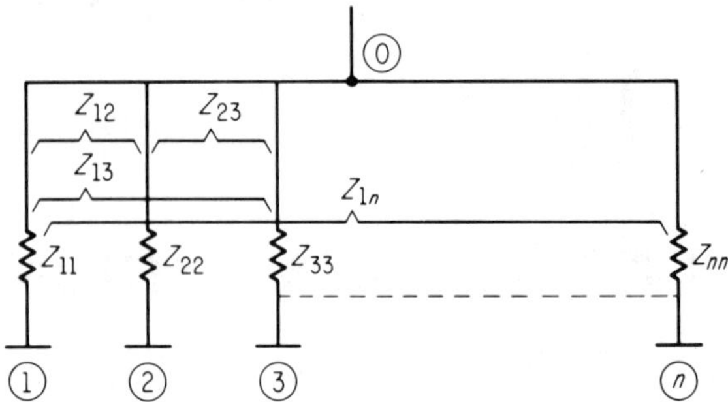

Fig. 10-1. The per phase short-circuit matrix equivalent or rake equivalent.

10-2. Basic Concepts and Assumptions

In order to insure a clear concept of the problem at hand, a few basic statements and assumptions will be given. Notice that these assumptions are geared for the simplified short-circuit studies, which is the primary objective of this chapter.

1. Our conception of the original network should be on a per phase basis, even though it is understood that a three-phase system is implied. A per phase circuit assumes balanced phase impedances of lines, generators and transformers, as well as balanced generated voltages in each phase of the generators. Refer to Fig. 10-2a, in which various generators (or stations) are all feeding from the neutral bus into the passive network.

2. In this chapter, equal generated voltage is assumed on all generators (both magnitude and phase angle). The rectangle of Fig. 10-2a, representing the passive network, may then be extended to include gen-

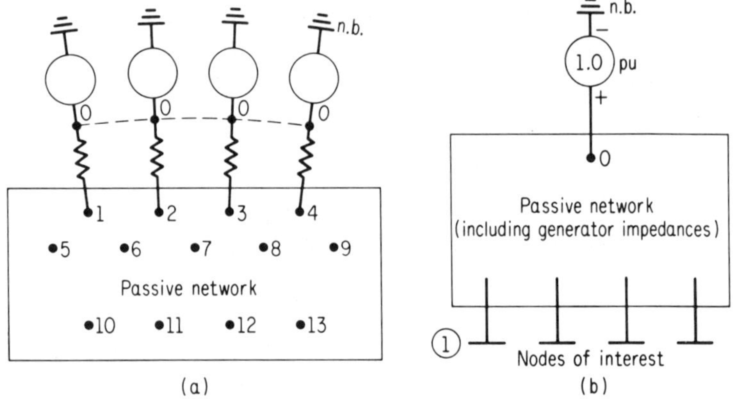

Fig. 10-2. Per phase representation of a power network. (Faults and loads excluded.)

erator impedances as shown by the dashed lines. Since the 0 terminals are at the same potential, they may be connected with an imaginary line, and all generation may now be replaced by one constant voltage generator. This is done in Fig. 10-2b. Usually the generator is assumed to generate 1.0 per unit volts for purposes of short-circuit studies.

3. While the passive network of Fig. 10-2b includes the impedances of generators, transformers, and lines, it does not normally include the fault or load impedances. In fact, the network as it stands exhibits no circulating currents of any kind. In order for currents to flow, an external fault or load path must exist back to the neutral bus. Effects of line capacitance could offer such a path, but line capacitance will be neglected here.

4. We will assume that any external load paths back to the neutral bus are of sufficiently high impedance so as to yield negligible load currents with respect to the fault currents. The per phase circuit for a symmetrical three-phase fault on bus 1 would then be complete if bus 1 of Fig. 10-2b were connected back around to the neutral bus of the generator (shown at ground potential). If unsymmetrical faults are to be studied, three separate passive networks could be needed, representing impedance networks to the positive-, negative- and zero-sequence currents.

5. Another simplification which will be made in the sample problems of this chapter is in regard to the resistive elements of the network. It will be assumed that the impedance elements are predominantly reactive and resistances will, therefore, be neglected. Recall that this same assumption was made in chapter four in the use of the d-c analyzer board for symmetrical short-circuit studies.

6. We may visualize buses (or nodes) of interest as being pulled outside the box for a closer look, leaving unwanted buses within the box. Later, when an equivalent is made of the network, unwanted nodes are said to be suppressed. Even though their identity has been lost, this in no way means that the effect of these suppressed nodes has not been considered. Figure 10-2b focuses special attention upon four buses. It may be that these are the buses to be faulted in the study or perhaps the voltages of these buses are of particular interest. In either case, it demands that, when an equivalent is made, these buses must be identified.

10-3. Formation of the Short-Circuit Matrix

All discussion of this section, unless otherwise specified, will apply only to *balanced* three-phase circuits and faults. This implies the use of only the *positive-sequence network*. Later the methods will be extended to cover the unbalanced faults as well.

The student might at first reason as follows: Why resort to this, another method, when the tried-and-true mesh and nodal matrix equations (of Chapter 5) will obviously yield a correct solution? The answer is that we resort to this method for much the same reason that we make our exit through a doorway in preference to the window—it is much easier. For example, while the conventional methods will require a complete and separate solution for each and every condition of fault, the short-circuit impedance matrix will in itself imply a complete solution regardless of which bus is faulted. This is but one of the several advantages of this method.

There are a number of ways of arriving at the short-circuit matrix and its corresponding equivalent. For the sake of better understanding we will first outline the method of forming the matrix by setting up the complete network and writing loop equations in a specific manner. The general procedure is as follows.

1. Determine the buses to be retained in the equivalent. Pull out these buses from the passive network and consider each bus to be provided with a fault switch connecting it to the neutral bus of the generator (see Fig. 10-3). In the writing of the loop equations, consider these

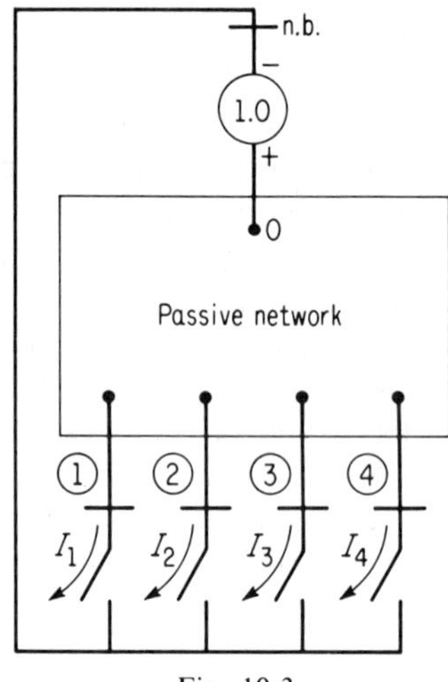

Fig. 10-3

fault switches closed, even though in a specific solution such may not be the case. (Later, certain of these fault currents can be set to zero without affecting our equivalent.)

2. Write loop equations in a specified manner. The tree-link criteria of Chapter 5 can be used to guarantee the sufficient number of independent loops. Loops are not chosen arbitrarily here, but several requirements should be met.

(a) Be sure that one and only one loop current passes through each fault switch. These particular loop currents also pass through the generator source.

(b) All other loop currents will flow only within the passive network.

3. Use matrix partitioning (as covered in Sec. 5-3) to eliminate the internal loop currents which flow within the passive network. The resulting impedance matrix is termed the "short circuit impedance matrix," or

$$\overline{Z}_{sc} = \overline{Z}_1 - \overline{Z}_2\overline{Z}_4^{-1}\overline{Z}_3 \tag{10-1}$$

4. The equivalent circuit is next drawn by inspection, realizing that the diagonal elements of the short-circuit matrix represent the self- or driving-point impedances of the equivalent of Fig. 10-1. The off-diagonal elements are the mutual or transfer impedances of the equivalent.

5. The short-circuit impedance matrix may now be applied readily to obtain currents and voltages throughout the system for a fault on any bus. This will be demonstrated later by Example 10-2.

While the foregoing procedure may not represent the best way for forming the desired matrix in actual practice, it does present a clearer concept at this stage of development. Example 10-1 should serve to answer questions regarding the general procedure.

Example 10-1. Given the network of Fig. 10-4a. The impedances of the system are all reactive, but the common j term has been omitted to simplify the procedure. Find the short-circuit matrix and its corresponding equivalent circuit. Retain buses 1, 2, and 3 plus the reference bus 0.

Solution. Following the procedure of the preceding section, buses 1,

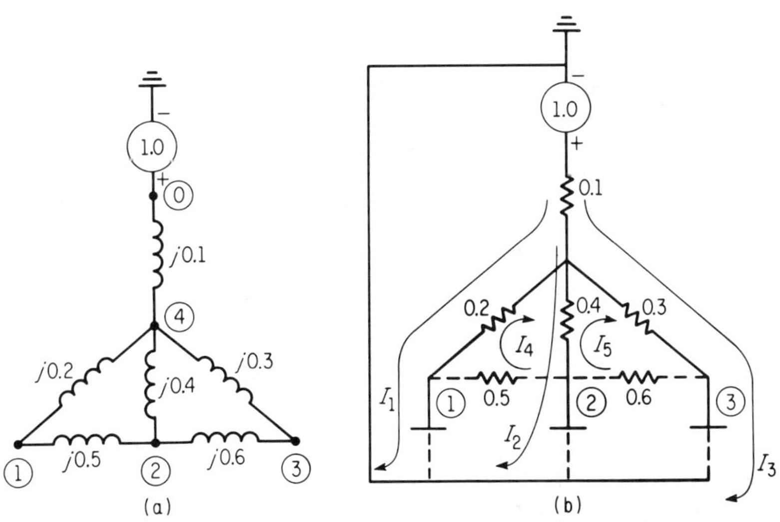

Fig. 10-4. (a) Circuit for Example 10-1. (b) Establishing loop currents using the tree-link approach. (Lines are dotted.)

2, and 3 are considered to have fault links to the neutral bus. The chosen tree is shown in Fig. 10-4b with links dashed. Recall from Sec. 5-2 that the addition of each link closes another loop and identifies another loop current. Voltage equations may be written in terms of the unknown loop (or link) currents. These equations can easily be written in matrix form as $[E] = [Z][I]$ as per Eq. 5-4. Actually, only the loop impedance matrix is needed but for clarity the complete equation is treated. Loop currents are arranged so that the fault link currents (to be retained) are placed first with the internal loop currents (to be eliminated) placed last. The loop matrix equation is

$$\left[\begin{array}{c} 1.0 \\ 1.0 \\ 1.0 \\ \hline 0 \\ 0 \end{array}\right]_{(\bar{E}_x,\ \bar{E}_y)} = \left[\begin{array}{ccc|cc} 0.3 & 0.1 & 0.1 & -0.2 & 0 \\ 0.1 & 0.5 & 0.1 & 0.4 & -0.4 \\ 0.1 & 0.1 & 0.4 & 0 & 0.3 \\ \hline -0.2 & 0.4 & 0 & 1.1 & -0.4 \\ 0 & -0.4 & 0.3 & -0.4 & 1.3 \end{array}\right]_{\left(\begin{smallmatrix}\bar{Z}_1 & \bar{Z}_2 \\ \bar{Z}_3 & \bar{Z}_4\end{smallmatrix}\right)} \left[\begin{array}{c} I_1 \\ I_2 \\ I_3 \\ \hline I_4 \\ I_5 \end{array}\right]_{(\bar{I}_x,\ \bar{I}_y)}$$

The loop currents I_4 and I_5 are eliminated by matrix partitioning according to Sec. 5-3. The matrix equation of 5-12 holds true if the partial arry $(\bar{E}_y)$ is zero. $\bar{E}_y$ contains the zero voltage sources of the loops 4 and 5 within the passive network. Then the short-circuit matrix is

$$\bar{Z}_{sc} = \bar{Z}_1 - \bar{Z}_2\bar{Z}_4^{-1}\bar{Z}_3 \qquad [10\text{-}1]$$

$$\bar{Z}_4^{-1} = \frac{1}{\Delta}\begin{bmatrix} 1.3 & 0.4 \\ 0.4 & 1.1 \end{bmatrix}$$

where $\Delta = (1.3)(1.1) - (-0.4)(-0.4)$
$\Delta = 1.27$

$$\bar{Z}_4^{-1} = \begin{bmatrix} 1.03 & 0.315 \\ 0.315 & 0.865 \end{bmatrix}$$

$$\bar{Z}_4^{-1}\bar{Z}_3 = \begin{bmatrix} 1.03 & 0.315 \\ 0.315 & 0.865 \end{bmatrix}\begin{bmatrix} -0.2 & 0.4 & 0 \\ 0 & -0.4 & 0.3 \end{bmatrix}$$

$$= \begin{bmatrix} -0.206 & 0.286 & 0.0945 \\ -0.063 & -0.22 & 0.2595 \end{bmatrix}$$

$$\bar{Z}_2\bar{Z}_4^{-1}\bar{Z}_3 = \begin{bmatrix} -0.2 & 0 \\ +0.4 & -0.4 \\ 0 & 0.3 \end{bmatrix}\begin{bmatrix} -0.206 & 0.286 & 0.0945 \\ -0.063 & -0.22 & 0.2595 \end{bmatrix}$$

$$= \begin{bmatrix} 0.0412 & -0.0572 & -0.0189 \\ -0.0572 & 0.2024 & -0.066 \\ -0.0189 & -0.066 & 0.0779 \end{bmatrix}$$

Finally,

$$\bar{Z}_1 - \bar{Z}_2\bar{Z}_4^{-1}\bar{Z}_3 = \begin{bmatrix} 0.3 & 0.1 & 0.1 \\ 0.1 & 0.5 & 0.1 \\ 0.1 & 0.1 & 0.4 \end{bmatrix} - \begin{bmatrix} 0.0412 & -0.0572 & -0.0189 \\ -0.0572 & 0.2024 & -0.066 \\ -0.0189 & -0.066 & 0.0779 \end{bmatrix}$$

or

$$\bar{Z}_{sc} = \begin{bmatrix} 0.259 & 0.157 & 0.119 \\ 0.157 & 0.298 & 0.166 \\ 0.119 & 0.166 & 0.322 \end{bmatrix}$$

The resulting short-circuit matrix equivalent (or rake equivalent) for the passive network is produced in Fig. 10-5. It should be realized that the common j term is a factor of each impedance and may be added at any time. It is simpler to treat the reactive network as if it were purely resistive.

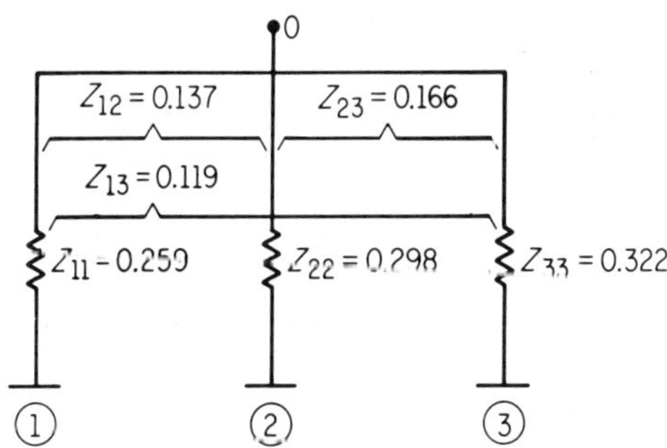

Fig. 10-5. Rake equivalent for Example 10-1.

10-4. Application of the Short-Circuit Impedance Matrix to Symmetrical Fault Studies

Before applying the matrix, we will examine its meaning more closely. Our resulting matrix equation after elimination of the internal loops example was $\bar{E}_x = \bar{Z}_{sc}\bar{I}_x$, or Example 10-1.

$$\begin{bmatrix} 1.0 \\ 1.0 \\ 1.0 \end{bmatrix} = \begin{bmatrix} Z_{11} & Z_{12} & Z_{13} \\ Z_{21} & Z_{22} & Z_{23} \\ Z_{31} & Z_{32} & Z_{33} \end{bmatrix} \begin{bmatrix} I_1 \\ I_2 \\ I_3 \end{bmatrix}$$

It was apparent that three loop equations were represented:

$$1.0 = Z_{11}I_1 + Z_{12}I_2 + Z_{13}I_3 \qquad (10\text{-}2)$$

$$1.0 = Z_{21}I_1 + Z_{22}I_2 + Z_{23}I_3 \qquad (10\text{-}2a)$$

$$1.0 = Z_{31}I_1 + Z_{32}I_2 + Z_{33}I_3 \qquad (10\text{-}2b)$$

These equations were based upon the closing of all three fault switches. However, we can easily alter our circuit by opening and closing these fault

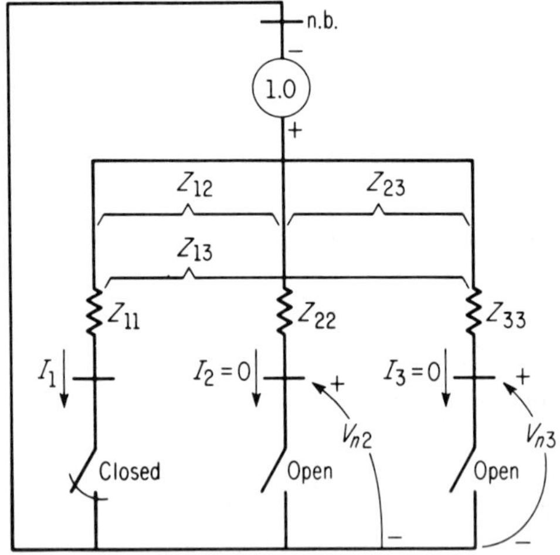

Fig. 10-6. Condition with bus 1 faulted.

links at will. For example, suppose that only bus 1 is faulted as in Fig. 10-6. Then $I_2 = I_3 = 0$ and Eq. 10-2 becomes

$$1.0 = Z_{11}I_1 + Z_{12}\cancelto{0}{I_2} + Z_{13}\cancelto{0}{I_3} \qquad (10\text{-}3)$$

or

$$I_1 = 1.0/Z_{11} \qquad \text{(bus 1 faulted)} \qquad (10\text{-}4)$$

Similarly with a fault only on bus 2, Eq. 10-2a becomes

$$1.0 = Z_{21}\cancelto{0}{I_1} + Z_{22}I_2 + Z_{23}\cancelto{0}{I_3}$$

or

$$I_2 = 1.0/Z_{22} \qquad \text{(bus 2 faulted)} \qquad (10\text{-}5)$$

Likewise with a fault on bus 3,

$$I_3 = 1.0/Z_{33} \tag{10-6}$$

It has previously been observed that the diagonal elements of the short-circuit impedance matrix are also the self-impedances of the "rake" equivalent. Equations 10-4 through 10-6 show that the *current flowing out of the fault* (*with bus* k *faulted is inversely proportional to the self impedance of that bus*, or

$$I_k = 1.0/Z_{kk} \qquad \text{(bus } k \text{ faulted)} \tag{10-7}$$

We turn our attention to the off-diagonal elements of the matrix $\overline{Z}_{sc}$. These elements were also observed to be the mutual impedances of the rake equivalent. Again assume that bus 1 is faulted in Fig. 10-6. · Equation 10-2a can be written as

$$1.0 - V_{n2} = Z_{21}I_1 + Z_{22}\cancelto{0}{I_2} + Z_{23}\cancelto{0}{I_3} \tag{10-8}$$

We see that I_2 and I_3 are both zero. Also in summing the voltage around loop 2, V_{n2} has taken on a value. Solving for V_{n2} from Eq. 10-8,

$$V_{n2} = 1.0 - Z_{21}I_1 = 1.0 - Z_{21}(1.0/Z_{11})$$

or

$$V_{n2} = 1.0 - Z_{21}/Z_{11}$$

Likewise

$$V_{n3} = 1.0 - Z_{31}/Z_{11} \qquad \text{(bus 1 faulted)} \tag{10 9}$$

The voltage matrix equation for a fault on bus 1 could have been written as

$$\begin{bmatrix} 1.0 \\ 1.0 - V_{n2} \\ 1.0 - V_{n3} \end{bmatrix} = \begin{bmatrix} Z_{11} & Z_{12} & Z_{13} \\ Z_{21} & Z_{22} & Z_{23} \\ Z_{31} & Z_{32} & Z_{33} \end{bmatrix} \begin{bmatrix} I_1 \\ 0 \\ 0 \end{bmatrix} \tag{10-10}$$

Similarly, for a fault on bus 2, the voltages on buses 1 and 3 would be

$$V_{n1} = 1.0 - Z_{12}/Z_{22}$$
$$V_{n3} = 1.0 - Z_{32}/Z_{22}$$

In general, the mutual impedances of the rake equivalent (obtained from the off-diagonal elements of the short-circuit impedance matrix) are related in a simple manner to the *m*th bus voltage of the network by the expression

$$V_{nm} = 1.0 - Z_{km}/Z_{kk} \qquad \text{(bus } k \text{ faulted)} \qquad (10\text{-}11)$$

The short-circuit impedance matrix is symmetrical with respect to its main diagonal, or

$$Z_{mk} = Z_{km}$$

Equations 10-7 and 10-11 serve as powerful tools in the network solution. It may be reasoned further that when a particular branch cur-

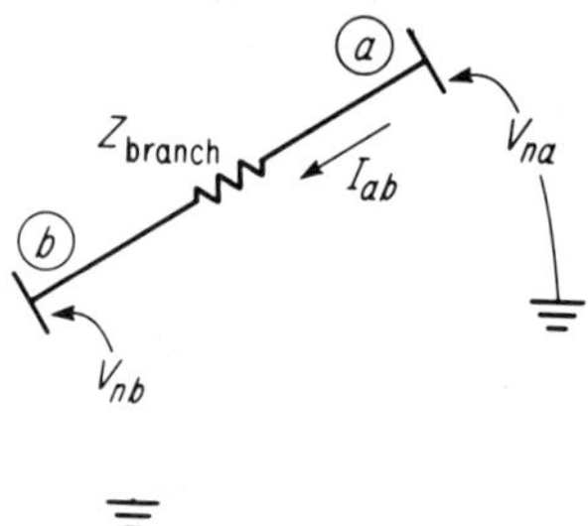

Fig. 10-7. Computing branch currents within the original network.

rent is needed within the original network, we need but to apply Ohm's law to the branch. See Fig. 10-7.

$$I_{ab} = \frac{V_{na} - V_{nb}}{Z_{\text{branch}}} = \frac{(1.0 - Z_{ak}/Z_{kk}) - (1.0 - Z_{bk}/Z_{kk})}{Z_{\text{branch}}}$$

$$I_{ab} = \frac{Z_{bk} - Z_{ak}}{Z_{kk} Z_{\text{branch}}} \qquad \text{(bus } k \text{ faulted)} \qquad (10\text{-}12)$$

Notice from Eq. 10-12 that any branch current which is to be found will necessarily require that both buses *a* and *b* be retained in the impedance matrix. Equation 10-12 will not, in general, hold true in case the branch in question has mutual coupling with some other branch in the original network.

Example 10-2. Given the per phase, positive-sequence network of Example 10-1, Fig. 10-4a. Using the resulting s-c matrix of Example 10-1 and the principles of the Sec. 10-4, determine the following:

(a) For a symmetrical fault on bus 1, find the *fault current* and the *voltages* from neutral to buses 2 and 3. Also find the *current* in the branch of the original network between terminals 2 and 3.

(b) For a symmetrical fault on bus 2, find the *fault current* and the voltages from neutral to buses 1 and 3.

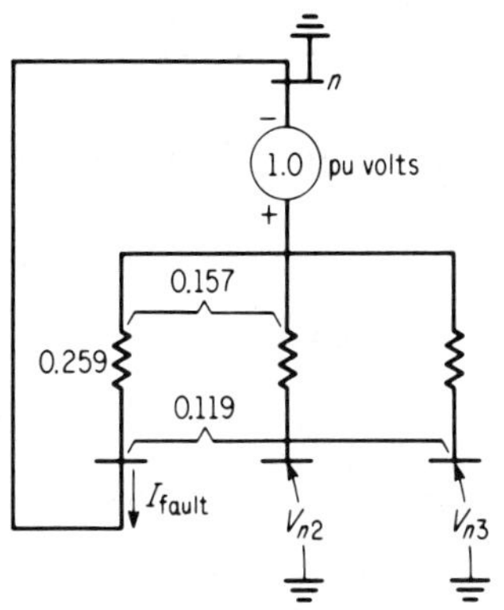

Fig. 10-8. Rake equivalent for Example 10-2a.

Solution (a). Refer to Fig. 10-8 which shows the fault condition on the rake equivalent.

$$I_{\text{fault}} = 1.0/Z_{11} = 1.0/0.259$$
$$= \underline{\underline{3.86}}^{*} \text{ pu amp}$$

$$V_{n2} = 1.0 - Z_{12}/Z_{11} = 1.0 - 0.157/0.259 = 1.0 - 0.606$$
$$= \underline{\underline{0.394}} \text{ pu volts}$$

$$V_{n3} = 1.0 - Z_{13}Z_{11} = 1.0 - 0.119/0.259$$
$$= \underline{\underline{0.46}} \text{ pu volt}$$

To obtain the branch current (I_{23}) flowing in the original network of Fig. 10-9, Eq. 10-12 can be used, or the more obvious relationship

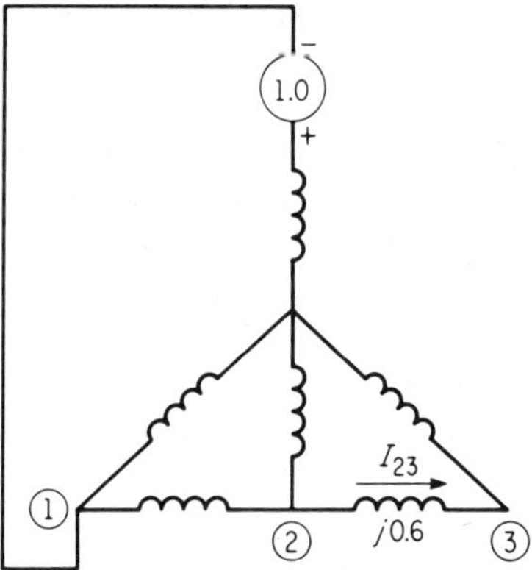

Fig. 10-9. Actual per phase network with bus 1 faulted.

$$I_{23} = \frac{V_{n2} - V_{n3}}{Z_{\text{branch}}} = \frac{0.394 - 0.46}{0.6} = \underline{\underline{-0.11}}\text{* pu amp}$$

From the sign on I_{23}, the current flows opposite from the assumed direction.

Solution (b)

$$I_{\text{fault}} = 1.0/Z_{22} = 1.0/0.298 = \underline{\underline{3.36}}\text{* pu amp}$$

$$V_{n1} = 1.0 - Z_{12}/Z_{22} = 1.0 - 0.157/0.298 = \underline{\underline{0.474}}\text{ pu volt}$$

$$V_{n3} = 1.0 - Z_{23}/Z_{22} = 1.0 - 0.166/0.298 = \underline{\underline{0.444}}\text{ pu volt}$$

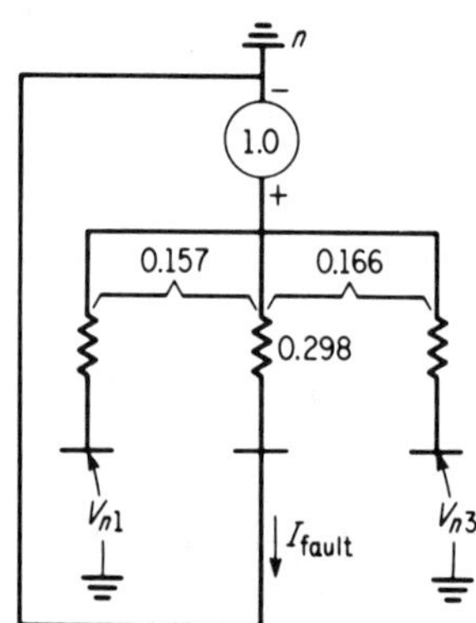

Fig. 10-10. Rake equivalent for Solution (b).

10-5. Application of the S-C Matrix to Line-to-Ground Faults

So far, any mention of the short-circuit impedance matrix has had reference to the positive-sequence system, being applicable to symmetrical three-phase faults. No such limitation is necessary however, since a s-c impedance matrix (and corresponding rake equivalent) is attainable for each of the positive-negative- and zero-sequence networks. In order to attain a solution for unsymmetrical faults, these network equivalents must be connected together as described in Chapter 8, according to the type of fault.

Assume a large network in which s-c impedance matrices have been obtained for each of the positive-, negative-, and zero-sequence systems. Assume four buses have been retained (plus reference bus) and a line-to-ground fault is placed on phase a of bus 1. The line-to-ground procedure

*Note: These currents are actually on an additional angle with regard to the generator voltage, of $-90°$, since the j term was dropped from all reactive elements from the beginning.

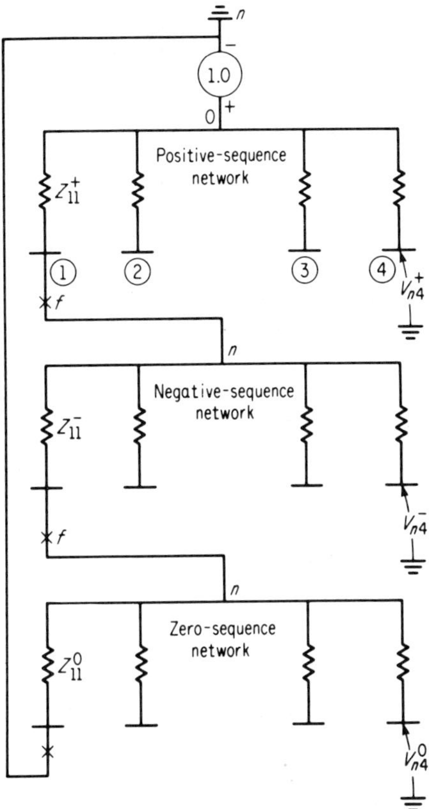

Fig. 10-11. Per phase sequence networks for *L-G* fault condition. (Transfer impedances not shown.)

of Chapter 8 requires that positive-, negative, and zero-sequence networks for phase *a* be placed in series in order to compute the zero-sequence fault current I_{af}^{0}. Using the rake equivalents obtained from the three s-c matrices, it is a relatively simple matter to visualize the fault condition and compute I_{af}^{0}. The L-G fault condition is shown in Fig. 10-11, where

$$I_{af}^{0} = \frac{1.0}{Z_{11}^{+} + Z_{11}^{-} + Z_{11}^{0}} \tag{10-13}$$

but

$$I_{af}^{0} = I_{af}^{+} = I_{af}^{-}$$

and

$$I_{af} = 3I_{af}^{0} \tag{10-14}$$

or

$$I_{af} = \frac{3.0}{Z_{11}^{+} + Z_{11}^{-} + Z_{11}^{0}} \tag{10-15}$$

The procedure is, of course, equally simple for a fault on any other bus. In general, for a fault on bus k,

$$I_{af} = \frac{3.0}{Z_{kk}^{+} + Z_{kk}^{-} + Z_{kk}^{0}} \tag{10-16}$$

Recall that the self-impedance Z_{kk} for any particular sequence is also the diagonal element of the kth row and the kth column in the s-c impedance matrix. Suppose the voltages on all three phases are needed at the bus 4 location. Again, it is a simple matter to determine these voltages with the aid of Fig. 10-11. The phase a relationships are

$$V_{n4}^{+} = 1.0 - I_{af}^{+} Z_{14}^{+} \tag{10-17}$$

$$V_{n4}^{-} = -I_{af}^{-} Z_{14}^{-} \tag{10-18}$$

$$V_{n4}^{0} = -I_{af}^{0} Z_{14}^{0} \tag{10-19}$$

now

$$\begin{aligned} V_{n4} &= V_{n4}^{+} + V_{n4}^{-} + V_{n4}^{0} \\ &= (1.0 - I_{af}^{+} Z_{14}^{+}) + (-I_{af}^{-} Z_{14}^{-}) + (-I_{af}^{0} Z_{14}^{0}) \end{aligned}$$

since

$$I_{af}^{+} = I_{af}^{-} = I_{af}^{0}$$

then

$$V_{n4} = 1.0 - I_{af}^{0}(Z_{14}^{+} + Z_{14}^{-} + Z_{14}^{0})$$

substituting I_{af}^{0} of Eq. 10-13,

$$V_{n4} = 1.0 - \frac{Z_{14}^{+} + Z_{14}^{-} + Z_{14}^{0}}{Z_{11}^{+} + Z_{11}^{-} + Z_{11}^{0}} \quad \text{(bus 1 faulted)} \tag{10-20}$$

Equation 10-20 yields the voltage from neutral to the bus 4 location for phase a with bus 1 faulted (L-G). A more general form of the phase a voltage at bus m with bus k faulted is

$$V_{nma} = 1.0 - \frac{Z_{km}^{+} + Z_{km}^{-} + Z_{km}^{0}}{Z_{kk}^{+} + Z_{kk}^{-} + Z_{kk}^{0}} \tag{10-21}$$

In order to obtain b and c phase voltages at bus m, we must be prepared to rotate the phase a sequence voltages by the proper phase angle:

$$V_{nmb} = V_{nmb}^{+} + V_{nmb}^{-} + V_{nmb}^{0}$$

or

$$V_{nmb} = V^{+}_{nma} \angle -120^\circ + V^{-}_{nma} \angle +120^\circ + V^{0}_{nma} \tag{10-22}$$

likewise

$$V_{nmc} = V^{+}_{nma} \angle +120^\circ + V^{-}_{nma} \angle -120^\circ + V^{0}_{nma} \tag{10-23}$$

Equations 10-22 and 10-23 will suffice for phases b and c if the sequence voltages are known. Another form is possible, similar to that of Eq. 10-21, where b and c phase voltages are written only in terms of sequence impedances. Rewriting Eq. 10-22,

$$\begin{aligned} V_{nmb} &= V^{+}_{nma} \angle -120^\circ + V^{-}_{nma} \angle +120^\circ + V^{0}_{nma} \\ &= (1.0 - I^{+}_{af} Z^{+}_{mk}) \angle -120^\circ + (-I^{-}_{af} Z^{-}_{mk}) \angle +120^\circ + (-I^{0}_{af} Z^{0}_{mk}) \end{aligned}$$

but

$$I^{+}_{af} = I^{-}_{af} = I^{0}_{af}$$

$$V_{nmb} = 1.0 \angle -120^\circ - I^{+}_{af}(Z^{+}_{mk} \angle -120^\circ + Z^{-}_{mk} \angle +120^\circ + Z^{0}_{mk})$$

Substituting for I^{+}_{af},

$$V_{nmb} = 1.0 \angle -120^\circ - \frac{Z^{+}_{mk} \angle -120^\circ + Z^{-}_{mk} \angle +120^\circ + Z^{0}_{mk}}{Z^{+}_{kk} + Z^{-}_{kk} + Z^{0}_{kk}} \tag{10-24}$$

Similarly, Eq. 10-23 becomes

$$V_{nmc} = 1.0 \angle +120^\circ - \frac{Z^{+}_{mk} \angle +120^\circ + Z^{-}_{mk} \angle -120^\circ + Z^{0}_{mk}}{Z^{+}_{kk} + Z^{-}_{kk} + Z^{0}_{kk}} \tag{10-25}$$

Again, k is the faulted bus with voltage taken at bus m. When the magnitude of E^{+}_{a} is other than 1.0, the preceding voltages and currents should be corrected by the ratio of $E^{+}_{a}/1.0$.

10-6. Application of the S-C Matrix to the Line-to-Line Fault

The phase a sequence network of Fig. 8-14 is still applicable for the L-L fault, where Z^{+} and Z^{-} are replaced by the positive and negative sequence rake equivalents. Figure 10-12 shows an example of this situation where four buses have been retained (plus reference bus) and a L-L fault exists on bus 1. We may proceed with calculations as follows:

$$I^{+}_{af} = -I^{-}_{af} = \frac{1.0}{Z^{+}_{11} + Z^{-}_{11}} \quad \text{(bus 1 faulted)} \tag{10-26}$$

With bus k faulted, merely exchange Z^{+}_{11} and Z^{-}_{11} with Z^{+}_{kk} and Z^{-}_{kk}. Of course the phase a fault current totals zero, since it is recalled from Sec.

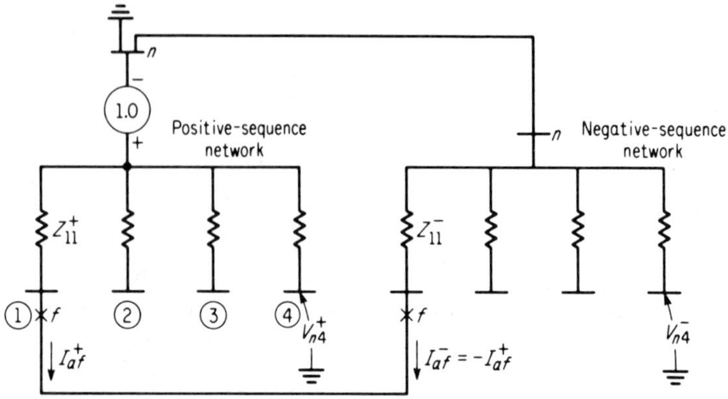

Fig. 10-12. Sequence network for phase a using rake equivalents. The condition is a line-to-line fault on bus 1. (Transfer impedances not shown.)

8-6 that phase a is the unfaulted phase. However, from Eq. 8-41,

$$I_{bf} = -I_{cf} = \sqrt{3}\, I_{af}^+ \angle -90^\circ \qquad [8\text{-}41]$$

The sequence voltages above neutral for bus 4 (phase a) are:

$$V_{n4}^+ = 1.0 - I_{af}^+ Z_{14}^+ \qquad (10\text{-}27)$$

$$V_{n4}^- = -I_{af}^- Z_{14}^- \qquad (10\text{-}28)$$

and

$$\begin{aligned} V_{n4} &= V_{n4}^+ + V_{n4}^- = (1.0 - I_{af}^+ Z_{14}) + (-I_{af}^- Z_{14}^-) \\ &= 1.0 - I_{af}^+(Z_{14}^+ - Z_{14}^-) \end{aligned}$$

Substituting I_{af}^+ of Eq. 10-26,

$$V_{n4} = 1.0 - \frac{(Z_{14}^+ - Z_{14}^-)}{(Z_{11}^+ + Z_{11}^-)} \quad \text{(bus 1 faulted)} \qquad (10\text{-}29)$$

In general, for an L-L fault on bus k, the phase a voltage on bus m is

$$V_{nma} = 1.0 - \frac{Z_{km}^+ - Z_{km}^-}{Z_{kk}^+ + Z_{kk}^-} \qquad (10\text{-}30)$$

To obtain phase b and c voltages on bus m (k faulted), rotate the a-phase sequence voltages by the appropriate angles, or

$$\begin{aligned} V_{nmb} &= V_{nmb}^+ + V_{nmb}^- \\ &= V_{nma}^+ \angle -120^\circ + V_{nma}^- \angle +120^\circ \end{aligned} \qquad (10\text{-}31)$$

and

$$V_{nmc} = V_{nma}^+ \angle +120^\circ + V_{nma}^+ \angle -120^\circ \qquad (10\text{-}32)$$

where

$$V^{+}_{nma} = 1.0 - I^{+}_{af} Z^{+}_{km} \tag{10-33}$$

and

$$V^{-}_{nma} = -I^{-}_{af} Z^{-}_{km} = +I^{+}_{af} Z^{-}_{km} \tag{10-34}$$

Substituting Eqs. 10-33 and 10-34 into Eq. 10-31,

$$\begin{aligned} V_{nmb} &= (1.0 - I^{+}_{af} Z^{+}_{km}) \underline{/-120^\circ} + (I^{+}_{af} Z^{-}_{km}) \underline{/+120^\circ} \\ &= 1.0 \underline{/-120^\circ} - I^{+}_{af}(Z^{+}_{km} \underline{/-120^\circ} - Z^{-}_{km} \underline{/+120^\circ} \end{aligned}$$

Substituting for I^{+}_{af},

$$V_{nmb} = 1.0 \underline{/-120^\circ} - \frac{Z^{+}_{km} \underline{/-120^\circ} - Z^{-}_{km} \underline{/+120^\circ}}{Z^{+}_{kk} + Z^{-}_{kk}} \tag{10-35}$$

Likewise, substituting Eqs. 10-33 and 10-34 into Eq. 10-32 would yield

$$V_{nmc} = 1.0 \underline{/+120^\circ} - \frac{Z^{+}_{km} \underline{/+120^\circ} - Z^{-}_{km} \underline{/-120^\circ}}{Z^{+}_{kk} + Z^{-}_{kk}} \tag{10-36}$$

Equations 10-30, 10-35, and 10-36 can all be further simplified if the positive- and negative-sequence rake equivalents are assumed equal for the network.

Again, if the magnitude of E^{+}_{a} is not 1.0, network currents and voltages should be corrected by the ratio of $E^{+}_{a}/1.0$.

10-7. Application of the S-C Matrix to Double Line-to-Ground Faults

The phase a sequence network of Fig. 8-19 is applicable here, where Z^{+}, Z^{-}, and Z^{0} are replaced by positive, negative, and zero-sequence rake equivalents respectively. Figure 10-13 shows this equivalent with a fault on bus 1.

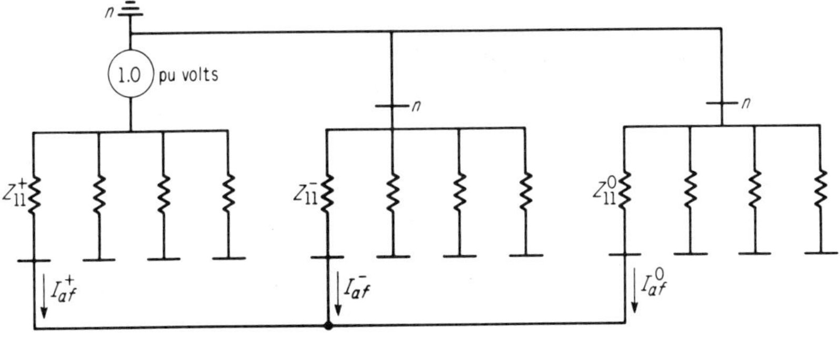

Fig. 10-13. Sequence network for double line-to-ground fault on bus 1 using rake equivalents. (Transfer impedances not shown.)

In general, for a fault on bus k, it is fairly obvious from Fig. 8-13 that

$$I_{af}^{+} = \frac{1.0}{Z_{kk}^{+} + (Z_{kk}^{-} Z_{kk}^{0})/(Z_{kk}^{-} + Z_{kk}^{0})} \tag{10-37}$$

or

$$I_{af}^{+} = \frac{Z_{kk}^{-} + Z_{kk}^{0}}{Z_{kk}^{+} Z_{kk}^{-} + Z_{kk}^{+} Z_{kk}^{0} + Z_{kk}^{-} Z_{kk}^{0}} \tag{10-38}$$

also

$$I_{af}^{-} = -V_{nfa}^{+}/Z_{kk}^{-}, \quad \text{and} \quad I_{af}^{0} = -V_{nfa}^{+}/Z_{kk}^{0} \tag{10-39}$$

where

$$V_{nfa}^{+} = 1.0 - I_{af}^{+} Z_{kk}^{+} \tag{10-40}$$

Of course V_{nfa}^{+} can also be written altogether in terms of impedances by substituting I_{af}^{+} from Eq. 10-38. This results in

$$V_{nfa}^{+} = \frac{Z_{kk}^{-} Z_{kk}^{0}}{Z_{kk}^{+} Z_{kk}^{-} + Z_{kk}^{+} Z_{kk}^{0} + Z_{kk}^{-} Z_{kk}^{0}} \tag{10-41}$$

While the fault current of the unfaulted a phase will be zero, the currents leaving the faulted b and c phases are needed. As before, they are found in terms of the phase a components. Referring back to Chapter 8,

$$I_{bf} = I_{af}^{+} \underline{/-120^\circ} + I_{af}^{-} \underline{/+120^\circ} + I_{af}^{0} \qquad [8\text{-}52]$$

$$I_{cf} = I_{af}^{+} \underline{/+120^\circ} + I_{af}^{-} \underline{/-120^\circ} + I_{af}^{0} \qquad [8\text{-}53]$$

Sequence voltages on phase a of bus m, for a fault on bus k are

$$\left.\begin{aligned} V_{nma}^{+} &= 1.0 - I_{af}^{+} Z_{km}^{+} \\ V_{nma}^{-} &= I_{af}^{-} Z_{km}^{-} \\ V_{nma}^{0} &= -I_{af}^{0} Z_{km}^{0} \end{aligned}\right\} \tag{10-42}$$

Finally, the voltages above neutral on bus m are

$$\left.\begin{aligned} V_{nma} &= V_{nma}^{+} + V_{nma}^{-} + V_{nma}^{0} \\ V_{nmb} &= V_{nma}^{+} \underline{/-120^\circ} + V_{nma}^{-} \underline{/+120^\circ} + V_{nma}^{0} \\ V_{nmc} &= V_{nma}^{+} \underline{/+120^\circ} + V_{nma}^{-} \underline{/-120^\circ} + V_{nma}^{0} \end{aligned}\right\} \tag{10-43}$$

As usual, individual branch currents throughout the network can be determined, once the voltages are known on the buses between which the branches terminate. Refer to Eq. 10-12.

10-8. Method of Developing the S-C Impedance Matrix One Step at a Time

The method of forming the s-c impedance matrix in Sec. 10-3 and Example 10-1 commenced with the complete network tree and closed all links simultaneously, yielding a set of loop-voltage equations. The corresponding loop matrix equation was partitioned and reduced in order to eliminate unwanted loop currents. It was reasoned that when only the fault-link currents remained, the final Z matrix would be the useful s-c impedance matrix from which the rake equivalent is taken. While the method of Sec. 10-3 serves well in gaining a clear concept of the s-c matrix, it is by no means the only method, nor is it considered to be the most advantageous one.

Section 10-8 will deal with a widely used approach, building up the s-c impedance matrix one step at a time by adding branches one at a time until the complete network has been formed. Each stage of the development will yield a new and modified short-circuit impedance matrix for the circuit as it stands, incomplete as it may be. It is most significant to note that, as branches are added, each addition must fall into one of four general types or categories, listed as follows:

1. Addition of a generator to a new bus.
2. Addition of a generator to an old bus.
3. Addition of a radial line or transformer from an old bus to a new bus.
4. Addition of a line or transformer between two old buses.

Actually, in the strictest sense, the additions are listed as they apply to the positive-sequence network, where generation is present. In order to revise the list to include the negative and zero-sequence networks, it is only to be realized that a path back to the neutral bus in these systems is treated as though it were a path through generation. For example, the grounded-Y generator may provide a path to the neutral bus for all three systems, even though generation is present only for the positive-sequence system. An even more contrasting example is found in the addition of a Y-Δ transformer with grounded Y. We have learned from Chapters 8 and 9 that this connection provides a low impedance path back to neutral bus for zero-sequence currents only. At first glance there appears to be no category to which this addition may apply. A closer look reveals that either type 1 or type 2 would serve this situation (with slight revision) since we are merely adding transformer impedance from the neutral bus to either a new or old bus.

One additional observation should be made before revising the list. It has been previously shown that the s-c impedance matrices (representing only the passive sequence networks) can be formed with only a knowledge of the system impedances. The revised list that follows is better adapted to include impedance additions of all three sequence networks.

I. Addition of an impedance (line, transformer, generator, etc.) from the neutral bus to a new bus.

II. Addition of an impedance (line, transformer, generator, etc.) from the neutral bus to an old bus.

III. Addition of an impedance (line, transformer, etc.) from an old bus to a new bus.

IV. Addition of an impedance (line, transformer, etc.) from an old bus to another old bus.

The four categories will be covered in order. Individual procedures will be explained for revising the old short-circuit impedance matrix to include a particular type of branch addition. For the sake of clarity, we will think specifically in terms of the *positive-sequence* network. The various loops involved are easier to visualize when identified with corresponding loop currents and source voltages of the positive-sequence system. Also in each case to follow, the s-c impedance matrix (before making the new branch addition) will be assumed to be a 3×3. The reader should not confuse the original three-bus system with the three phases. Again it is emphasized that the s-c impedance matrix and corresponding rake equivalent are on a *per-phase* basis, representing only phase "*a*" of the system.

Type 1. Addition of a generator to a new bus. This type merely extends the s-c impedance matrix from an order of $n \times n$ to that of an $(n + 1) \times (n + 1)$. Its new rake equivalent is likewise modified in a very simple manner as in Fig. 10-14. A new bus is retained and a new loop is formed (shown dotted) through a fault link. Any fault current I_4 which might flow through Z_{gen} is not likely to be mutually coupled to the other branches, thus accounting for the zero elements of the new 4×4 matrix, which is

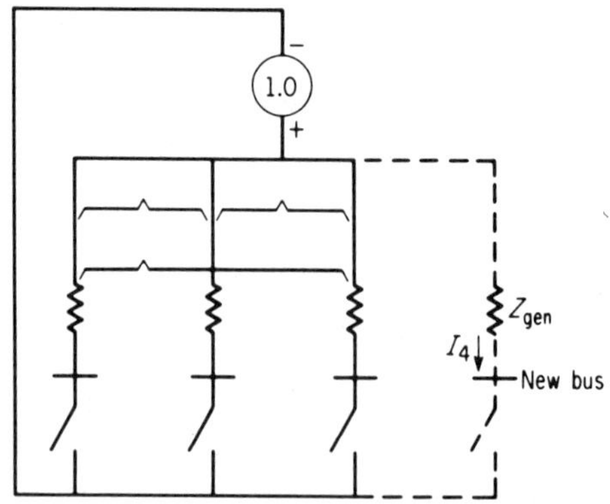

Fig. 10-14. Final rake equivalent for type 1 addition.

old s-c matrix

$$\overline{Z}_{\text{s-c new}} = \left[\begin{array}{ccc|c} Z_{11} & Z_{12} & Z_{13} & 0 \\ Z_{21} & Z_{22} & Z_{23} & 0 \\ Z_{31} & Z_{32} & Z_{33} & 0 \\ \hline 0 & 0 & 0 & Z_{\text{gen}} \end{array}\right] \tag{10-44}$$

The dashed portion of this matrix is simply the old 3 × 3 s-c matrix.

Type 2. Addition of a generator to an old bus. An internal loop has been formed, totally within the passive network as indicated in Fig. 10-15a. Here the new generator is shown connected to bus 3, forming the

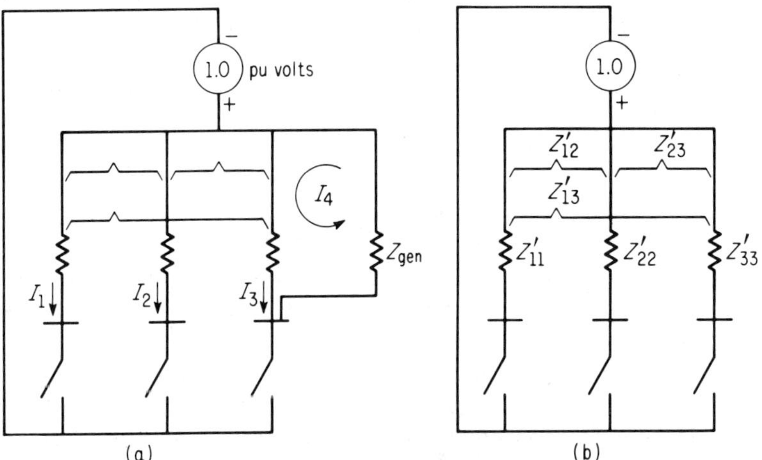

Fig. 10-15. (a) Example of type 2 modification. (b) Final rake equivalent after eliminating I_4.

loop for I_4. The procedure for our example then entails (a) writing the loop-impedance matrix $\overline{Z}_{\text{loop}}$ and (b) eliminating the internal loop by matrix partitioning to obtain $\overline{Z}_{\text{s-c new}}$ and the new 3-bus equivalent of Fig. 10-15b.

(a) The loop impedance matrix is

$$\overline{Z}_{\text{loop}} = \left[\begin{array}{ccc|c} Z_{11} & Z_{12} & Z_{13} & Z_{13} \\ Z_{21} & Z_{22} \;\textcircled{\scriptsize $\overline{Z}_1$} & Z_{23} & Z_{23} \;\textcircled{\scriptsize $\overline{Z}_2$} \\ Z_{31} & Z_{32} & Z_{33} & Z_{33} \\ \hline Z_{31} & Z_{32} \;\textcircled{\scriptsize $\overline{Z}_3$} & Z_{33} & Z_{33} + Z_{\text{gen}} \;\textcircled{\scriptsize $\overline{Z}_4$} \end{array}\right] \tag{10-45}$$

Once again the 3 × 3 portion of the portion identified as $\overline{Z}_1$ is the old short-circuit impedance matrix. Notice that $\overline{Z}_2$ and $\overline{Z}_3$ are simply repeats of the preceding column or row, respectively, of the old matrix. This last statement seems logical, since the loop currents I_1, I_2, and I_3 see the same mutual impedance in common with loop 4 as they did with loop 3. It might be argued here that the sign on the elements of $\overline{Z}_2$ and $\overline{Z}_3$ would all be negative had I_4 been chosen in the reverse direction. This is true, but since loop 4 and I_4 are to be eliminated, it could also be reasoned that the assumed current direction should have no bearing upon the end result. Step (b) will show that it makes no difference whether the positive or negative sign is used for the elements in $\overline{Z}_2$ and $\overline{Z}_3$ as long as one is consistent. We will assume the counterclockwise direction of I_4 here and the corresponding positive signs. The loop 4 self-impedance is the sum of the generator impedance and the self-impedance of loop 3 as indicated in Eq. 10-45. This fact is apparent from inspection of Fig. 10-15a.

(b) To eliminate the new internal loop 4, again reduce the matrix by the method of Sec. 5-3, which yields a new 3 × 3 short-circuit impedance matrix, where

$$\overline{Z}_{sc} = \overline{Z}_1 - \overline{Z}_2\overline{Z}_4^{-1}\overline{Z}_3 \tag{10-1}$$

Note: By the time the product $\overline{Z}_2\overline{Z}_4^{-1}\overline{Z}_3$ is taken, it will have been immaterial whether elements of both $\overline{Z}_2$ and $\overline{Z}_3$ were taken as positive or negative, since the product of two negatives yield a positive sign. Recall that the choice of signs from step (a) was dependent upon the assumed direction of I_4.

Type 3. Addition of an impedance between an old (already identified) bus and a new bus. Figure 10-16a shows the addition of a line impedance between bus 2 and the new bus 4. Assume no mutual coupling

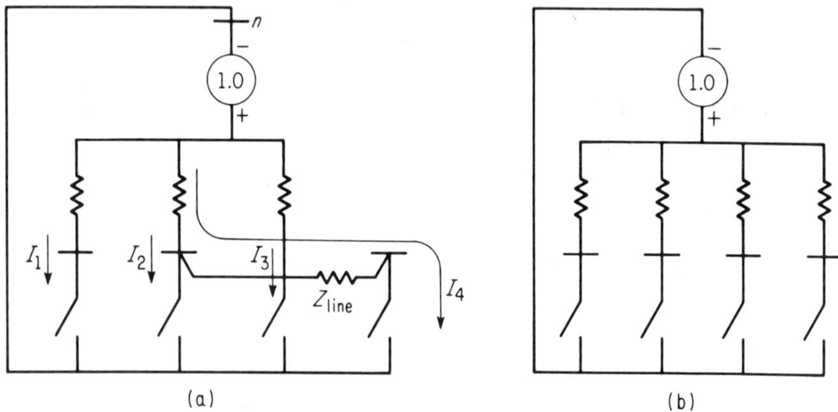

Fig. 10-16. (a) Adding a line from old bus 2 to new bus 4. (b) Final rake equivalent of Fig. 10-16a. (Transfer impedances not shown.)

between the new line and the remaining network. Notice that the new bus is provided with a fault switch as prescribed by Sec. 10-3. The self- and mutual-impedances of loops 1,2, and 3 remain unchanged. However, we have expanded to a four-bus system and will expect a new 4×4 s-c matrix, where

$$\overline{Z}_{\text{s-c new}} = \overbrace{\begin{bmatrix} Z_{11} & Z_{12} & Z_{13} & Z_{14} = Z_{12} \\ Z_{21} & Z_{22} & Z_{33} & Z_{24} = Z_{22} \\ Z_{31} & Z_{32} & Z_{33} & Z_{34} = Z_{32} \\ Z_{21} & Z_{22} & Z_{23} & (Z_{22} + Z_{\text{line}}) \end{bmatrix}}^{\text{old s-c matrix}} \tag{10-46}$$

The dashed 3×3 portion of the matrix is the old s-c matrix. The corner element in the 4th row and 4th column is the sum of Z_{line} plus the self-impedance of the old bus to which the line is attached. Note also that the new mutual impedances Z_{14}, Z_{24}, and Z_{34} of the new matrix are identical with Z_{12}, Z_{22}, and Z_{32} respectively. In other words, loop currents I_1, I_2, and I_3 will reflect the same voltage upon loop 4 as they reflect upon loop 2 because of the path which loops 2 and 4 have in common. Of course, as before, the matrix is symmetrical and $Z_{km} = Z_{mk}$.

While the impedance matrix of Eq. 10-46 is satisfied by Fig. 10-16a, it is also satisfied by the new four-bus rake equivalent of Fig. 10-16b.

Type 4. Addition of a line, transformer, etc., between two old buses. This time the rake equivalent of Fig. 10-17 has been modified to include a line impedance between buses 2 and 3. This forms an internal loop with I_4 circulating within. A 4×4 loop-impedance matrix may be written by

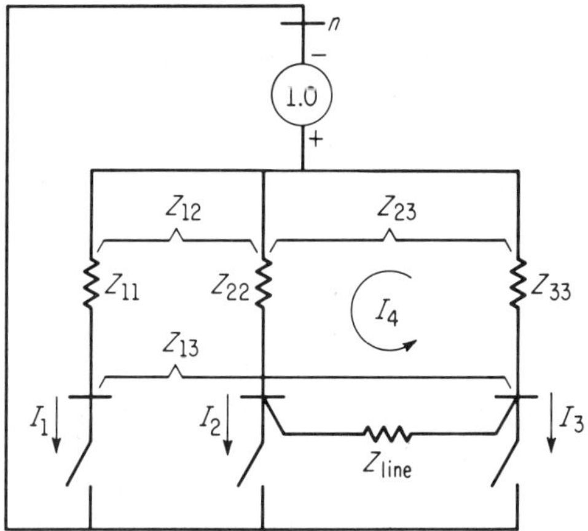

Fig. 10-17. Type 4 addition to 3-bus system.

inspection and then reduced to obtain the new 3 × 3 s-c impedance matrix.

$$\overline{Z}_{\text{loop}} = \left[\begin{array}{ccc|c} Z_{11} & Z_{12} & Z_{13} & Z_{14} \\ Z_{21} & Z_{22} \;\textcircled{\small $\overline{Z}_1$} & Z_{23} & Z_{24} \;\textcircled{\small $\overline{Z}_2$} \\ Z_{31} & Z_{32} & Z_{33} & Z_{34} \\ \hline Z_{41} & Z_{42} \;\textcircled{\small $\overline{Z}_3$} & Z_{43} & \textcircled{\small $\overline{Z}_4$}\; Z_{44} \end{array}\right] \tag{10-47}$$

The old 3 × 3 matrix is the $\overline{Z}_1$ of Eq. 10-47. In order to obtain row 4 and column 4 of $\overline{Z}_{\text{loop}}$, observe the loop currents of Fig. 10-17.

The Z_{44} element should represent all of the self-impedance encountered by the current I_4. This includes $Z_{22} + Z_{33} + Z_{\text{line}}$ plus any mutual effect that I_4 has upon its own loop. Notice that as I_4 moves downward through Z_{22} it reflects a voltage upon the Z_{33} branch which is proportional to Z_{23}. The sign of this mutual voltage will be opposite to the voltage created by I_4 moving upward through Z_{33} and therefore Z_{23} is included with a negative sign. It must be included twice, since as I_4 moves upward through Z_{33} it also reflects a voltage upon the Z_{22} branch of its own loop, proportional again to Z_{23}. To summarize,

$$Z_{44} = Z_{22} + Z_{33} + Z_{\text{line}} - 2Z_{23} \tag{10-48}$$

A more general equation can be written for this corner element ($\overline{Z}_4$) of $\overline{Z}_{\text{loop}}$. If k and m are the buses between which Z_{line} is connected, then

$$\overline{Z}_4 = Z_{kk} + Z_{mm} + Z_{\text{line}} - 2Z_{km} \tag{10-49}$$

where Z_{kk} and Z_{mm} are the old self-impedances of buses k and m before modification and Z_{km} is the mutual impedance between buses k and m of the old rake equivalent.

Next, look at the remaining elements in column 4 of Eq. 10-47; Z_{14}, Z_{24}, and Z_{34}. These are equal to the row elements Z_{41}, Z_{42}, and Z_{43} respectively. To determine these three values we will investigate Z_{41} in particular. As I_4 traverses loop 4, it will reflect a voltage into loop 1 because of both Z_{12} and Z_{13}, or

$$Z_{41} = Z_{14} = Z_{12} - Z_{13} \tag{10-50}$$

A more general expression may be written for the mutual between the internal loop current I_p and bus i with a line added between bus k and m:

$$Z_{pi} = Z_{ip} = Z_{ki} - Z_{mi} \qquad p \neq i \tag{10-51}$$

If I_p is assumed downward through bus k and upward through bus m, the signs on Z_{ki} and Z_{mi} are as shown in Eq. 10-51. If the reverse direction were assumed on I_4, then the signs on Z_{ki} and Z_{mi} would be interchanged.

However since loop p is to be eliminated then it should not matter in which direction I_p was assumed, as long as one is consistent regarding the use of signs in Eq. 10-51. These loop currents are, in fact, only mentioned to aid in a better understanding of Eqs. 10-49 and 10-51.

In the elimination of the internal loop, we again resort to the elimination equation

$$\bar{Z}_{\text{s-c new}} = \bar{Z}_1 - \bar{Z}_2\bar{Z}_4^{-1}\bar{Z}_3 \qquad [10\text{-}1]$$

This will, in our three-bus example, take the 4 × 4 Z_{loop} matrix Eq. 10-47 and reduce it to the new 3 × 3 s-c impedance matrix.

As a practical consideration, a single loop *elimination* may be accomplished on an *individual element basis*. For example, it can be proven that individual elements of the 3 × 3 $\bar{Z}_{\text{s-c new}}$ matrix can be found directly in terms of the individual elements of the 4 × 4 $\bar{Z}_{\text{loop}}$ matrix. Let Z'_{mn} be an element (mth row and nth column) of the new matrix and Z_{mn} be an element in $\bar{Z}_{\text{loop}}$. Then

$$Z'_{mn} = Z_{mn} - Z_{m4}Z_{44}^{-1}Z_{4n} \qquad (10\text{-}52)$$

Example 10-2. Show how to obtain the element Z'_{23} for $Z_{\text{s-cnew}}$ from the elements of Z_{loop}, where loop 4 is to be eliminated.

$$\bar{Z}_{\text{loop}} = \left[\begin{array}{ccc|c} Z_{11} & Z_{12} & Z_{13} & Z_{14} \\ Z_{21} & Z_{22} & \textcircled{Z_{23}} & \textcircled{Z_{24}} \\ Z_{31} & Z_{32} & Z_{33} & Z_{34} \\ \hline Z_{41} & Z_{42} & \textcircled{Z_{43}} & \textcircled{Z_{44}} \end{array}\right]; \qquad \bar{Z}_{\text{s-c new}} = \begin{bmatrix} Z'_{11} & Z'_{12} & Z'_{13} \\ Z_{21} & Z'_{22} & \textcircled{Z'_{23}} \\ Z'_{31} & Z'_{32} & Z'_{33} \end{bmatrix}$$

The value of Z'_{23} is obtained from Eq. 10-52 as

$$Z'_{23} = Z_{23} - Z_{24}Z_{44}^{-1}Z_{43}$$

The elements involved are circled in the matrix equations.

10-9. Modifying the S-C Matrix to Accommodate System Changes

Suppose one were provided with a s-c matrix and the system is subsequently altered. The matrix may generally be modified without too much difficulty. The addition of new lines, generators etc. may be treated as in Sec. 10-8. Other changes are also possible.

REMOVING LINES

This procedure is identical to that of adding lines, except that the removed line or branch is considered as negative impedance, in order to cancel the effect of the branch. This principle also applies to the removal of a generator where a type 1 or 2 addition is made, only with negative generator impedance.

OPENING A BUS BREAKER

Suppose a bus breaker is opened so as to divide the bus into two parts. We will assume that the original network had two connections (Z_x and Z_y) between bus 2 and bus 3 as in Fig. 10-18a. Opening the

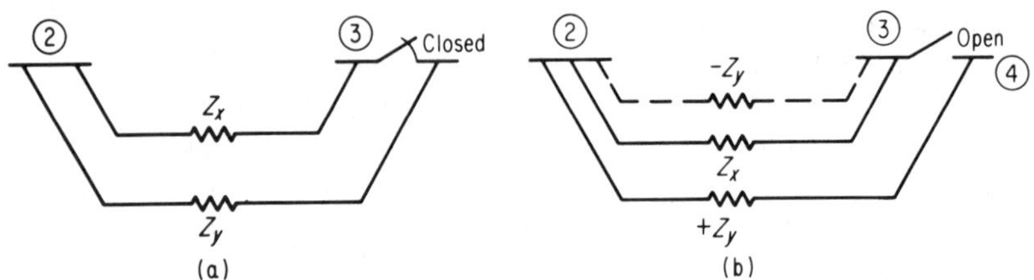

Fig. 10-18. (a) Original connection between buses 2 and 3. (b) Concept of opening a bus breaker.

breaker effectively creates a new bus 4 with a different voltage level than bus 3. At the same time we can consider that we have removed the line Z_y from bus 2 to bus 3 (type 4 addition of $-Z_y$) while adding Z_y between buses 2 and 4, a type 3 addition. This example is not intended to exhaust the subject of opening bus breakers, but rather to demonstrate that most modifications can be worked out through an understanding and application of the foregoing principles.

10-10. A General Computer Procedure for Developing the S-C Matrix

It has already been determined that, once the s-c impedance matrices are obtained for the sequence networks, fault calculations are a relatively simple matter. One general procedure will be indicated here for arriving at these matrices by computer. We will refer to Sec. 10-8, where the network is built up one step at a time.

A punched card of information could be associated with each branch in the network. A separate card for each generator transformer, line, etc., would contain such information as positive-, negative-, and zero-sequence impedance per phase, and identification of the nodes or buses on which these impedances terminate. We have learned from Sec. 10-8 that only four types of branch additions will be made. A logical selection process is demonstrated in Fig. 10-19, given in flow-chart form. A card is read in to determine whether the impedance goes back to the neutral bus through generation. The presence of reference node 0 on the card as one of the branch terminating nodes could easily establish this fact. If generation is

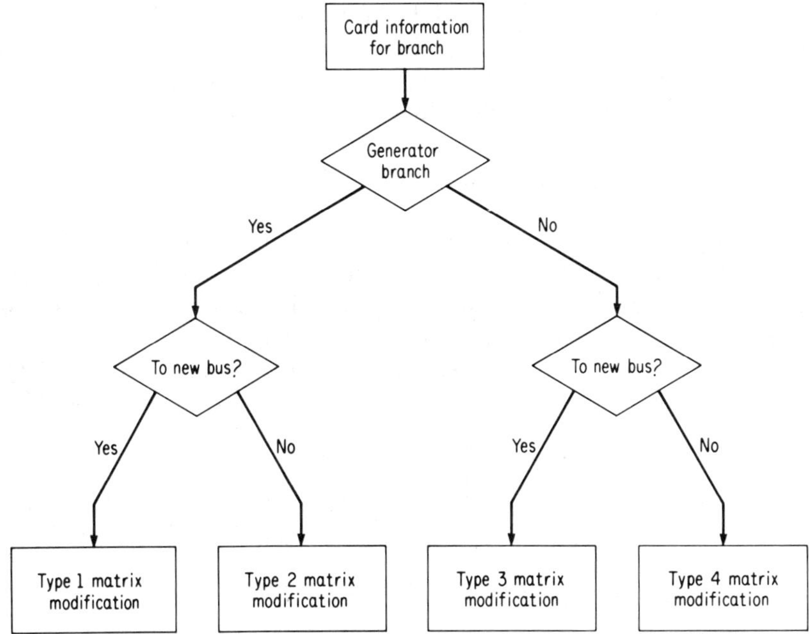

Fig. 10-19. Partial flow chart for building up the positive-sequence network.

present another question is asked—namely, does this new generator branch impedance connect to a new bus which has not yet been identified in a previous step or does it connect with an old (already identified) bus? The answer to this question will differentiate between the type 1 and type 2 additions. When this has been determined, instructions are given to proceed with the necessary matrix modification.

The cards could previously have been ordered or arranged so as to guarantee that each branch has at least one node to be connected to a bus previously identified. If a generator is not present in the branch it is still necessary to determine whether the other terminating bus is an old or new bus. This distinguishes between the type 3 or type 4 addition.

After information on all cards has been processed and all matrix modifications are completed the final matrix will be the s-c impedance matrix for the positive-sequence system (for our particular example). Negative- and zero-sequence systems are treated similarly except for the absence of generation. Impedance paths still exist back to the neutral bus for these systems, which allows for the type 1 and 2 additions. To make Fig. 10-19 applicable to the negative- and zero-sequence networks we should replace the words "Generator branch?" in the first decision box with the question "Does this branch have a path back to the neutral bus?" Also to be more general, the four modifications could be written as Roman numeral type I, II, III, and IV, to take in the negative-, and zero-sequence systems.

The actual program may or may not be extended to include the complete fault calculations in addition to the formation of the matrix.

10-11. Solving a Large Network by Sections

It is possible for a system to be so large that a complete step-by-step matrix development would exceed the capabilities of the computer. This section will describe a procedure for accomplishing this development by dividing the large system of many nodes into sections.

First, the concept of *suppressed nodes* in the s-c matrix and corresponding rake equivalent should be mentioned. Refer to Fig. 10-20 in

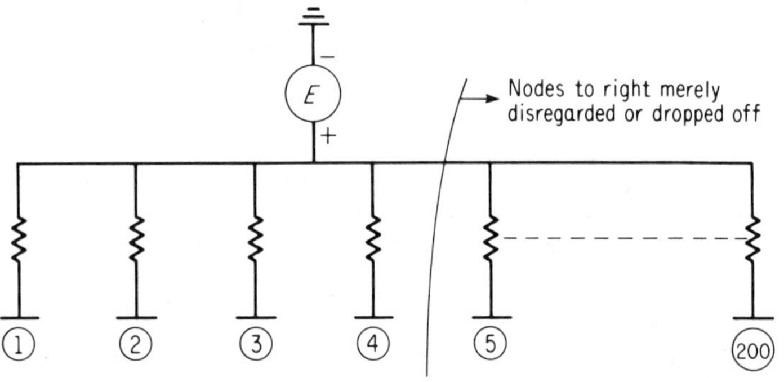

Fig. 10-20. Simple procedure of suppressing nodes in a rake equivalent. (Transfer impedances not shown.)

which 200 buses were developed in the rake equivalent. We will assume that only 4 of these buses are of real interest in the study. If the other 196 buses are not to be faulted in the study, they will carry no current and therefore have no effect upon the four buses of interest. Therefore, this means of node suppression is a mere matter of dropping off unwanted branches from the rake equivalent. Corresponding rows and columns are likewise dropped out of the 200 × 200 matrix, leaving a simple 4 × 4 matrix.

Now consider the example of a positive-sequence twelve-bus network which has been divided into three sections (a, b, and c). These sections are connected by certain tie lines. Refer to Fig. 10-21. One procedure for handling this problem is to (1) consider all tie lines open and develop the s-c matrices for the three sections separately; (2) next suppress unwanted nodes in each section as explained in the preceding paragraph, retaining only nodes of interest, including tie-line nodes; (3) now join the three

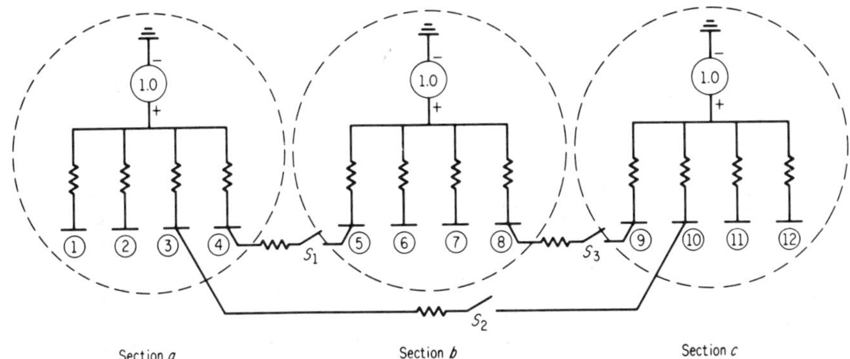

Fig. 10-21. Example of dividing a 12-bus system into three areas or sections. (Mutual or transfer impedances not shown.)

reduced sections into one short-circuit matrix with tie lines still open; and (4) modify this matrix by closing the tie-line switches one step at a time. This modification falls under the heading of a type 4 addition, covered in Sec. 10-8.

Suppose we are interested in all of the buses in section *a* (for later fault studies in the *a* section) but only tie buses 5, 8, 9 and 10 from the *b* and *c* sections. Other unwanted nodes (6, 7, 11 and 12) are merely dropped off as shown in Fig. 10-22.

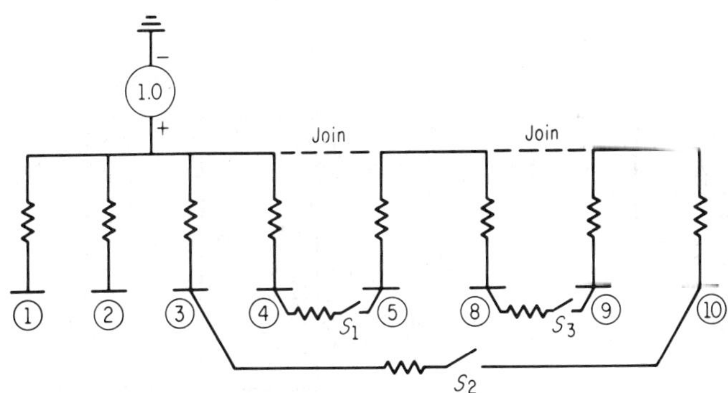

Fig. 10-22. Joining the section equivalents (tie lines still open). Unwanted nodes (6, 7, 11, and 12) are dropped off. (Transfer impedances not shown.)

The three equivalents are now joined together (see dotted lines) and one reduced 8 × 8 s-c matrix results. Now S_1 is closed and the internal loop which it creates is eliminated according to the type 4 modification procedure. A new 8 × 8 matrix results. The modification procedure is

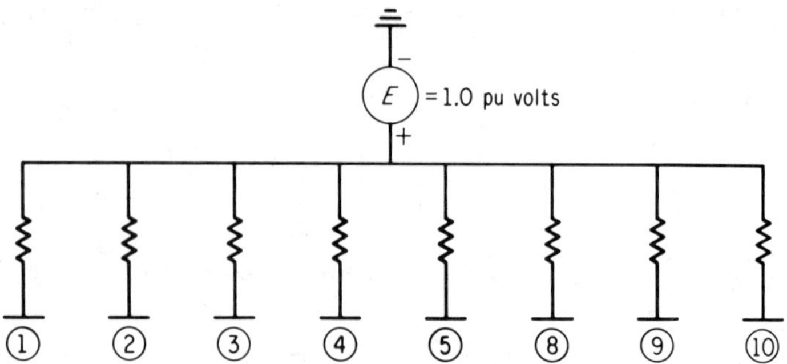

Fig. 10-23. Final rake equivalent for circuit of Fig. 10-21 with nodes 6, 7, 11, and 12 suppressed. (Transfer impedances not shown.)

repeated for S_2 and S_3 in turn, and finally we are left with the eight-bus (plus reference bus 0) rake equivalent of Fig. 10-23.

10-12. Alternate Concept of the S-C Matrix Method

In some of the literature, the voltage matrix equation for the s-c equivalent is treated in a different manner. Compare Fig. 10-24a with that of Fig. 10-24b in which we visualize a symmetrical three-phase fault on a

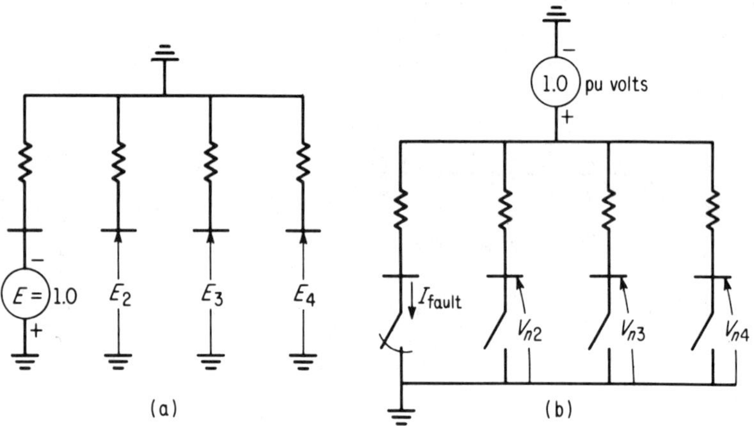

Fig. 10-24. (a) Alternate concept for symmetrical fault. (b) Fault concept used in this chapter. (Transfer impedances not shown.)

four-bus system. Instead of placing generation between the neutral bus (or ground) and the reference bus 0, it is placed between ground and the faulted bus. However, from strictly a conceptual point of view the author has chosen the circuit of Fig. 10-24b (throughout Chapter 10) along with the corresponding matrix equation of the form of Eq. 10-10.

10-13. Handling Mutuals* Between Branch Lines

Section 10-3 formed the s-c impedance matrix as a complete network, arriving at $\overline{Z}_{\text{loop}}$ and eliminating internal unwanted loops to yield loop $\overline{Z}_{\text{s-c}}$. $\overline{Z}_{\text{loop}}$ was formed by inspection of the loop currents although it could also have been formed (as in Chapter 6) by transforming a branch impedance matrix into $\overline{Z}_{\text{loop}}$ by

$$\overline{Z}_{\text{loop}} = C_t Z_{br} C \qquad [6\text{-}7]$$

In either case, mutual effects can be included in Z_{loop}. Yet in Sec. 10-8 it was pointed out that it is often more convenient to form the s-c matrix one step at a time, adding one branch at a time. This text has omitted the treatment of mutual couplings for this latter method. By doing so, it is felt that the complexity of the problem and the space required for its treatment have been considerably reduced in favor of a clearer understanding of the basic problem. Mutual impedance is often considered negligible between sets of transmission lines, insofar as the positive- and negative-sequence systems are concerned, but some line mutuals may be significant in the zero-sequence network as explained in Sec. 9-4. In case of such mutual coupling it should be realized that methods exist to include mutuals, even while building the s-c matrix one step at a time. The reader may find an extensive coverage of the treatment of mutuals in a paper entitled "Digital Calculation of L-G Short Circuits by Matrix Method," by A. H. El-Abiad, *AIEE Transactions*, Vol. 79, Part III (1960), pp. 323–332.

Even though the effect of mutuals is included within the resulting s-c matrix and corresponding rake equivalent, original branch currents (for a particular faulted bus) could still be a problem where branches are mutually coupled. For example, suppose two original branch lines X and y (with mutual Z_{xy} between them) exist as in Fig. 10-25a, where the

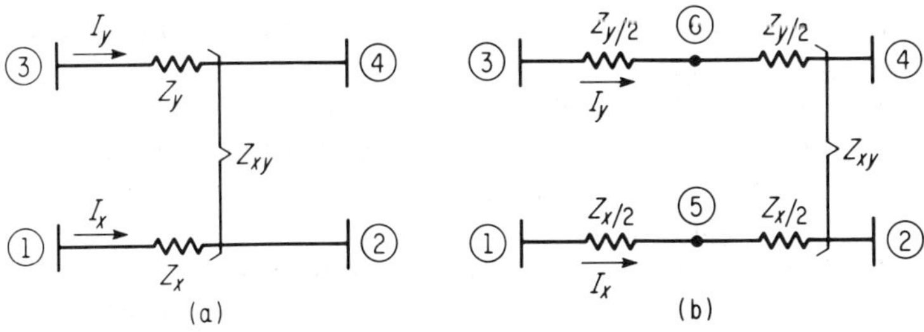

Fig. 10-25. (a) Two original branch lines with mutual coupling. (b) Use of fictitious nodes to simplify branch current solution.

*When speaking of "mutuals between branches," recall that the system is on a per phase basis and reference here is made to mutuals (per phase) between two or more sets of three-phase lines running together.

lines do not terminate on the same buses. The s-c matrix makes possible a quick calculation of voltages above ground on all four buses as per Eq. 10-11. Furthermore, when no coupling exists in the original branches, Eq. 10-12 holds. With the mutual coupling present this current equation obviously no longer holds true. Actually, two equations and two unknowns (I_x and I_y) would be necessary to obtain these branch currents. They are

$$(V_{n3} - V_{n4}) = I_y Z_y + I_x Z_{xy}$$
$$(V_{n1} - V_{n2}) = I_x Z_x + I_y Z_{xy}$$

Of course these equations become increasingly difficult as a branch is coupled with more and more other branches.

One technique for permitting the use of the single Eq. 10-12 involves the creation of fictitious nodes in the original network. For example, Fig. 10-25b has broken up the branch self impedances into two parts, creating nodes 5 and 6. Now all of the coupling is assumed to exist on one end of each line, while the other ends contain only self-impedance ($Z_{y/2}$ and $Z_{x/2}$). Once again the currents I_y and I_x can be found as

$$I_y = \frac{V_{n3} - V_{n6}}{Z_y/2}, \qquad \text{and} \qquad I_x = \frac{V_{n1} - V_{n5}}{Z_x/2}$$

Of course Eq. 10-12 may be applied in place of the above form. One disadvantage of this method is seen in that new buses must necessarily be retained. This requires a larger s-c impedance matrix.

10-14. Summary

The short-circuit impedance matrix, with its corresponding rake equivalent has proven to be a most powerful tool in the solution of short-circuit problems. This is not to say that it is limited to these problems, but this chapter was primarily concerned with the short-circuit application, given the assumptions of Sec. 10-2.

It was demonstrated in Sec. 10-4 that the self- (or driving-point) impedances of the rake equivalent correspond to the diagonal elements of the s-c impedance matrix and they make possible an immediate fault current solution for a symmetrical fault on any bus. Refer to Eq. 10-7. The transfer or mutual impedances of the rake equivalent correspond to the off-diagonal elements of the matrix and are related to the voltages on the unfaulted buses by Eq. 10-11. One s-c matrix suffices for a fault on any bus.

Unsymmetrical faults (L-G, L-L, L-L-G) have a ready solution by applying the principles of symmetrical components to the rake equivalents for the positive-, negative-, and zero-sequence networks. This application was made in Secs. 10-5 through 10-7.

Various methods are available for the formation of the s-c matrix. Perhaps the most obvious procedure involves the writing of loop equations in a specific manner as was done in Sec. 10-3 then eliminating unwanted loops until the s-c matrix was the result. However, probably a more acceptable method (from the computer solution point of view) builds the matrix gradually by adding one branch at a time. The matrix is thus continually modified and expanded to accommodate each new addition. Only four basic additions are possible as listed in Sec. 10-8. As regards the positive-sequence network, these additions are (1) adding a generator and a new bus, (2) adding a generator to an old bus, (3) adding a branch (line, transformer, etc.) from an old bus to a new bus, and (4) adding a branch (line, transformer, etc.) between two old buses. By proper numbering of the buses it is not difficult for the computer to determine which type of addition is being made, at which point the appropriate matrix modification procedure is carried out.

Of course, if the s-c matrix can be built up one step at a time, it can likewise be modified to accommodate system changes—the addition or removal of lines and the like.

Not the least of the advantages of the s-c impedance matrix is the realization that a large system of many buses can be reduced simply by dropping off rows and columns corresponding to unwanted buses. These nodes or buses are thus suppressed by compressing the matrix, which is the equivalent of dropping off branches of the rake equivalent. If necessary, very large systems can be (1) broken up into smaller areas with an s-c impedance matrix for each area, tie lines between areas being open, (2) unwanted nodes may be suppressed, (3) area matrices may be joined together into one s-c matrix, and (4) area tie lines are included through a type 4 modification procedure. The resulting s-c matrix now includes the entire system. It is also noteworthy that these area and/or system matrices can be kept on file for future use.

Problems

10-1. Find Z_{bus} for Fig. 10-4a using the method of Sec. 10-8. (Answer given in result of example Prob. 10-1.)

10-2. Referring to Figs. 10-4a and 10-5, it is desired to add one radial line of $j0.7$ to bus 3, establishing a new bus 4. Use the method of Sec. 10-8 to obtain a new 4×4 Z_{bus} matrix.

10-3. Referring to Figs. 10-4a and 10-5, a line of $j0.4$ ohm is added from node 1 to node 2. Find the new Z_{bus} using the method of Sec. 10-8.

10-4. Referring to Figs. 10-4a and 10-5, a generator (with transformer) is to be added to node 2. The generator and transformer impedance is given as $j0.2$. Find the new Z_{bus}.

10-5. Given the system (four-bus + neutral bus as reference) of Prob. 8-12. Find the positive-sequence bus impedance matrix, using the method of Sec. 10-3. (Answer given in Prob. 10-12.)

10-6. Repeat Prob. 10-5, only using the method of Sec. 10-8.

10-7. Suppose the Z_{bus} of Prob. 10-5 were to be solved by inversion of an admittance matrix. (The inversion method is not recommended.) Write by inspection, the nodal admittance matrix which would be required by this method. Don't bother to invert the matrix.

10-8. Repeat Prob. 10-5 by the method given in Chapter 4, Sec. 13. Even though an analyzer is not available, the methods of that section can be applied to longhand calculation. Merely apply an emf to the reference bus, fault the individual buses one at a time and make calculations to determine driving-point and transfer impedances.

10-9. Refer to Prob. 10-5. Find the Z_{bus} matrix for the zero-sequence system. (The answer is given in Prob. 10-12.) Use the method of Sec. 10-3.

10-10. Repeat Prob. 10-9 using the method of Sec. 10-8.

10-11. Repeat Prob. 10-9 using the method of Sec. 4-13, but substituting hand calculation for the analyzer-board determination of driving-point and transfer impedances.

10-12. The $Z_{s\text{-}c}$ matrices for Prob. 8-12 are calculated to be

$$\bar{Z}^{+}_{bus} = \bar{Z}^{-}_{bus} = j\begin{array}{c} \begin{array}{cccc} d & e & g & f \end{array} \\ \begin{bmatrix} 0.355 & 0.245 & 0.300 & 0.300 \\ 0.245 & 0.355 & 0.300 & 0.300 \\ 0.300 & 0.300 & 0.450 & 0.450 \\ 0.300 & 0.300 & 0.450 & 0.600 \end{bmatrix} \end{array}$$

$$\bar{Z}^{0}_{bus} = j\begin{array}{c} \begin{array}{cccc} d & e & g & f \end{array} \\ \begin{bmatrix} 0.2 & 0.2 & 0.2 & 0.2 \\ 0.2 & 1.2 & 0.7 & 0.7 \\ 0.2 & 0.7 & 0.95 & 0.95 \\ 0.2 & 0.7 & 0.95 & 1.10 \end{bmatrix} \end{array}$$

Given a symmetrical three-phase fault at *f*. Using the short-circuit matrix method find the fault current, the voltages (*L-n*) at nodes *d* and *g*, and the current in line *d-g*.

10-13. Repeat Prob. 10-12 for a symmetrical fault at point *d*. Use the bus impedance matrix method.

10-14. Given the matrices of Prob. 10-12, applying to Prob. 8-12. Apply a line-to-ground fault at point *f*. Determine the fault current and the voltages to neutral on all three phases of bus *d*, using the bus impedance method.

10-15. Repeat Prob. 10-14 only with a line-to-line fault on bus *f*.

10-16. Repeat Prob. 10-14 only with a double line-to-ground fault at point *f*.

chapter 11

CONCEPTS IN THE CONTROL OF VOLTAGE, WATTS, AND VARS

11-1. Introduction

Before looking into the system calculations of load flow and economic dispatch, it seems more appropriate to gain a conceptual understanding of the factors that influence the flow of real and reactive power. This chapter will stress the relationship of bus voltage magnitudes to var flow and bus voltage phase angles to watts flow. An analogy is made between the frequency-load curves of generator prime movers and the voltage-var curves of the generator terminals. The effect of tap changes upon reactive power flow is also included. Some of the problems encountered in the paralleling of automatic load-tap-changing (LTC) transformers are mentioned, as well as methods in which these problems may be overcome.

A method for determining all bus voltage changes with the switching of shunt capacitors is set forth in this chapter. Again the bus impedance matrix (and corresponding "rake" equivalent) is a useful tool for this application. The method is also extended to include the effect upon all bus voltages of injecting vars at generation through field excitation. In addition, the voltage changes at all buses can be readily determined for a tap change at any point in the network. Since generators, shunt capacitors, and tap-changing transformers are the primary means of controlling vars and voltage, the "rake" equivalent method may be used as a device for dispatching vars from a voltage-profile standpoint.

Network solutions (such as the load flow studies of Chapter 12) may be oversimplified unless one is aware of certain variables in the actual system. It is often not sufficient to assume a static system of given loads, and known impedances, when in reality the system is dynamic, constantly responding automatically to change. For example, there are responses of automatic generator voltage regulation, automatic tap changes, and automatic switching of shunt capacitor banks, all of which may be sensitive to voltage and/or vars. It will be seen later that power input to generators may be sensitive to system frequency and/or load changes. It is important that certain of these basic concepts be understood before proceeding with the studies of total system load flows or economic dispatch.

It is also important, as a background to this chapter, for the reader to be familiar with the fundamentals of Sec. 3.3, in which the concepts of the power triangle are reviewed.

11-2. Circulating Vars Between Transformers of Unequal Turns Ratio

Paralleled transformers of unequal turns ratio are said to circulate lagging amperes from the transformer of higher induced secondary voltage back around to the other transformer. Figure 11-1 is a per phase

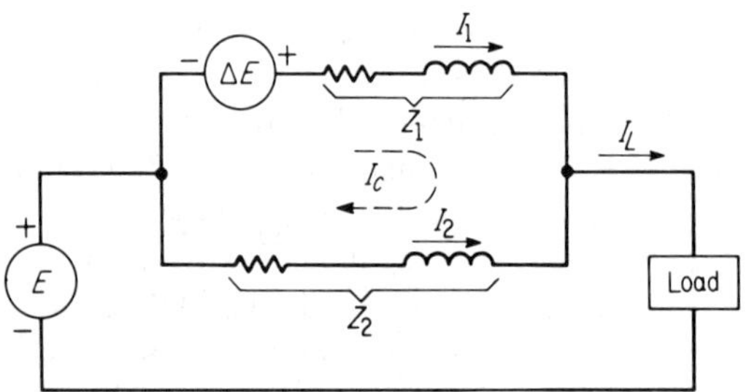

Fig. 11-1. Paralleled transformers feeding a common load.

representation of two paralleled transformers. Normally the load current would divide through the units inversely as their impedances, or

$$I_1 = I_L \frac{Z_2}{Z_1 + Z_2}; \qquad \text{and} \qquad I_2 = I_1 \frac{Z_1}{Z_2} \tag{11-1}$$

However, a boost in the tap setting of transformer 1 has the effect of superimposing a ΔE into the circuit as shown by the dashed circle. Consider I_1 and I_2 as the load current components due to E. In addition to these load components, the voltage of ΔE circulates I_c between the transformers. The load impedance is high with respect to the transformer impedances in series, and therefore

$$I_c = \frac{\Delta E}{Z_1 + Z_2} \tag{11-2}$$

If Z_1 and Z_2 are essentially reactive, then I_c is lagging ΔE (and E) by almost 90°. Vars are said to circulate within the loop for this reason. Normally the currents I_1, I_2, and I_L are nearly in phase with each other, and their phase angles depend, for the most part, upon the power factor of the load. An example is given to demonstrate the effect of I_c in adding to I_1 and subtracting from I_2.

Example 11-1. Two transformers are paralleled to a load of 0.8 + j0.6 pu ohms. Transformer 1 has a 5 percent boost in its tap setting, while transformer 2 is set "flat" on its nominal voltage tap. The transformer reactances are j0.08 and j0.10 pu ohms respectively, both being based on the same mva as the load impedance. Find the approximate currents (pu) in both transformers. Consider input volts as 100 percent.

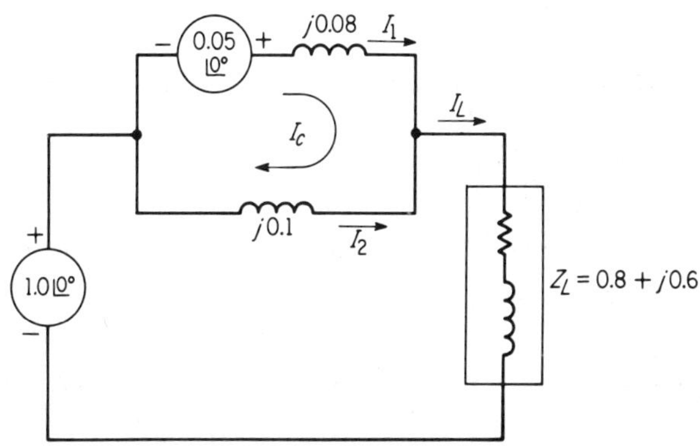

Fig. 11-2

First consider the source voltage of 1.0 pu:

$$I_L = \frac{1.0}{Z_L + \dfrac{Z_1 Z_2}{Z_1 + Z_2}} = \frac{1.0}{0.8 + j0.6 + \dfrac{(j0.08)(j0.1)}{j0.08 + j0.1}}$$

$$= \frac{1.0}{0.8 + j0.645} = 0.970 \angle -38.9^\circ$$

$$I_1 = I_L \frac{Z_2}{Z_1 + Z_2} = (0.97 \angle -38.9^\circ) \left(\frac{j0.1}{j0.08 + j0.1} \right)$$

$$= 0.538 \angle -38.9^\circ$$

$$I_2 = I_1 \frac{Z_1}{Z_2} = 0.538 \angle -38.9^\circ \left(\frac{j0.08}{j0.1} \right) = 0.432 \angle -38.9^\circ$$

$$I_c = \frac{\Delta E}{Z_1 + Z_2} = \frac{0.05}{j0.1 + j0.08} = \frac{0.05}{j0.18} = -j0.278$$

Total pu transformer currents $I_{\text{tot }1}$ and $I_{\text{tot }2}$ are found as

$$\begin{aligned} I_{\text{tot }1} &= I_1 + I_c = 0.538 \angle -38.9^\circ - j0.278 \\ &= 0.419 - j0.337 - j0.278 = 0.419 - j0.615 \\ &= \underline{\underline{0.744 \angle -55.7^\circ}} \text{ per unit} \end{aligned}$$

$$
\begin{aligned}
I_{\text{tot}\,2} &= I_2 - I_c = 0.432 \,\underline{/-38.9^\circ} - (-j0.278) \\
&= 0.337 - j0.271 + j0.278 = 0.337 + j0.007 \\
&= \underline{\underline{0.337 \,\underline{/1.1^\circ}\ \text{per unit}}}
\end{aligned}
$$

Note that boosting transformer 1 taps by 5 percent has forced it to assume most of the vars to the load.

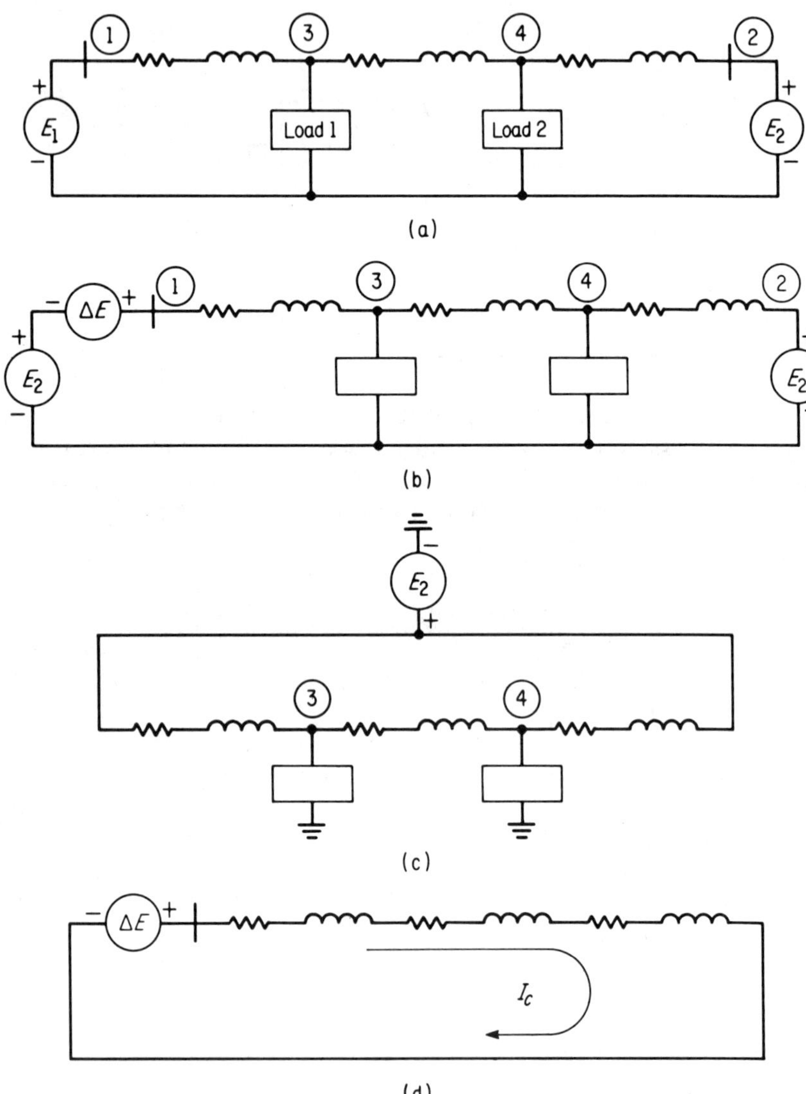

Fig. 11-3. (a) Per phase diagram with balanced three-phase loads. (b) Per phase diagram with superimposed voltage. (c) Circuit for approximating load current components. (d) Circuit for approximating circulating current.

11-3. Circulating Vars Between Buses of Unequal Voltage Magnitude

This more general consideration is almost identical with that of Sec. 11-2. Refer to the per phase diagram of Fig. 11-3a in which several balanced, three-phase loads were taken from lines between the two source buses (E_1 and E_2) where $|E_1| > |E_2|$, phase angles being equal. Let $E_1 = E_2 + \Delta E$. Fig. 11-3a may be replaced with that of Fig. 11-3b where ΔE is a superimposed voltage upon the circuit. The problem might be solved in two parts as indicated in Figs. 11-3c and d, superimposing the circulating components of current upon the load components. Again, in determining the circulating current I_c, the load impedances are considered

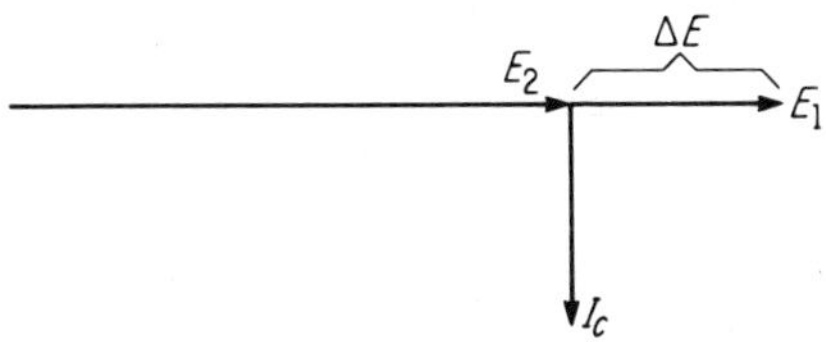

Fig. 11-4. Phasor diagram for Fig. 11-3d where $X_{\text{line}} >> R_{\text{line}}$ and voltage magnitudes are unequal.

high with respect to line (or series) impedances. ΔE will drive I_c around the circuit from bus 1 to bus 2. The value of I_c is approximately

$$I_c = \frac{\Delta E}{Z_{\text{line}}} \tag{11-3}$$

If the line $X \gg R$, then circulating current is of a lagging nature as seen from the phasor diagram of Fig. 11-4 and

$$\text{pu vars}_{\text{circ}} \doteq \frac{\text{pu } \Delta E}{\text{pu } Z_{\text{line}}} \tag{11-4}$$

11-4. Circulating Watts Between Buses of Differing Voltage Phase Angles

Again refer to Fig. 11-3a. However, now assume equal voltage magnitudes, where E_1 leads E_2 as indicated in the phasor diagram of Fig. 11-5. Again ΔE drives circulating current from bus 1 to bus 2.

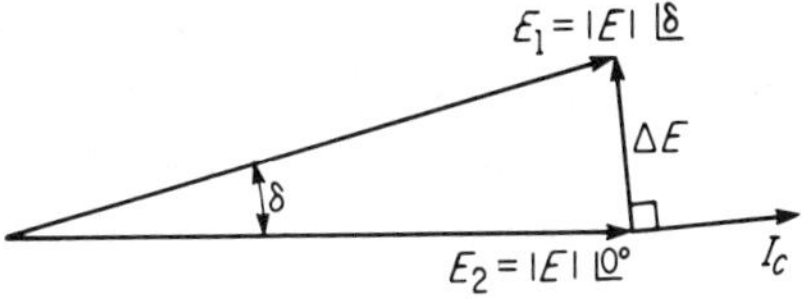

Fig. 11-5. Voltage phase angles unequal.

I_c is seen to lag ΔE by 90° ($X \gg R$) but is nearly in phase with the bus voltages E_1 and E_2. For this reason, watts are said to circulate from the bus of higher phase angle to the bus of lower angle. This fact is kept in mind when closing ties or interconnections between systems, or before closing a line between any two buses in a given system.

Notice in Fig. 11-5 that

$$|\Delta E| \doteq |E| \sin \delta \tag{11-5}$$

Now, from Eq. 11-3

$$|I_c| \doteq \frac{\Delta E}{X_{\text{line}}}$$

Then

$$|I_c| \doteq \frac{|E| \sin \delta}{X_{\text{line}}} \tag{11-6}$$

Then the watts circulated in each phase between the buses is

$$\text{watts}_{\text{circ}} \doteq |E| \times |I_c|$$

$$\doteq |E| \times \frac{|E| \sin \delta}{X_{\text{line}}}$$

$$\text{watts}_{\text{circ}} \doteq \frac{|E|^2 \sin \delta}{X_{\text{line}}} \tag{11-7}$$

Note the similarity of the approximate expression of Eq. 11-7 to the power-transfer Eq. 4-1.

Several facts should be apparent from Secs. 11-2 through 11-4. First of all, system var flow can be regulated by altering the voltage magnitudes between the buses in question. This is largely accomplished by

1. Adjustment of field excitation of generators and synchronous condensers. Switching of shunt-capacitor banks as a source of vars will be treated later.
2. Changing taps of transformers in parallel. On the other hand, watt flow through the network is primarily determined by the voltage phase angles on the buses. These angles are regulated to a great extent by the relative shaft input from the generator prime movers.

11-5. A Two-Generator Experiment

In this section, the author will attempt to provide the reader with a better concept for the subject of circulating watts and vars between generators, as well as their effect upon system voltage and frequency. It is felt that this can be accomplished with the aid of a model of two generators of equal ratings paralleled together and serving one common load of P_L and Q_L. Generator synchronous reactances are included with generator resistances neglected. Manual voltage controls are available for the generator fields and manual speed controls available for both prime movers. See Fig. 11-6 for the per phase diagram of the model, and consider both prime movers (M_1 and M_2) as separately excited d-c machines with simple speed control from field rheostats. It is understood that before the three-phase paralleling switch (represented on a per phase basis by S_p) was closed, that the requirements for paralleling alternators were met. It should be recalled that these paralleling requirements are (1) same phase sequence on both generators, (2) same frequency of generation, (3) same magnitude of voltage, and (4) same voltage phase angle at instant of closing the switch.

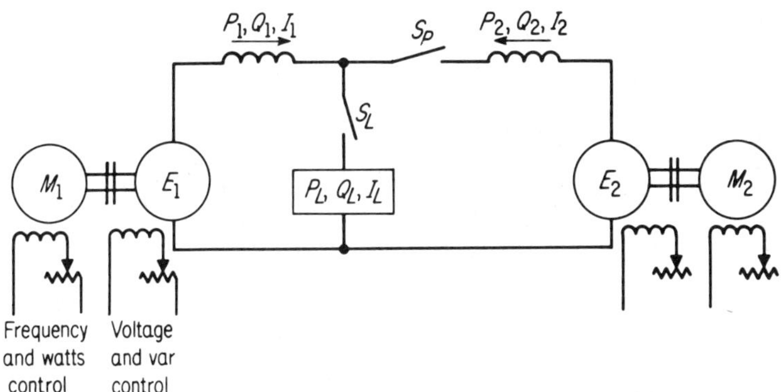

Fig. 11-6

Significant factors in this model experiment will be taken in two parts.

(a) *The effect of prime-mover control upon frequency and watts distribution.* Associated with each prime mover is a curve of frequency (or speed) vs. real load in watts. The load impedance will determine the total load where $P_L = P_1 + P_2$. Both generators have frequency in common. Of course they also have speed in common if both units have the same number of poles, since speed (rpm) = $120f \div$ number of poles. Note from Fig. 11-7a how the load divides between the units. Unit 1 is "hogging" the load in this figure.

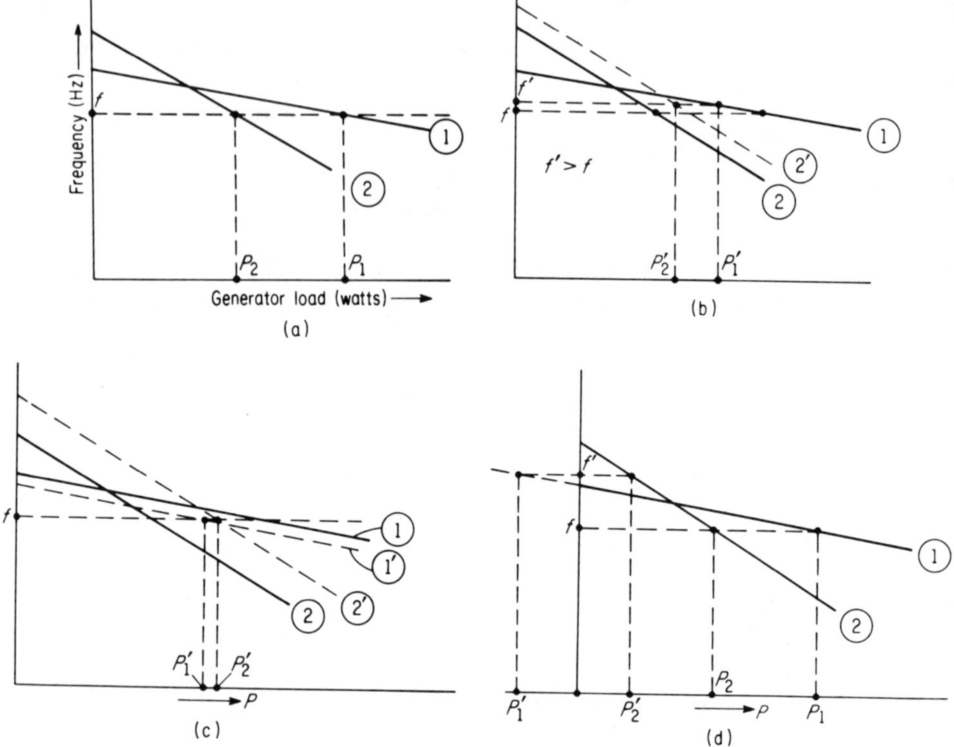

Fig. 11-7. (a) Load divided between units with unit 1 "hogging" the load. (b) Demonstrating the effect of adjusting speed control on unit 2. (c) Redistributing load and maintaining frequency. (d) Effect of opening load switch.

Now suppose the prime mover speed control of unit 2 is readjusted in order for unit 2 to take on more load. Altering this control has the effect of moving the f-load curve of unit 2 to a new position, parallel to the old curve. See Fig. 11-7b. With the sum of $P_1 + P_2$ as constant (determined by P_L) the frequency is forced higher to f'.

In order to redistribute the load and still maintain frequency, the field current of M_2 is reduced (moving toward a speed increase) while the field current of M_1 is increased (moving toward a speed decrease). This dual adjustment permits a transfer of load from unit 1 to unit 2. In general, to transfer load from one unit to another, speed controls must be moved in *opposite* directions if frequency is to be maintained. In our case, we have advanced the electrical angle on E_2 and retarded the angle on E_1. The effect of resetting both speed controls is seen in Fig. 11-7c.

Next refer to Fig. 11-7d in which the load is distributed as shown (P_1 and P_2). Now suppose the load switch S_L is suddenly opened. Since $P_1' + P_2' = P_L' = 0$, then $P_1' = -P_2'$. Generator 2 circulates power to unit

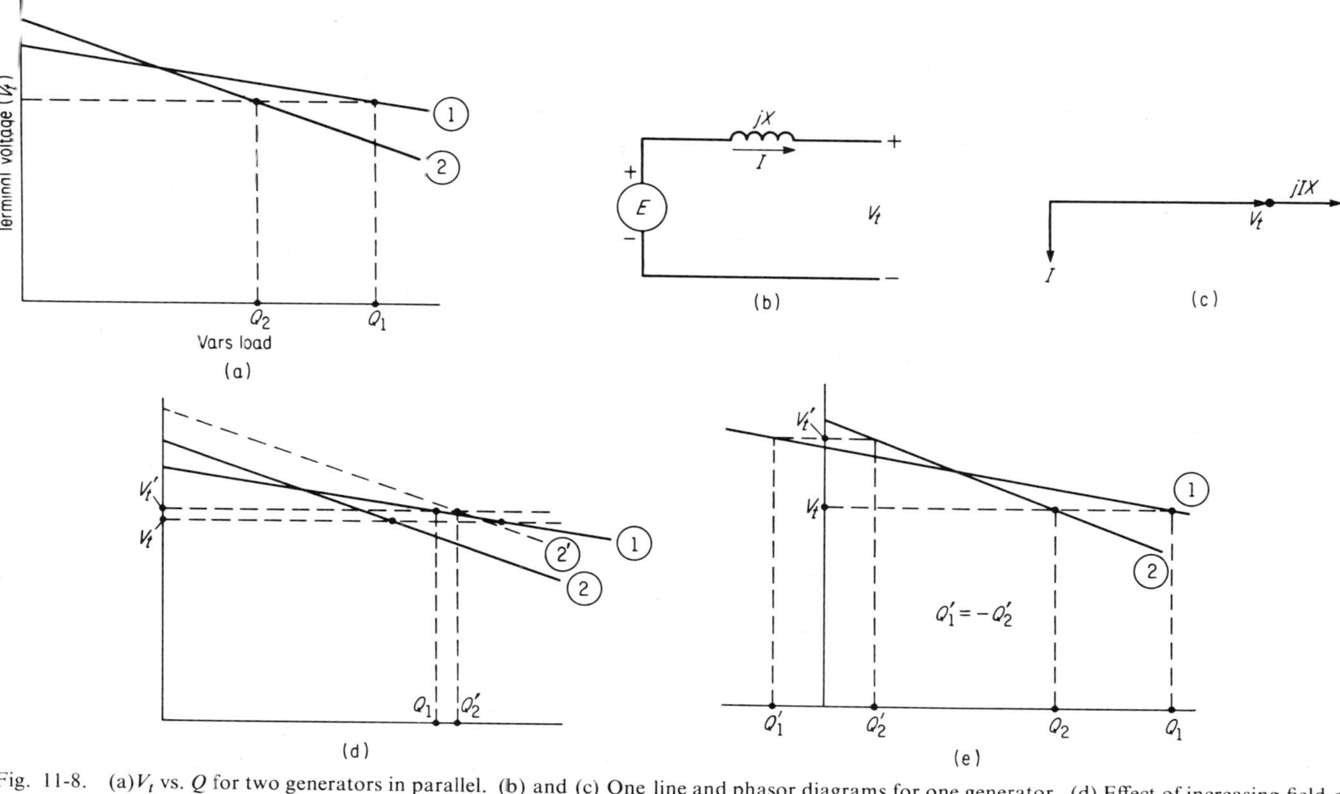

Fig. 11-8. (a) V_t vs. Q for two generators in parallel. (b) and (c) One line and phasor diagrams for one generator. (d) Effect of increasing field excitation on unit 2. (e) Effect of opening load switch.

1. The no-load operating frequency rises to f' (assuming no speed adjustment is made). Notice also that when load is removed, the unit with the least droop will drop its load more rapidly.

(b) *The effect of field excitation control upon terminal voltage and var distribution.* In order to determine the var distribution between parallel generators feeding a common load, refer to the generator curves of Fig. 11-8a. A curve of terminal voltage V_t vs. var loading is plotted for each unit. These linear curves are justified by realizing that, for each unit, $E = V_t + jIX$. For var flow (corresponding to lagging amperes) the IX drop of Figs. 11-8b and c is proportioned to I and in phase with V_t and E. Terminal voltage is common to both generators, so vars will split as shown in Fig. 11-8a, where $Q_L = Q_1 + Q_2$.

For a given var load Q_L it will be necessary to change *both* generator field excitations in *opposite* directions in order to shift between units. Otherwise, the terminal voltage would change. For example, in order to permit unit 2 to assume more vars output, we must *increase* excitation on generator 2 and decrease excitation on unit 1. This will move both units to new curves which are parallel to the original curves.

If only the excitation of unit 2 is altered, the load voltage would necessarily rise as shown in Fig. 11-8d in order for $Q_1 + Q_2$ to remain equal to the load Q_L.

If the load switch S_L were opened on Fig. 11-6, the no-load terminal voltage would rise according to Fig. 11-8e and some vars will circulate from unit 2 to unit 1. This is to be expected since vars do circulate from a bus of higher no-load voltage magnitude (E_2 in this case) to the other bus E_1, as demonstrated in Sec. 11-3.

A useful parallel is apparent in the analysis of Secs. 11-5a and b. Dual concepts are tabulated for convenience in Table 11-1.

Table 11-1

DUAL CONCEPTS APPLYING TO SECS. 11-5a AND 11-5b

Section 11-5a	Section 11-5b
Prime mover speed control	Generator field excitation
Frequency (f)	Terminal voltage (V_t)
$P_{load} = P_1 + P_2$	$Q_{load} = Q_1 + Q_2$
f vs. watts curves	V_t vs. vars curves
To redistribute watts, holding f constant: move speed controls in opposite directions	To redistribute vars, holding V_t constant: move generator excitation controls in opposite directions
Increasing speed control on one prime mover only: f rises	Increasing excitation on one generator only: V_t rises
Reducing P_{load}: 1. f rises 2. Unit with least slope drops watts more rapidly	Reducing Q_{load}: 1. V_t rises 2. Unit with least slope drops vars more rapidly

11-6. Controlling System Voltage with Appropriate Var Injections Using Bus Impedance Matrix

The majority of loads served by a power utility will draw current at a lagging power factor. In addition to this load var requirement, considerable additional vars (I^2X) may be required by the transmission and distribution network itself. A number of problems are connected with inductive var loading. Included among them are:

1. Added I^2R losses in lines and equipment.
2. Increased MVA investment for lines and equipment.
3. Poor voltage regulation, or falling voltage profile from generation toward the load.

The third consideration is of primary concern in this section. Much is being done in order to dispatch vars from an economic standpoint (to minimize I^2R losses). However, the approach to be set forth here is that of injecting vars where necessary to improve the system voltage profile. The methods of this section may be used either as a tool for dispatching vars or merely as a means of determining the effect of a given var insertion upon all system voltages. In the improvement of voltage regulation or the flattening of the voltage profile, one will, for the most part, be improving the economic picture as well.

The effect upon all steady-state bus voltages will be investigated due to the injection of vars from (a) shunt capacitor banks, (b) generators and synchronous condensers, and (c) transformer tap changes. While the last item listed is not, in itself, a source of vars, Sec. 11-2 has demonstrated the effectiveness of a transformer tap change in altering the course of circulating vars within a loop of the network. The same might be said for line regulators. The three considerations will be treated in order.

BUS VOLTAGE CHANGES DUE TO SWITCHING OF SHUNT CAPACITOR BANKS

Section 3-3 reviews the general subject of power-factor correction by use of shunt capacitors. Actually, several advantages are to be gained through the use of shunt capacitors. They may be looked upon as a source of vars to release generator capacity. More important to this section is the use of the capacitors to raise voltage, both on the bus at which the injection is made and at surrounding or nearby buses (electrically speaking). The standard method of determining the change in voltage (ΔV) experienced at the bus being switched was to use Thévenin's equivalent E_0 and Z_0 at the bus being switched. To determine the amount of the voltage change, refer to Fig. 11-9a which shows the capacitor current I_c which exists after the switch is closed. E_0 is the Thévenin's open-circuit voltage at the bus while $R_0 + jX_0$ is the through or Thévenin's impedance from the bus back to the neutral bus. The phasor diagram

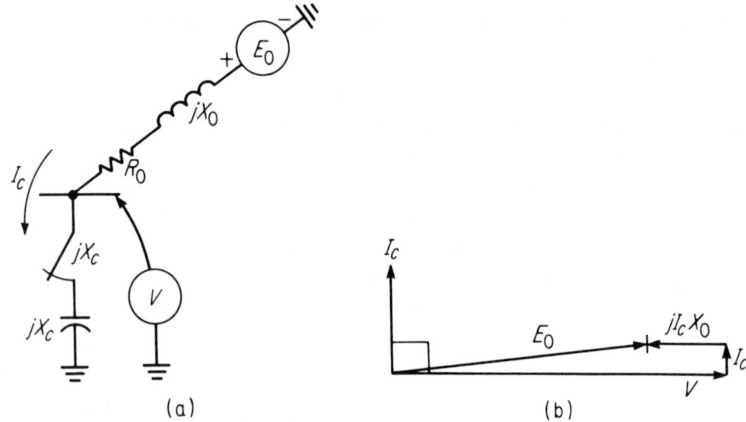

Fig. 11-9. Effect of capacitor switching upon bus voltages.

of Fig. 11-9b shows I_c to lead V by 90°. The difference between $|E_0|$ and $|V|$ is often referred to as the *capacitor flicker voltage*, ΔV, occurring at that bus. The value of ΔV may be approximated as

$$|\Delta V| \doteq |I_c| X_0 \tag{11-8}$$

For a given bus voltage, capacitor vars are proportional to I_c, and therefore proportional to ΔV. Again, this points up the direct correlation between var flow and voltage level.

While Thévenin's method is useful in determining the ΔV at the particular capacitor bus being switched, it is also of importance that the ΔV's be known on other buses in the network. Prior knowledge of this will not only place a handle upon this aspect of voltage regulation but may also prevent a hunting situation which could otherwise occur in the automatic on-and-off switching of adjacent voltage-controlled capacitor banks. The bus impedance matrix and its corresponding rake equivalent will be used to obtain this ΔV information. Figure 11-10 is just such an equivalent, showing a shunt capacitor on bus 1. Again, Z_{11}, $Z_{22}, \ldots, Z_{mm}$, are the self- or driving-point impedances of the buses. They are also identical, in this case, to the "through impedance" or "Thévenin's impedance" back to the neutral bus. It is to be recalled from Chapter 10 that these impedances are the diagonal elements of the bus impedance matrix, where the neutral bus is taken as reference. Generator impedances are included in the equivalent. The transfer impedances are the off-diagonal elements of the bus impedance matrix. One may consider all relatively high-impedance elements to neutral as being excluded from the equivalent—such as loads, transformer shunt impedance, or line-charging capacitance. If one is concerned with their effect, they can be added externally to the appropriate buses. The fact of the existence of load does

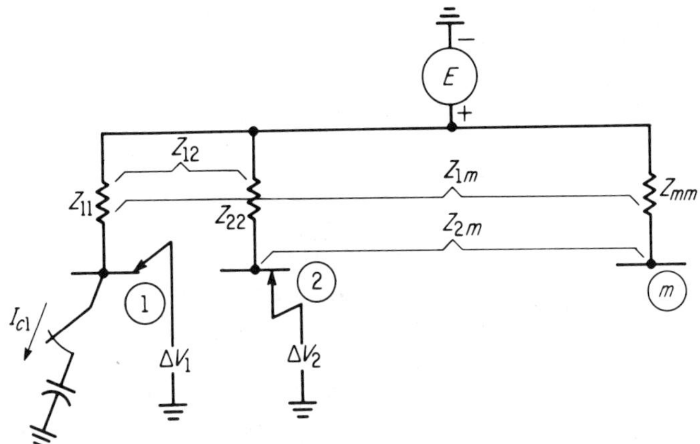

Fig. 11-10. Rake equivalent applied to capacitor switching.

not have a great effect upon the voltage changes when switching capacitors, as this load is present in nearly the same amount both before and after the switching takes place.

Suppose now that the capacitor bank of bus 1 is to be switched on. Adding generation E behind the passive network, the phasor diagram for bus 1 due to switching at bus 1 is similar to that of Fig. 11-9. The phasor value of ΔV_{11} is

$$\Delta \overline{V}_{11} = \overline{I}_{c1} \cdot \overline{Z}_{11} \tag{11-9}$$

For purposes of regulation, the magnitude difference in V_1 before and after switching is $|E_0| - (V_1)$ and will be referred to as ΔV_{11}. Its approximate value is

$$\Delta V_{11} \doteq |I_{c1}|\,|X_{11}| \tag{11-10}$$

Similarly, the phasor $\overline{I}\overline{Z}$ induced upon bus 2 is

$$\Delta \overline{V}_{12} = \overline{I}_{c1} \overline{Z}_{12} \tag{11-11}$$

If one were to neglect the phase-angle difference between the voltages $V_1, V_2, \ldots, V_m$, then the change of voltage magnitudes occurring on the various buses due to the switching of bus 1 capacitor would be approximated as

$$\left.\begin{aligned} \Delta V_{11} &\doteq |I_{c1}|\,|X_{11}| \\ \Delta V_{12} &\doteq |I_{c1}|\,|X_{12}| \\ &\vdots \\ \Delta V_{1m} &\doteq |I_{c1}|\,|X_{1m}| \end{aligned}\right\} \tag{11-12}$$

If X values are in percent, and I_c in per unit, then ΔV values will be in percent.

Example 11-2. A practical example in the use of these approximations is taken from a 72-bus reactance matrix used at the Detroit Edison Company. Each of the 72 buses had capacitor banks ranging in size from 4.8 mvar to 18 mvar, and they were located in the 24- and 40-kv subtransmission system. Only six buses of the matrix are retained here as the sample matrix of Table 11-2. Reactances are in percent on a 50-mva base.

Table 11-2
A SAMPLE REACTANCE MATRIX (FROM BUS IMPEDANCE MATRIX) OF DETROIT EDISON SYSTEM
(Values are in percent ohms on a 50-mva base, matrix is symmetrical)

	Chilson	Cody	Fowler-ville	Howell	Pinckney	Webber-ville
Chilson	21.1	5.19	12.7	14.2	8.30	11.9
Cody........................		6.95	4.4	4.54	4.05	4.35
Fowlerville.................			32.2	16.8	12.6	29.2
Howell......................				18.5	10.2	15.4
Pinckney					16.1	12.4
Webberville						36.0

Find the effect of switching "on" a capacitor bank at Fowlerville upon the voltages at Fowlerville, Webberville, and Howell. The Fowlerville capacitor is rated at 4.8 mvar. Refer to Fig. 11-11.

Solution. The percent ΔV at Fowlerville for switching at Fowlerville is

$$\%\Delta V_{FF} = \text{pu } I_c \cdot \%X_{FF}$$

where

$$\text{pu } I_{cf} \doteq \frac{\text{pu mvar}}{\text{pu } V_f} = \frac{4.8/50}{1.0} = 0.096$$

$$\%V_{FF} = 0.096 \times 32.2$$

$$= \underline{\underline{3.09\%}} \text{ volts rise}$$

The percent ΔV at Webberville and Howell for switching the 4.8-mvar bank at Fowlerville is

$$\%\Delta V_{WF} = \text{pu } I_c \cdot \%X_{WF}$$

$$= 0.096 \times 29.2$$

$$= \underline{\underline{2.81\%}} \text{ volts rise}$$

Similarly,

$$\%V_{HF} = \text{pu } I_c \cdot \%X_{HF}$$

$$= 0.096 \times 16.8$$

$$= \underline{\underline{1.61\%}} \text{ volts rise at Howell}$$

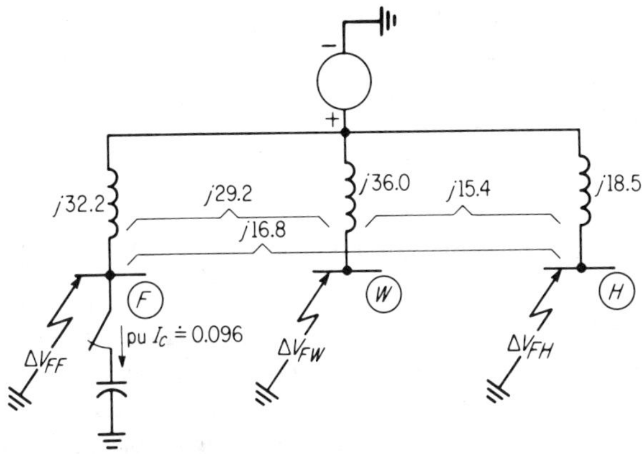

Fig. 11-11. Rake equivalent for Example 11-2.

EFFECT OF GENERATOR VAR INJECTION UPON SYSTEM BUS VOLTAGES

In Fig. 10-2, generator voltages were assumed equal and their points of equal polarity were joined, lumping all generation together. Generator impedances then became a part of the passive network. Suppose for the present that we start with this same assumption of equal generator emf's in showing one source feeding the rake equivalent, then alter the equivalent to include the effect of unequal generator voltage magnitudes.

For example, suppose bus 1 is the terminal bus of generator 1, which is to be given an increase in field excitation, thereby increasing its voltage magnitude and var output. In order to deal with this change, bus 1 must be retained in the equivalent. Keep in mind that this generator impedance is already included in the system matrix and theoretically should be removed from the equivalent before externally injecting vars from generator 1. From Fig. 11-12 the negative of Z_{gen} is added to the rake equivalent with the use of a type 2 matrix modification as outlined in Sec. 10-9.

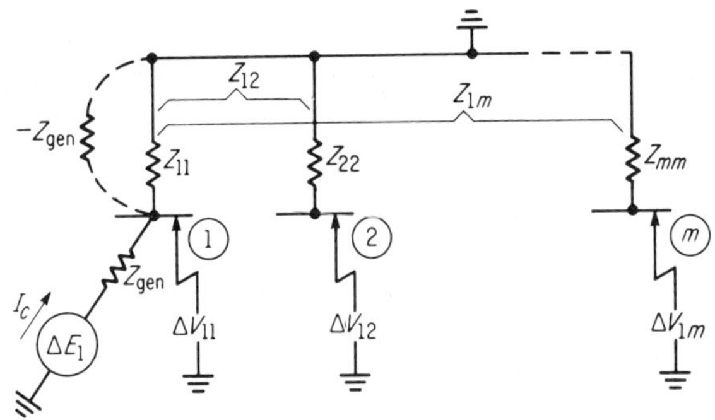

Fig. 11-12. Injecting vars into a system from generator 1 where the effect of ΔE is superimposed upon the system.

Current (I_c) will circulate primarily in the path from ΔE_1 through Z'_{11} (corrected to exclude Z_{gen}). As this current moves through leg 1, it will induce a ΔV upon the other buses. The phasor value of voltage change on bus 1 due to a var injection on bus 1 is nearly identical to that of Eq. 11-9, since this injection is not basically different from a capacitor var injection, or

$$\Delta \bar{V}_{11} = \bar{I}_c \bar{Z}'_{11} \tag{11-13}$$

The only difference between Eqs. 11-9 and 11-13 is in the corrected value of Z_{11}. From a practical standpoint it may even prove unnecessary to modify Z_{11} with the removal of Z_{gen}, wherever the synchronous impedance of the generator is much greater than the through impedance of bus 1.

Again, for the sake of regulation calculation, change in magnitude of bus voltage levels (before and after the var injection is made) is approximated by Eqs. 11-11 and 11-12. If a particular bus voltage is down (say 2 percent), a reactance table—with capacitor buses and generator buses both retained—will give a quick indication of just where to go for the most effective var injection. In fact, a reactance matrix (in percent X) can easily be converted to a voltage table with the units of "percent volts per mvar injection" by simply multiplying all reactance values by the appropriate per unit current (I_c) corresponding to a 1-mvar injection. In other words, use the factor of

$$K_I = \frac{1 \text{ mvar}}{\text{Base mvar}} \tag{11-14}$$

We have then converted the bus impedance (reactance) matrix to an incremental voltage matrix by

$$[\Delta V/\text{mvar}] = K_I[X_{bus}] \tag{11-15}$$

DETERMINING EFFECT OF TRANSFORMER TAP CHANGES UPON SYSTEM VOLTAGES

Again, look at the system rake equivalent of Fig. 11-13a. Assume a transformer branch exists between buses 1 and 2 and a tap change is to be made in such a way as to boost voltage by some percent ΔE in the

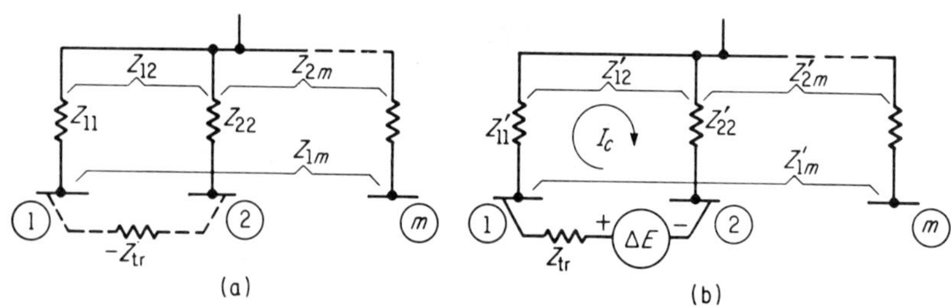

Fig. 11-13

direction of bus 1. First, the system matrix is modified to remove the transformer impedance branch, using the type 4 modification of Sec. 10-9. Next, consider the transformer branch as inserted back into the modified equivalent with the $\%\Delta E$ included in order to represent the tap boost from bus 2 to bus 1. The circulating current (I_c) flowing in the loop will see the following impedance:

$$Z_{\text{loop}} = Z_{tr} + Z'_{11} + Z'_{22} - 2Z'_{12} \tag{11-16}$$

Then

$$I_c = \frac{\Delta E}{Z_{\text{loop}}}, \qquad \text{all values in pu} \tag{11-17}$$

Now the phasor voltage increases seen by the buses due to the tap change are equal to

$$\Delta \overline{V}_1 = \overline{I}_c(\overline{Z}'_{11} - \overline{Z}'_{12}) \tag{11-18}$$

Substituting Eq. 11-17 for I_c,

$$\left.\begin{aligned} \Delta V_1 &= \frac{\Delta E}{Z_{\text{loop}}}(Z'_{11} - Z'_{21}) \\ \Delta V_2 &= \frac{\Delta E}{Z_{\text{loop}}}(Z'_{12} - Z'_{22}) \\ &\vdots \\ \Delta V_m &= \frac{\Delta E}{Z_{\text{loop}}}(Z'_{1m} - Z'_{2m}) \end{aligned}\right\} \tag{11-19}$$

The above equations are complex. Theoretically, the value of ΔV_m does not necessarily represent the difference in the voltage $|\Delta V_m|$ before and after the tap change. Refer to the phasor diagram of Fig. 11-14. How-

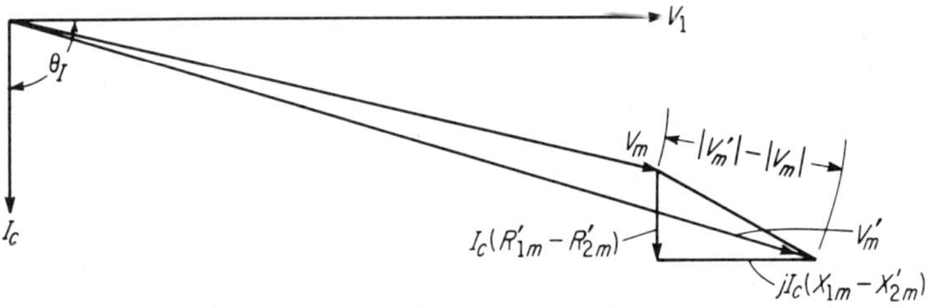

Fig. 11-14. Phasor diagram showing how the phasor ΔV_m can differ from the magnitude change of V_m.

ever, if the Z's of Eq. 11-19 have a sufficiently large X/R ratio and if the angles on the bus voltages (V_1 and V_m) themselves are close together, then the magnitude change of the m-bus voltage ($|V'_m| - |V_m|$) for regulating purposes can be approximated as the $I_c(X'_{1m} - X'_{2m})$ component, or

$$\Delta V_m \doteq \frac{\Delta E}{|X_{\text{loop}}|}(X'_{1m} - X'_{2m}) \tag{11-20}$$

In general, Eq. 11-20 would apply as well to any voltage regulator or autotransformer in the network (if no appreciable transformer phase shift is present).

11-7. Paralleling Load-Tap-Changing (LTC) Transformers with Automatic Controls

A typical power transformer today might be equipped with both fixed taps (where turns ratio is varied manually at no load) and with the additional feature of automatic tap changing under load (TCUL). For example, the high-voltage winding might be equipped with a nominal voltage turns ratio plus four 2½ percent fixed tap settings to yield ±5 percent buck or boost voltage. In addition to this, there could be provision, on the low-voltage winding, for 32 incremental steps of ⅝ percent each, giving an automatic range of ±10 percent.

The voltage-control feature is made to regulate the voltage automatically either to the transformer secondary, or more likely to some predetermined load feed point. To demonstrate this, Fig. 11-15 is given. The

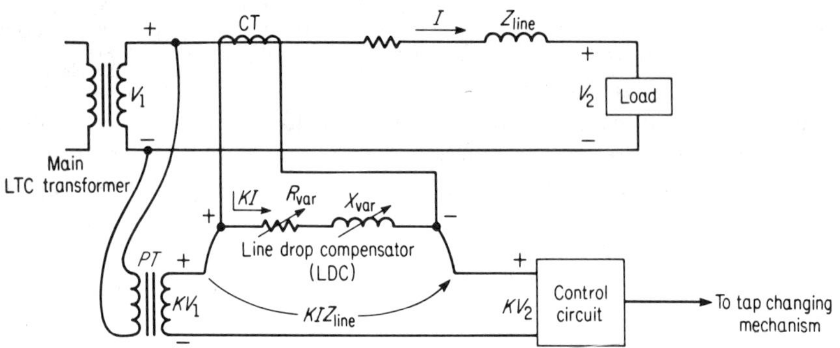

Fig. 11-15. Simplified schematic of an LTC transformer with automatic voltage control.

control circuit will be located physically on the main transformer itself, while the load voltage to be controlled (V_2) may be at a considerable distance away. For this reason, the more convenient voltage V_1 will be taken from the potential transformer (*PT*) while the *IZ* drop of the line will be represented artificially with the use of a line-drop-compensator (LDC). The LDC is fed from the current transformer (CT) of Fig. 11-15. The appropriate *R* and *X* of the line is represented by adjusting the variable *R* and *X* values. The *PT* secondary uses 120 volts as base voltage for the control circuit. It is seen that the voltage to the control circuit (KV_2) is proportioned to V_2 where $V_2 = V_1 - IZ_{\text{line}}$. Of course, there is often a question as to exactly where the load point is located, especially in a complex network, where ties and loads may be tapped off in various places. This is always an individual operational problem, but for our purposes, it will be assumed that the load location is clearly defined. The control circuit will be sensitive to any over- or under-voltage beyond a given set point. However, the control will not be permitted to initiate a tap change on a sudden or short-time voltage deviation from the set point. Rather, a change will be initiated only when the summation of periods of over- or undervoltage exceeds a predetermined time period (say 30 seconds).

Two problems are inherent with the paralleling of automatic LTC transformers. Refer to Fig. 11-16, keeping in mind that both control

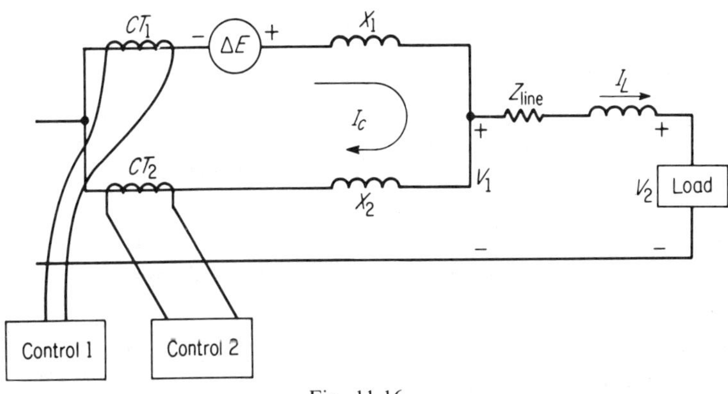

Fig. 11-16

circuits are attempting to regulate to V_2. Now, suppose that the load current I_L increases, dropping V_2. If control 1 is faster than control 2, a tap change on transformer 1 will circulate current (I_c) between the units. The line-drop-compensator for unit 1 will indicate a further I_cX drop and will attempt to raise its voltage even higher, further increasing I_c. At the same time, unit 2 will see a higher V_2 than before and attempt to lower voltage on transformer 2. See the phasor diagrams of Figs. 11-17a and b.

Unless this problem were solved, transformer 1 would run to the all raise position and transformer 2 would run all lower. Circulating current could also overload the transformers.

Another problem would arise, should one transformer be removed from service. The remaining unit would pick up total current and would give a false (high) indication of the line IZ drop to the line drop compensator.

Because of the above problems inherent in paralleling automatic LTC transformers, some paralleling method must be employed. Five of these methods are:

1. Reverse-reactance method.
2. Cross-current method.
3. Circulating-current method (also called the *current balance method*).
4. The mechanical interlock method.
5. The electrical interlock method (sometimes called *leader-follower*).

Space will not permit a detailed coverage of the above methods. In brief, the *reverse reactance method* permits the reversal of currents through the variable reactance of both line-drop compensators, which has the effect of interchanging the phasor diagrams of Fig. 11-17. Thus the KV_2 voltages

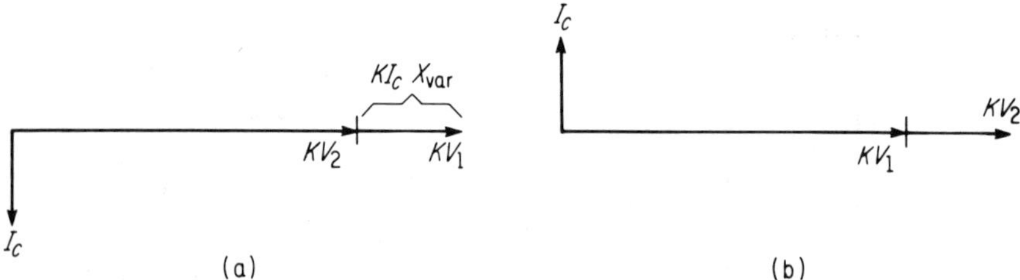

Fig. 11-17. (a) Voltage drop as seen by unit 1 due to I_c only. (b) Voltage rise as seen by unit 2 due to I_c only.

seen by the control circuits are brought closer together. Of course, the true representation of the line *IX* drop is destroyed with some sacrifice in regulation, especially in the case of poor power factor loads.

The *cross-current method* merely uses the secondary of CT 1 to activate the LDC of control 2, while CT 2 secondary is connected to LDC 1. This method is good for only 2 units in parallel.

The *circulating current method* can be used for any number of units in parallel. It is a more sophisticated scheme and more satisfactory as well. All circulating current is "blocked" out of the line-drop-compensator circuits with auxiliary blocking CT's, leaving only the true line *IZ* drop represented. At the same time if one unit is removed from service, auxiliary equalizing CT's will prevent the line-drop compensator(s) of the unit(s) remaining in service from picking up total current.

The interlock methods are especially useful where it is necessary to hold the paralleled transformers on identical taps. Such an application is found when paralleling two autotransformers with low loop impedance, where even one tap step apart would circulate excessive current.

11-8. Overexcitation Protection for the Generator Transformer

In recent years considerable attention has been given to the overexcitation protection of large generator step-up transformers. If a core is excited with flux (ϕ) to a sufficient degree beyond saturation, stray fluxes entering the structural and bracing members, tank, etc., can do irreparable damage in very short order. Steel members can be heated red hot and to the melting point. Blistering of paint on the exterior of the tank has been known to occur. Even if direct failure does not result, years may be taken from the normal life of the transformer. It is possible that the unit could be subjected unknowingly to this condition of overexcitation without direct failure. However, subsequent failure could occur at the first sign of mechanical stress due to short circuit, synchronizing out of step, etc. Weakened steel members which snap under this stress might leave the impression that the transformer failed due to the short circuit, out-of-step synchronizing, and so on.

Recall from transformer theory that the voltage equation for a transformer is

$$E_{rms} = Kf\phi_{max}$$

or

$$\phi_{max} \propto \frac{E_{rms}}{f} \tag{11-21}$$

Equation 11-21 states simply that the excitation level can be expressed as proportional to volts per Hertz or percent volts per Hertz. ASA standards specify in general that an unloaded transformer should not be excited beyond 110 percent on its primary side while a fully loaded transformer should not exceed 105 percent on its secondary side.

Figure 11-18 demonstrates a typical method for matching generator

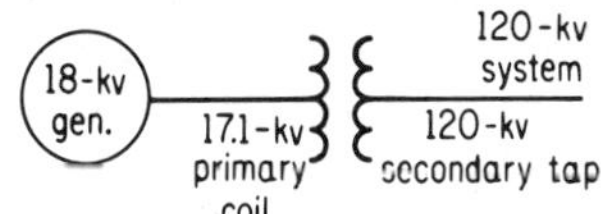

Fig. 11-18. Example of matching generator and transformer with a fixed 5 percent boost toward the secondary.

and transformer by placing a 5 percent fixed boost on the primary coil. This will to some extent compensate for IZ drop in the transformer under load conditions. At unity power factor the IZ drop is almost negligible. However, a 50 percent reactive component of current flowing through a transformer Z of 10 percent would drop approximately 5 percent across the transformer. The common voltage range for generator operation is 95 percent to 105 percent. However, 105 percent for the 18-kv generator of Fig. 11-18 is about 110 percent of the rating of the

17.1-kv winding. The upper end of the generator range then places the transformer close to its upper excitation limit.

The generator transformer is also particularly susceptible to overexcitation for other reasons. Some of the ways in which this condition might occur are:

1. Bringing a generator up to speed where f is low and automatic regulators attempt to hold V_t constant (V/f is high). Shutting down the unit could have the same effect.
2. Sudden removal of load while generator terminal voltage rises.
3. Unintentional overexcitation of the generator field (either manual or automatic) for whatever reason. This could be due to failure of a regulator, an attempt of an operator to pick up more reactive output to the system, etc.

Permissible durations for transformer overexcitation have been established and accepted by the industry. Refer to Fig. 11-19. Note that

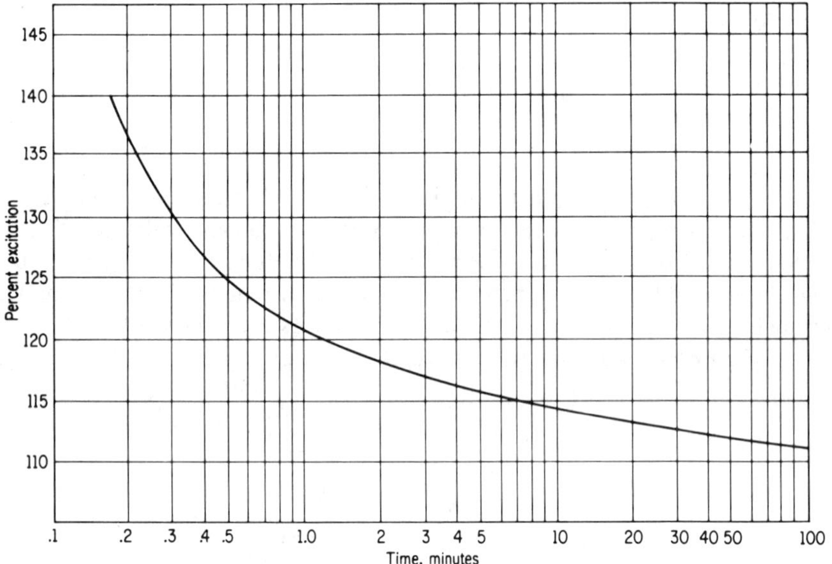

Fig. 11-19. General guide for permissible short-time overexcitation of power transformers.

130 percent excitation would permit only 18 seconds if damage is to be prevented. Overexcitation relays are available which are calibrated in percent volts per Hertz and can be incorporated (with timers) to initiate alarms, field tripping circuits, generator shutdown, or whatever protection scheme is desired. Sometimes two devices are used, one to alarm the operator at perhaps 110 percent, and the other to trip the machine or field off automatically at a higher setting, say 114 percent.

11-9. Modeling Transformers with Off-Nominal Turns Ratio

In load flows of large systems the simple transformer model of Sec. 4-4 may not always suffice. Recall that this model represented the transformer as a simple series branch impedance. Our new model will be an equivalent π. Now consider a transformer with an off-nominal tap setting. In the per unit system, the transformer turns ratio will be considered 1:1 if this ratio equals the system nominal voltage ratio. Otherwise, in arriving at the equivalent, the transformer can be first considered as two two-port networks in cascade as shown in Fig. 11-20.

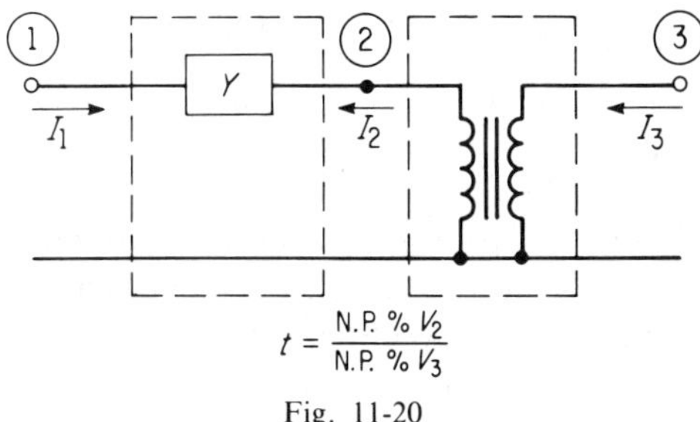

Fig. 11-20

The nodal equation for the two-port between nodes 1 and 2 is

$$\begin{bmatrix} I_1 \\ I_2 \end{bmatrix} = \begin{bmatrix} Y & -Y \\ -Y & Y \end{bmatrix} \begin{bmatrix} V_1 \\ V_2 \end{bmatrix} \tag{11-22}$$

The second two-port on the right in Fig. 11-20 is an ideal transformer. When in the per unit system, this ratio can be expressed as a ratio of two per unit quantities, where rated nameplate voltages are expressed as a percent of nominal system voltages. Then

$$t = \frac{\text{N.P. \%}\, V_2}{\text{N.P. \%}\, V_3} \tag{11-23}$$

where, for example,

$$\text{N.P. \%}\, V_3 = \frac{\text{Nameplate rating of secondary} \times 100}{\text{Nominal system voltage on secondary side}} \tag{11-24}$$

Voltages and currents of the left-hand two-port can be expressed in terms of the end terminals of Fig. 11-20 as

$$\begin{bmatrix} V_1 \\ V_2 \end{bmatrix} = \begin{bmatrix} 1 & 0 \\ 0 & t \end{bmatrix} \begin{bmatrix} V_1 \\ V_3 \end{bmatrix} \tag{11-25}$$

and

$$\begin{bmatrix} I_1 \\ I_3 \end{bmatrix} = \begin{bmatrix} 1 & 0 \\ 0 & t \end{bmatrix} \begin{bmatrix} I_1 \\ I_2 \end{bmatrix} \tag{11-26}$$

Substituting expressions of Eqs. 11-22 and 11-25 into Eq. 11-26 yields

$$\begin{bmatrix} I_1 \\ I_3 \end{bmatrix} = \begin{bmatrix} 1 & 0 \\ 0 & t \end{bmatrix} \begin{bmatrix} Y & -Y \\ -Y & Y \end{bmatrix} \begin{bmatrix} 1 & 0 \\ 0 & t \end{bmatrix} \begin{bmatrix} V_1 \\ V_3 \end{bmatrix}$$

$$= \begin{bmatrix} 1 & 0 \\ 0 & t \end{bmatrix} \begin{bmatrix} Y & -Yt \\ -Y & Yt \end{bmatrix} \begin{bmatrix} V_1 \\ V_3 \end{bmatrix}$$

$$\begin{bmatrix} I_1 \\ I_3 \end{bmatrix} = \begin{bmatrix} Y & -Yt \\ -Yt & Yt^2 \end{bmatrix} \begin{bmatrix} V_1 \\ V_3 \end{bmatrix} \tag{11-27}$$

Recall that in an admittance matrix such as is found in Eq. 11-27, the diagonal elements Y and Yt^2 are the summation of all admittances around nodes 1 and 3 respectively, while the off-diagonal elements corresponding to the 1–3 nodes are the negatives of branch admittances between nodes 1 and 3. The π values are then found as follows:

$$Y = Y_{01} + Y_{13} \tag{11-28}$$

$$Yt^2 = Y_{03} + Y_{13} \tag{11-29}$$

where

$$Y_{13} = Yt \tag{11-30}$$

Substituting Y_{13} of Eq. 11-30 into Eq. 11-28 and Eq. 11-29 yields

$$Y_{01} = Y - Yt$$

$$Y_{01} = Y(1 - t) \tag{11-31}$$

$$Y_{03} = Yt^2 - Yt$$

$$Y_{03} = Yt(t - 1) \tag{11-32}$$

The final π equivalent for the off-nominal turns ratio transformer is shown in Fig. 11-21 with admittance added. The parameters of this equivalent can now be incorporated into a system study just as any other network. If, for example, such an off-nominal tapped transformer is within a network where it tends to circulate current within this network, the use of this model in a load-flow study will automatically account for such circulation.

As a test of this model, the circulating current will be determined for the transformers of Example 11-1.

Example 11-3. Given the two transformers of Example 11-1. Open

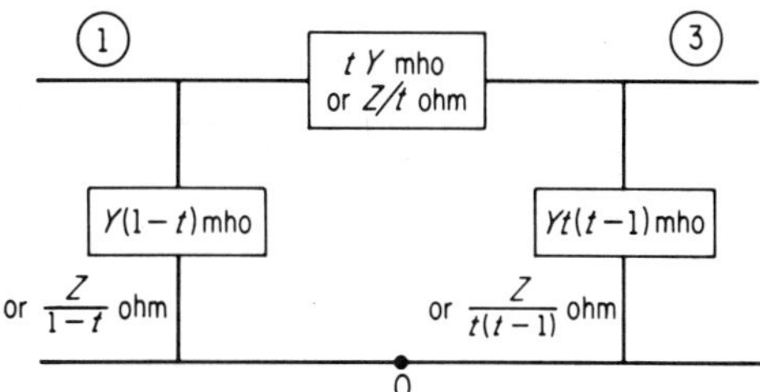

Fig. 11-21. π equivalent for off-nominal turns ratio transformer.

the load circuit and determine the circulating current between the units using the model of Fig. 11-21. Again, transformer 1 has a 5 percent boost in direction of the load,

$$t = \frac{1}{1.05} = 0.952$$

$$\text{pu } Z_{ab} = \frac{1}{tY} = \frac{j0.08}{0.952}$$

$$= \underline{\underline{j0.0840}}$$

$$\text{pu } Z_{0a} = \frac{1}{Y(1 - t)} = \frac{j0.08}{(1 - 0.952)}$$

$$= \frac{j0.08}{0.048} = \underline{\underline{j1.667}}$$

$$\text{pu } Z_{0b} = \frac{1}{Yt(t - 1)} = \frac{j0.08}{0.952(0.952 - 1)}$$

$$= \underline{-j1.75}$$

Note the 0-*b* branch of Fig. 11-22 is in series with the parallel

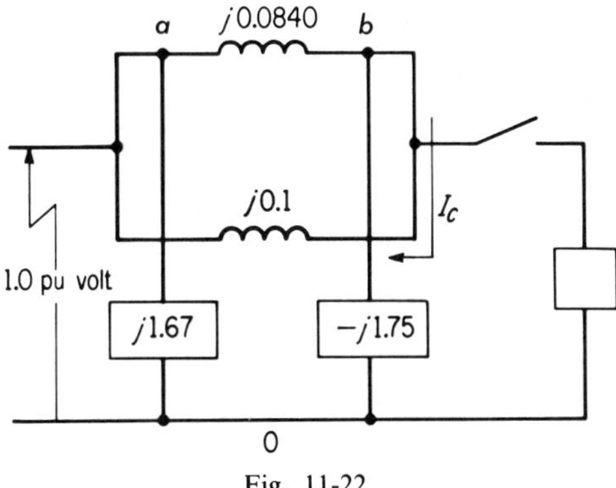

Fig. 11-22

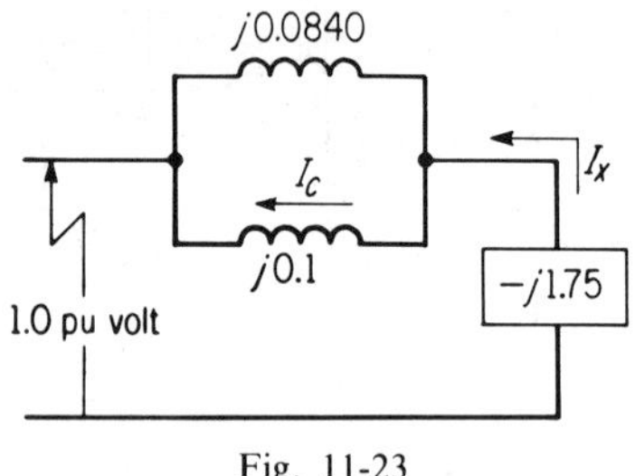

Fig. 11-23

combination of $j0.1$ ohm and $j0.0840$ ohm. The circulating current is that part flowing in the $j0.1$-ohm branch. Refer to Fig. 11-23.

$$Z_p = \frac{j0.0840 \times j0.1}{j0.0840 + j0.1} = j0.0457$$

$$I_x = \frac{1.0}{j1.75 - j0.0457} = -j0.586$$

$$I_c = I_x\left(\frac{j0.0840}{j(0.0840 + 0.1)}\right) = -j0.586 \times \frac{0.0840}{0.1840}$$

$$= \underline{\underline{-j0.267}} \text{ pu amp}$$

The value of 0.267 pu amp compares favorably with the value of 0.278 pu amp of circulating current as found by the approximate method of Example 11-1.

Notice in Fig. 11-21 where $t \neq 1$, one shunt branch will yield a negative quantity. Current will normally flow upward through this negative admittance and downward through the other. At no load and where the transformer is not paralleled with other units or with any other part of the network, these currents add to zero. This is to be expected, since under such conditions no line current would be required. The fact that these two shunt currents add to zero (at no load, with no parallel network branches) is easily verified by noting that the left-hand impedance from nodes 0 to 1 of Fig. 11-21 is the negative of the impedance sum from nodes 0 to 1 by the other path, or

$$\frac{Z}{1 - t} = -\left[\frac{Z}{t} + \frac{Z}{t(t-1)}\right] \tag{11-33}$$

Problems

11-1. Two transformers are in parallel, sharing a common load. Unit 1 is rated at 45 mva, 120 kv to 24 kv, Δ-Y, with an equivalent impedance of $j0.08$ pu ohms on its own base. Unit 2 is rated at 30 mva, 120 kv to 24 kv, Δ-Y, with $Z_e = j0.07$ pu on its own base. To what total mva load would you limit the transformers to prevent either unit from being overloaded? Give the load split in each unit.

11-2. Given the transformers of Prob. 11-1. Assume taps of the larger unit are set with a 2½ percent boost toward the low side (117 kv to 24 kv tap), leaving the smaller unit at 120 kv to 24 kv. They are, together, sharing a load of 65 mva at 0.9 p.f. lagging. Using the method of Sec. 11-2, calculate, (a) the circulating current in the transformers in per unit. Specify the base used, and (b) the load component of current in each transformer in per unit, and (c) the total secondary current in each unit in amperes.

11-3. The two transformers of Prob. 11-1 are to be paralleled together with taps of unit 1 set at 123 kv to 24 kv and unit 2 set at 120 kv to 24 kv. The load has been removed. Find the circulating current in per unit which flows using (a) method of Sec. 11-2, and (b) method of Sec. 11-9.

11-4. In the circuit of Fig. 11-3a, the following data are given: $Z_{13} = j1.0$, $Z_{34} = j1.5$, $Z_{24} = j2.0$, $E_2 = 100 + j0$, $E_1 = 110 + j0$, $I_{\text{load 1}} = 5.0 - j2.0$, $I_{\text{load 2}} = 6.0 - j3.0$. Using the approximate methods of Secs. 11-3 and 11-4, determine the following values in each of the lines: circulating current component, load current components, and total line currents.

11-5. Repeat Prob. 11-4, after revising E_1 to $100 + j10$.

11-6. Refer to Table 11-2 in Example 11-2. Given an 18 mvar capacitor bank at Cody station. Determine the percent voltage change at all six buses on the table if the Cody bank is switched on.

11-7. In the system of Example 11-2, a line is to be removed from Howell to Fowlerville. Consider its impedance to be $j25$ percent on a 50-mva base. Modify the matrix given in example Prob. 11-2 by the methods of Sec. 10-8 and rework Example 11-2.

11-8. Refer to Example 11-2. As an aid in operation of the portion of the system described in the example, it is desired to provide the dispatcher with a table which indicates "percent volts change per 10 mvar injected."

(a) Show the resulting table.

(b) Determine where (besides Chilson) you would inject vars (in form of capacitors or synchronous condensors) to give the maximum voltage rise at Chilson.

(c) How many vars would you inject at the location of part (b) to raise the Chilson voltage 3 percent?

(d) How are the other bus voltages affected by the injection of part (c)? Give your answers in percent.

11-9. A bus impedance matrix which retains only three buses of a system is given in percent ohms on a 100-mva base as follows:

$$Z_{\text{bus}} = j \begin{bmatrix} 2.0 & 1.2 & 0.6 \\ 1.2 & 7.0 & 0.5 \\ 0.6 & 0.5 & 1.5 \end{bmatrix}$$

(columns labelled ① ② ③)

Bus 1 is on the high side of a transformer and bus 2 is the low bus of the same unit. Bus 3 is a generator bus. The transformer and generator impedances have been included in arriving at the matrix. The neutral bus is reference. Taps are to be reset on the transformer which exists in the network between buses 1 and 2. An additional 2 ½ percent boost is to be given the unit in the direction of the low bus. The transformer is rated at 45 mva with an 8 percent reactance on its own base. Determine the loop impedance, the per unit circulating current picked up by the transformer and the voltage change on all three buses. Use the method of Sec. 11-6. Modify the matrix by the method of Sec. 10-8.

11-10. Given the three-bus system of Prob. 11-9. A 100-mva generator with a reactance of 12 percent (on its own base) is to deliver an additional 40 mvar to the system. Find the $\%\Delta V$ (voltage change) on each bus as a result. (Modify your matrix by the method of Sec. 10-8.)

chapter 12

SELECTED NUMERICAL METHODS AND THE LOAD-FLOW PROBLEM

12-1. Introduction

Power utilities must plan far into the future to keep pace with load growth. New loads, generators, system interconnections and lines will all have their effect upon system voltages and currents, as well as real and reactive power flow. These effects must be studied beforehand and with some considerable degree of accuracy. The d-c analyzer board, which has been used so successfully for short-circuit studies, does not qualify for use on normal load-flow problems. Both the d-c board of Sec. 4-11 and the short-circuit methods of Chapters 7–10 make about the same simplifying assumptions. Certain of these assumptions are not justifiable for load-flow studies. For example, it has been previously explained that *short-circuit studies* often assume (among other things) that all generator voltage magnitudes (per unit) are equal as well as voltage phase angles, that resistance of all branches is insignificant with respect to inductance and can therefore be neglected, and that load currents themselves are often negligible with respect to fault currents. In the load-flow problem such simplifying assumptions are unjustified. This means that complex impedances, phase-angle relationships, power factor of loads, etc., become an important part of the problem. The a-c analyzer board was developed for the purpose of including such factors and is described in Sec. 4-11. It is taken for granted that the reader realizes the futility of attempting extensive load studies by longhand method. In fact, an a-c board study requires considerable time and expense; thus the computer methods have come to the forefront.

While the a-c board is regarded as a miniature system in replica of the actual system, the computer must rely for the most part upon direct or indirect solution of simultaneous system equations. A complete coverage of all methods is beyond the scope of this book and would better fall under the heading of numerical analysis. At the same time, certain selected methods are included as space permits. For the benefit of those without a basic understanding of determinants and matrices, the appendix is intended to provide this background.

There are many ways of obtaining a direct solution to n simultaneous linear equations in unknowns. Probably among the very first methods introduced to the reader entails the simple algebraic substitution of one variable in terms of another, until one variable stands alone. This procedure works fine for two or three unknowns but beyond that the student might next have been taught to write the equations in a more orderly fashion and utilize Cramer's rule for a conventional determinant solution. This latter method (using minors, cofactors, etc.) is covered in Appendix C-1. Next in the learning process the student writes the equation in matrix form, such as $\bar{E} = \bar{Z}\bar{I}$, where $\bar{I}$ is unknown. The solution for this column array of currents can be obtained by a conventional matrix inversion of $\bar{Z}$ (adjoint method) and $\bar{I} = \bar{Z}^{-1}\bar{E}$. This adjoint method is also covered in the appendix and takes its place with the many other numerical methods which are available. A partial listing of *direct* methods for solving simultaneous linear equations is given as follows:

1. Determinant method using Cramer's rule, cofactors, etc.—Appendix C-1.
2. Conventional or adjoint method of matrix inversion—Appendix C-2.
3. Pivotal condensation for determinants—Sec. 12-2.
4. Gauss reduction method—Sec. 12-3.
5. Crout reduction method—Sec. 12-4.
6. Matrix inversion and solution with elementary row operations—Sec. 12-5.

Simultaneous linear equations may also be solved by *iterative* techniques which employ successive approximations, eventually converging upon a solution. These iterative methods are readily adapted to computer use. The *Gauss iterative* and the *Gauss-Seidel iterative* methods will be emphasized because of their widespread application to modern load-flow studies. The *Relaxation iterative* approach is also included in Sec. 12-12.

Certain of the load-flow techniques involve the solution of a set of nonlinear equations. For this purpose, the *Newton-Raphson* method is presented.

12-2. Pivotal Condensation Applied to Determinants

This method is used in evaluating a determinant and the principle is also found useful in certain matrix inversions. Given the determinant

$$D = \begin{vmatrix} a_{11} & a_{12} & a_{13} & a_{14} \\ a_{21} & a_{22} & a_{23} & a_{24} \\ a_{31} & a_{32} & a_{33} & a_{34} \\ a_{41} & a_{42} & a_{43} & a_{44} \end{vmatrix} \tag{12-1}$$

There is a rule that states that the value of a determinant is unchanged if the elements of a row (or column) are changed by adding to them a constant multiple of the corresponding elements in any other row (or column). Suppose that through such row transformations, that the positions indicated in the triangle of Eq. 12-1 are all reduced to zero, yielding

$$D = \begin{vmatrix} a_{11} & a_{12} & a_{13} & a_{14} \\ 0 & b_{22} & b_{23} & b_{24} \\ 0 & 0 & c_{33} & c_{34} \\ 0 & 0 & 0 & d_{44} \end{vmatrix} \tag{12-2}$$

We know from the expansion of a determinant by its minors that the value of D would be equal to the product of the diagonal elements, or

$$D = a_{11} b_{22} c_{33} d_{44} \tag{12-3}$$

One could think of this process as first reducing all elements below a_{11} to zero, giving a value of the determinant $D = a_{11} \mid 3 \times 3 \mid$. Next column 1 of the 3×3 is treated in the same manner yielding $D = a_{11} b_{22} \mid 2 \times 2 \mid$. Finally the 2×2 is similarly treated until Eq. 12-3 results.

Now in order to obtain zeros in column 1 under a_{11}, we can subtract appropriate multiples of row 1 from each of rows 2, 3, and 4, without changing the value of D. This would yield

$$D = \begin{vmatrix} a_{11} & a_{12} & a_{13} & a_{14} \\ 0 & (a_{22} - a_{21}a_{12}/a_{11}) & (a_{23} - a_{21}a_{13}/a_{11}) & (a_{24} - a_{21}a_{14}/a_{11}) \\ 0 & (a_{32} - a_{31}a_{12}/a_{11}) & (a_{33} - a_{31}a_{13}/a_{11}) & (a_{34} - a_{31}a_{14}/a_{11}) \\ 0 & (a_{42} - a_{41}a_{12}/a_{11}) & (a_{43} - a_{41}a_{13}/a_{11}) & (a_{44} - a_{41}a_{14}/a_{11}) \end{vmatrix} \tag{12-4}$$

or

$$D = a_{11} \begin{vmatrix} b_{22} & b_{23} & b_{24} \\ b_{32} & b_{33} & b_{34} \\ b_{42} & b_{43} & b_{44} \end{vmatrix} \tag{12-5}$$

where the bracketed terms of Eq. 12-4 are replaced with the b terms of

Eq. 12-5. In a similar fashion, b_{32} and b_{42} positions are made zero and

$$D = a_{11}b_{22}\begin{vmatrix} c_{33} & c_{34} \\ c_{43} & c_{44} \end{vmatrix} \tag{12-6}$$

This process is continued, making the c_{43} position zero and the full triangle of zeros is complete as shown in Eq. 12-2.

The pivotal points for the above example were a_{11}, b_{22}, and c_{33}. The computer will perform these operations in a direct and orderly manner. However, there are conditions to be met. For example, if a_{11} is zero it could not serve as a pivot point for the first operation, and in any subsequent operation this test for zero must be made. Also it is desirable for the sake of better accuracy that the pivotal element chosen be the largest element of the column from which it is taken. Should the upper left-hand element not be the largest in any particular step, rows may be interchanged by changing the sign on the determinant. The reasoning behind this last requirement (of larger element for pivotal purposes) is as follows. Use of the smallest pivot would make factors of the first reduction (a_{21}/a_{11}, a_{31}/a_{11}, etc.) greater than one, causing loss of accuracy especially where the number of significant figures is fixed as in floating point arithmetic.

12-3. Gauss Reduction Method

Given a set of linear, algebraic equations:

$$a_{11}x_1 + a_{12}x_2 + a_{13}x_3 = a_{14} \tag{12-7}$$

$$a_{21}x_1 + a_{22}x_2 + a_{23}x_3 = a_{24} \tag{12-8}$$

$$a_{31}x_1 + a_{32}x_2 + a_{33}x_3 = a_{34} \tag{12-9}$$

The procedure is orderly and straightforward. For the simple set given, it involves the reduction of three equations and three unknowns (in x_1, x_2, and x_3) to two equations and two unknowns (x_2 and x_3) and finally one equation which is solved for x_3. Substitution of x_3 back into the two-equation form yields x_2 and so forth. To be more specific,

1. (a) Divide Eq. 12-7 by a_{11} yielding

$$x_1 + (a_{12}/a_{11})x_2 + (a_{13}a_{11})x_3 = a_{14}/a_{11} \tag{12-10}$$

(b) Multiply Eq. 12-10 by a_{21} and subtract the result from Eq. 12-8, eliminating x_1.

(c) Multiply Eq. 12-10 by a_{31} and subtract the result from 12-9, eliminating x_1. An auxiliary set of equations has been formed. They are

$$b_{22}x_2 + b_{23}x_3 = b_{24} \tag{12-11}$$

$$b_{32}x_2 + b_{33}x_3 = b_{34} \tag{12-12}$$

2. Repeat the general procedure of part 1 on Eqs. 12-11 and 12-12 in order to eliminate x_2. This yields

$$c_{33}x_3 = c_{34} \tag{12-13}$$

or

3. Substitute the value of x_3 back into Eqs. 12-11 or 12-12 and solve for x_2. Then substitute x_2 and x_3 into one of the original equations to obtain x_1. In general we work backward through the equations, making substitutions as we go.

One obvious restriction on the procedure is that none of the leading coefficients (a_{11} and b_{11} in our example) may equal zero. If a_{11} equals zero, we must simply renumber our equations and/or variables to prevent dividing the equation in step 1(a) by zero. This renumbering may even be desirable if the leading coefficient is small with respect to other coefficients.

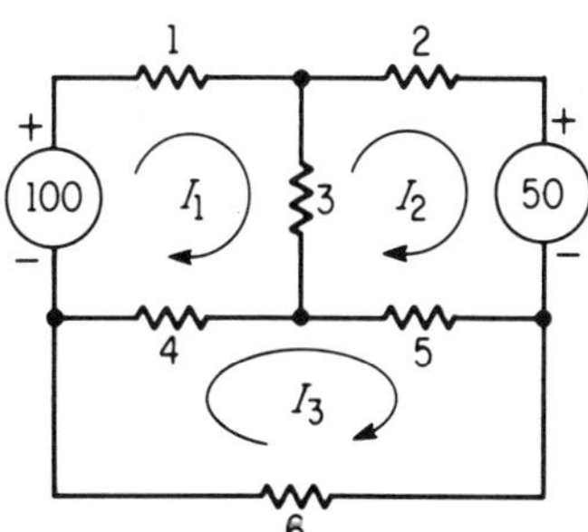

Fig. 12-1. Circuit for Example 12-1.

Example 12-1. Given the circuit of Fig. 12-1 and three mesh (open-loop) currents I_1, I_2, and I_3. Write the voltage equations and solve by the Gauss reduction method. The voltage equations are:

$$8I_1 - 3I_2 - 4I_3 = 100 \tag{12-14}$$

$$-3I_1 + 10I_2 - 5I_3 = -50 \tag{12-15}$$

$$-4I_1 - 5I_2 + 15I_3 = 0 \tag{12-16}$$

An augmented array of the coefficients is tabulated for each mesh equation. An additional summation column is added as a continuous error check. A row summation is the sum total of the row coefficients plus the constant. If we perform the same operations upon the summation column as the others, each successive value of the check column should agree with the sum of the coefficients plus the constant in that same row.

	Row Number	I_1	I_2	I_3	K	Σ	Comments
Original Equations	(1)	8	−3	−4	100	101	
	(2)	−3	10	−5	−50	−48	
	(3)	−4	−5	15	0	6	
First Reduction	(4)	1	−.375	−.5	12.5	12.625	row (1) ÷ 8
	(5)	−3	1.125	1.5	−37.5	−37.875	row (4) × (−3)
	(6)	−4	1.5	2.0	−50	−50.5	row (4) × (−4)
	(7)	0	8.875	−6.5	−12.5	−10.125	row (2) − row (5)
	(8)	0	−6.5	13	50	56.5	row (3) − row (6)
Final Reduction	(9)	0	1.0	−.732	−1.409	−1.141	row (7) ÷ 8.875
	(10)	0	−6.5	4.761	9.155	7.415	row (9) × (−6.5)
	→ (11)		0	8.239	40.845	49.045	row (8) − row (10)

From the final reduction,

$$8.239 I_3 = 40.845, \qquad \text{or} \qquad I_3 = \underline{\underline{4.96}}$$

Working backwards, using row 9,

$$I_2 - (0.732)(4.96) = -1.409$$

$$I_2 = \underline{\underline{2.22}}$$

Again working back into row 4,

$$I_1 - (0.375)(2.22) - (0.5)(4.96) = 12.5$$

$$I_1 = \underline{\underline{15.81}}$$

While the system of this example may seem trivial (involving only three unknowns), it serves to exemplify the method. Also, for solution on the computer or hand calculator, more significant figures would be retained.

One variation of the Gauss method is called the *Gauss-Jordan reduction* method. This approach would eliminate the need for back substitution. At the same time it does require other additional work. Suppose we take the first equation of one set in the Gaussian reduction and use it to eliminate the first unknown of that set from the first equation of each preceding set. If this is done repeatedly, the resulting equations would have all coefficients but one equal to zero. This technique will be demonstrated by using the simple Eqs. 12-7 through 12-9. The first equations for each step of the Gaussian reduction are

$$a_{11}x_1 + a_{12}x_2 + a_{13}x_3 = a_{14} \qquad [12\text{-}7]$$

$$b_{22}x_2 + b_{23}x_3 = b_{24} \qquad [12\text{-}11]$$

$$c_{33}x_3 = c_{34} \qquad [12\text{-}13]$$

Now Eq. 12-13 can be used to eliminate x_3 from both Eqs. 12-11 and 12-7. Equation 12-11 can be used to eliminate x_2 from Eq. 12-7. Three resulting equations are

$$\left.\begin{aligned} a_{11}x_1 \qquad\qquad\qquad &= a'_{14} \\ b_{22}x_2 \qquad\quad &= b'_{24} \\ c_{33}x_3 &= c_{34} \end{aligned}\right\} \qquad (12\text{-}17)$$

From Eq. 12-17, the complete solution is

$$x_1 = a'_{14}/a_{11}; \qquad x_2 = b'_{24}/b_{22}; \qquad x_3 = c_{34}/c_{33} \qquad (12\text{-}18)$$

12-4. The Crout Reduction

This method again modifies the Gauss reduction, and is better adapted to computer use. Through this reduction a compact triangular form of the equations is directly obtained. Auxiliary sets of equations are deleted. In other words, we move directly from the original set of equations (or corresponding coefficients) to a final array. This method demands less rewriting of coefficients and less storage space as well. Again, look at the sample equations:

$$a_{11}x_1 + a_{12}x_2 + a_{13}x_3 = a_{14} \qquad [12\text{-}7]$$

$$a_{21}x_1 + a_{22}x_2 + a_{23}x_3 = a_{24} \qquad [12\text{-}8]$$

$$a_{31}x_1 + a_{32}x_2 + a_{33}x_3 = a_{34} \qquad [12\text{-}9]$$

It will be recalled that the Gauss reduction of these equations yielded a triangular set of equations, which were solved for x_3, x_2 and x_1 in that order. These were Eqs. 12-7, 12-11, and 12-13, or when normalized to allow a unit coefficient on the first terms, they can be written as

$$x_1 + e_{12}x_2 + e_{13}x_3 = e_{14} \qquad (12\text{-}19)$$

$$x_2 + e_{23}x_3 = e_{24} \qquad (12\text{-}20)$$

$$x_3 = e_{34} \qquad (12\text{-}21)$$

We will first set up an augmented (expanded) matrix array ($\bar{A}$) representing the original equations.

$$\bar{A} = \begin{bmatrix} a_{11} & a_{12} & a_{13} & a_{14} \\ a_{21} & a_{22} & a_{23} & a_{24} \\ a_{31} & a_{32} & a_{33} & a_{34} \end{bmatrix} \qquad (12\text{-}22)$$

The final array ($\bar{B}$) will contain the e coefficients needed in the solution of x_1 x_2 and x_3. The d values are important auxiliary coefficients, retained for the purpose of calculating the triangular array of e coefficients. The final $\bar{B}$ array is

$$\bar{B} = \begin{bmatrix} d_{11} & e_{12} & e_{13} & e_{14} \\ d_{21} & d_{22} & e_{23} & e_{24} \\ d_{31} & d_{32} & d_{33} & e_{34} \end{bmatrix} \tag{12-23}$$

The first column of d coefficients is merely equal to the first column of $\bar{A}$, or

$$d_{11} = a_{11}; \qquad d_{21} = a_{21}; \qquad d_{31} = a_{31} \tag{12-24}$$

The e coefficients of row 1 are determined (from Gauss discussion) as

$$e_{12} = a_{12}/a_{11}; \qquad e_{13} = a_{13}/a_{11}; \qquad e_{14} = a_{14}/a_{11} \tag{12-25}$$

The remaining coefficients are not so obvious. The second column of d coefficients results from the first auxiliary reduction, and the third column of d coefficients results from the second auxillary reduction, etc. It is possible to follow through the Gauss procedure and justify the Crout reduction formulas. However, for purposes of this book, the general reduction equations will merely be given, along with a numerical example. The equations are

$$d_{ij} = a_{ij} - \sum_{k=1}^{k=j-1} d_{ik} e_{kj}; \qquad i \geq j \tag{12-26}$$

$$e_{ij} = \frac{1}{d_{ii}} \left(a_{ij} - \sum_{k=1}^{k=i-1} d_{ik} e_{kj} \right); \qquad i < j \tag{12-27}$$

$$x_i = e_{in} - \sum_{k=i+1}^{k=n} e_{ik} x_k \tag{12-28}$$

Example 12-2. Solve the equations of Example 12-1 by the Crout reduction method. Again they are

$$8I_1 - 3I_2 - 4I_3 = 100 \qquad [12\text{-}14]$$

$$-3I_1 + 10I_2 - 5I_3 = -50 \qquad [12\text{-}15]$$

$$-4I_1 - 5I_2 + 15I_3 = 0 \qquad [12\text{-}16]$$

The original augmented array is

$$\bar{A} = \begin{bmatrix} a_{11} & a_{12} & a_{13} & a_{14} \\ a_{21} & a_{22} & a_{23} & a_{24} \\ a_{31} & a_{32} & a_{33} & a_{34} \end{bmatrix} = \begin{bmatrix} 8 & -3 & -4 & 100 \\ -3 & 10 & -5 & -50 \\ -4 & -5 & 15 & 0 \end{bmatrix}$$

The $\bar{B}$ array is to be found as

$$\bar{B} = \begin{bmatrix} d_{11} & e_{12} & e_{13} & e_{14} \\ d_{21} & d_{22} & e_{23} & e_{24} \\ d_{31} & d_{32} & d_{33} & e_{34} \end{bmatrix}$$

Applying the Crout reduction Eqs. 12-26 and 12-27

$$d_{11} = a_{11} = 8; \qquad e_{12} = a_{12}/a_{11} = -3/8 = -0.375$$

$$d_{21} = a_{21} = -3; \qquad e_{13} = a_{13}/a_{11} = -4/8 = -0.5$$

$$d_{31} = a_{31} = -4; \qquad e_{14} = a_{14}/a_{11} = 100/8 = 12.5$$

$$d_{22} = a_{22} - a_{21}e_{12} = 10 - (-3)(-0.375) = 8.875$$

$$d_{32} = a_{32} - a_{31}e_{12} = -5 - (-4)(-0.375) = -6.5$$

$$e_{23} = \frac{1}{d_{22}}(a_{23} - d_{21}e_{13}) = \frac{1}{8.875}[-5-(-3)(-0.5)] = -0.732$$

$$e_{24} = \frac{1}{d_{22}}(a_{24} - d_{21}e_{14}) = \frac{1}{8.875}[(-50) - (-3)(12.5)] = -1.41$$

$$d_{33} = a_{33} - d_{31}e_{13} - d_{32}e_{23}$$

$$= 15 - (-4)(-0.5) - (-6.5)(-0.732) = 8.24$$

$$e_{34} = \frac{1}{d_{33}}[a_{34} - d_{31}e_{14} - d_{32}e_{24}]$$

$$= \frac{1}{8.24}[0 - (-4)(12.5) - (-6.5)(-1.41)] = 4.96$$

$$\bar{B} = \begin{bmatrix} 8 & -0.375 & -0.5 & 12.5 \\ -3 & 8.875 & -0.732 & -1.41 \\ -4 & -6.5 & 8.24 & 4.96 \end{bmatrix}$$

The unknown currents can now be found directly through the application of Eq. 12-28. Perhaps it is better here to realize that the triangular e coefficients of $\bar{B}$ represent equations similar to Eq. 12-19 through 12-21. In this way I_1, I_2 and I_3 may be found in reverse order:

$$I_1 - 0.375 I_2 - 0.5 I_3 = 12.5$$

$$I_2 - 0.732 I_3 = -1.41$$

$$I_3 = \underline{\underline{4.96}}$$

$$I_2 = -1.41 + (0.732)(4.96) = \underline{\underline{2.22}}$$

$$I_1 = 12.5 + (0.375)(2.22) + (0.5)(4.96)$$

$$= \underline{\underline{15.8}}$$

12-5. Achieving a Matrix Solution and Inverse with Elementary Row Operations

In solving linear algebraic equations at the level of introductory algebra, it is recalled that an equation is not basically altered by multiplying both sides by a constant $c \neq 0$. Also equations of a set can be added or subtracted to one another by the simple rules of algebra. In fact, these simple rules were applied in the solution of algebraic equations. The author refers to solutions of the type shown.

(1) $3x + 4y = 5$

(2) $x - 2y = 7$

Multiply (2) by 3 and subtract from (1):

(1) $3x + 4y = 5$
(2) $3x - 6y = 21$

(3) $10y = -16$, or $y = -1.6$

In basic algebra we might find x by substitution or even a further elimination process such as multiplying Eq. 3 by the quantity $(-4/10)$ and adding to Eq. (1).

While justification for the above elementary operations seems almost obvious, these operations are closely akin to the subject of row operations for matrices. After all, the matrices with which we are involved are representative of sets of just such equations. A matrix $\bar{B}$ is said to be *row equivalent* to a matrix $\bar{A}$ if $\bar{B}$ can be obtained from $\bar{A}$ by a finite number of row operations.

Let us examine the voltage matrix equation $\bar{E} = \bar{Z}\bar{I}$, where the column matrix $\bar{E}$ and the square matrix $\bar{Z}$ are both known. It is required to find the column matrix $\bar{I}$ which is the solution for all unknown currents. The matrix equation is

$$\begin{bmatrix} Z_{11} & Z_{12} & Z_{13} \\ Z_{21} & Z_{22} & Z_{23} \\ Z_{31} & Z_{32} & Z_{33} \end{bmatrix} \begin{bmatrix} I_1 \\ I_2 \\ I_3 \end{bmatrix} = \begin{bmatrix} E_1 \\ E_2 \\ E_3 \end{bmatrix} \tag{12-29}$$

One procedure is to write an expanded or augmented matrix array including Z and E, as

$$\left[\begin{array}{ccc|c} Z_{11} & Z_{12} & Z_{13} & E_1 \\ Z_{21} & Z_{22} & Z_{23} & E_2 \\ Z_{31} & Z_{32} & Z_{33} & E_3 \end{array}\right] \tag{12-30}$$

Row operations can be performed to convert the portion previously occupied by the Z elements into the unit matrix which automatically converts the E column into the desired I column, or

$$\left[\begin{array}{ccc|c} 1 & 0 & 0 & I_1 \\ 0 & 1 & 0 & I_2 \\ 0 & 0 & 1 & I_3 \end{array}\right] \tag{12-31}$$

Actually this procedure is similar to the Gauss-Jordan technique explained in Sec. 12-3 where all of the off-diagonal coefficients of the Eq. 12-17 were reduced to zero. There can also be considerable overlap between the concept of reduction by row operations and the method of pivotal condensation of Sec. 12-2.

One further modification will be made upon the augmented array of Eq. 12-30 and the resulting matrix of Eq. 12-31. Apparently the row operations had the effect of multiplying the $\bar{E}$ column by the $\bar{Z}$ inverse to obtain the $\bar{I}$ column. We will, with our modification, be able not only to find the $\bar{I}$ solution but also to retain the $\bar{Z}$ inverse in the process. We proceed as follows. The matrix of Eq. 12-30 is expanded to include a unit matrix $\bar{U}$ between the Z and E elements. The new augmented matrix is

$$\left[\begin{array}{ccc|ccc|c} Z_{11} & Z_{12} & Z_{13} & 1 & 0 & 0 & E_1 \\ Z_{21} & Z_{22} & Z_{23} & 0 & 1 & 0 & E_2 \\ \underbrace{Z_{31} \quad Z_{32} \quad Z_{33}}_{\bar{Z}} & & & \underbrace{0 \quad 0 \quad 1}_{\bar{U}} & & & \underbrace{E_3}_{\bar{E}} \end{array}\right] \tag{12-32}$$

Next, row operations are performed upon the matrix of Eq. 12-32 until the $\bar{Z}$ matrix portion is converted to the unit matrix. Now the position which was occupied by $\bar{U}$ in Eq. 12-32 has been transformed into Z^{-1} while the position originally occupied by $\bar{E}$ has been converted to the column $\bar{I}$, which is the solution. In other words the matrix of Eq. 12-32 or

$$[\bar{Z} \mid \bar{U} \mid \bar{E}] \tag{12-33}$$

is transformed to

$$[U \mid \bar{Z}^{-1} \mid \bar{I}] \tag{12-34}$$

It does seem reasonable that the same operations that transform $\bar{Z}$ into $\bar{U}$ should at the same time transform $\bar{U}$ into $\bar{Z}^{-1}$. To demonstrate this fact, suppose T represents all of the operations performed upon the matrix of Eq. 12-34 until

$$T\bar{Z} = \bar{U} \tag{12-35}$$

But $\bar{Z}^{-1}\bar{Z}$ also equals $\bar{U}$, therefore T has the same effect as multiplying by $\bar{Z}^{-1}$. Therefore

$$T\bar{U} = \bar{Z}^{-1} \tag{12-36}$$

Example 12-3. Once again we will solve the equations of Example 12-1 using row operations upon the augmented matrix (similar to Eq. 12-32). In addition to the solution of I, we will obtain the inverse of Z. The equations are

$$\begin{aligned} 8I_1 - 3I_2 - 4I_3 &= 100 \\ -3I_1 + 10I_2 - 5I_3 &= -50 \\ -4I_1 - 5I_2 + 15I_3 &= 0 \end{aligned}$$

The augmented matrix is

$$\left[\begin{array}{ccc|ccc|c} 8 & -3 & -4 & 1 & 0 & 0 & 100 \\ -3 & 10 & -5 & 0 & 1 & 0 & -50 \\ -4 & -5 & 15 & 0 & 0 & 1 & 0 \end{array}\right]$$

The row operation for each step will be indicated on the left of each matrix where R_1, R_2, and R_3 stand for rows 1, 2, and 3 respectively.

$$R_3 + 0.5R_1 \left[\begin{array}{ccc|ccc|c} 8 & -3 & -4 & 1 & 0 & 0 & 100 \\ -3 & 10 & -5 & 0 & 1 & 0 & -50 \\ \textcircled{0} & -6.5 & 13 & 0.5 & 0 & 1 & 50 \end{array}\right]$$

$$R_2 + 0.375R_1 \left[\begin{array}{ccc|ccc|c} 8 & -3 & -4 & 1 & 0 & 0 & 100 \\ \textcircled{0} & 8.875 & -6.5 & 0.375 & 1 & 0 & -12.5 \\ 0 & -6.5 & 13 & 0.5 & 0 & 1 & 50 \end{array}\right]$$

$$R_3 + \frac{6.5}{8.875}R_2 \left[\begin{array}{ccc|ccc|c} 8 & -3 & -4 & 1 & 0 & 0 & 100 \\ 0 & 8.875 & -6.5 & 0.375 & 1 & 0 & -12.5 \\ 0 & \textcircled{0} & 8.24 & 0.775 & 0.733 & 1 & 40.85 \end{array}\right]$$

$$R_1 + \frac{4}{8.24}R_3 \left[\begin{array}{ccc|ccc|c} 8 & -3 & \textcircled{0} & 1.375 & 0.355 & 0.485 & 119.8 \\ 0 & 8.875 & -6.5 & 0.375 & 1 & 0 & -12.5 \\ 0 & 0 & 8.24 & 0.775 & 0.733 & 1 & 40.85 \end{array}\right]$$

$$R_2 + \frac{6.5}{8.24}R_3 \left[\begin{array}{ccc|ccc|c} 8 & -3 & 0 & 1.376 & 0.355 & 0.485 & 119.8 \\ 0 & 8.875 & \textcircled{0} & 0.987 & 1.578 & 0.79 & 19.7 \\ 0 & 0 & 8.24 & 0.775 & 0.733 & 1 & 40.85 \end{array}\right]$$

$$R_1 + \frac{3}{8.875}R_2 \left[\begin{array}{ccc|ccc|c} 8 & \textcircled{0} & 0 & 1.708 & 0.888 & 0.752 & 126.45 \\ 0 & 8.875 & 0 & 0.987 & 1.578 & 0.79 & 19.7 \\ 0 & 0 & 8.24 & 0.775 & 0.733 & 1 & 40.85 \end{array}\right]$$

$$\frac{R_1}{8};\ \frac{R_2}{8.875};\ \frac{R_3}{8.24} \left[\begin{array}{ccc|ccc|c} 1 & 0 & 0 & 0.213 & 0.111 & 0.094 & 15.8 \\ 0 & 1 & 0 & 0.111 & 0.178 & 0.089 & 2.22 \\ \multicolumn{3}{c|}{\underbrace{0 \quad 0 \quad 1}_{\bar{U}}} & \multicolumn{3}{c|}{\underbrace{0.094 \quad 0.089 \quad 0.1215}_{\bar{Z}^{-1}}} & \underbrace{4.96}_{\bar{I}} \end{array}\right]$$

$$I_1 = \underline{\underline{15.8}}, \quad I_2 = \underline{\underline{2.22}}, \quad I_3 = \underline{\underline{4.96}}$$

$$\bar{Z}^{-1} = \begin{bmatrix} 0.213 & 0.111 & 0.094 \\ 0.111 & 0.178 & 0.089 \\ 0.094 & 0.089 & 0.1215 \end{bmatrix}$$

12-6. The Gauss Iterative Technique

This method of successive approximations is also used in the solution of n linear equations in n unknowns. In years past, before the advent of high-speed computers, such "trial-and-error" procedures were likely to be frowned upon as tedious and inexact. However, today they have proven to be of the most practical nature, often taking preference to the direct methods.

Given n linear equations in n unknowns $(x_1, x_2, \ldots, x_n)$ where the a coefficients and the "y" dependent variables are known.

$$\left.\begin{array}{l} a_{11}x_1 + a_{12}x_2 + a_{13}x_3 + \cdots + a_n x_n = y_1 \\ a_{21}x_1 + a_{22}x_2 + a_{23}x_3 + \cdots + a_{2n}x_n = y_2 \\ \cdots\cdots\cdots\cdots\cdots\cdots\cdots\cdots\cdots\cdots \\ a_{n1}x_1 + a_{n2}x_2 + a_{n3}x_3 + \cdots + a_{nn}x_n = y_n \end{array}\right\} \quad (12\text{-}37)$$

The above equations can be rewritten in a different form as

$$\left.\begin{array}{l} x_1 = \dfrac{1}{a_{11}}[y_1 - a_{12}x_2 - a_{13}x_3 - \cdots - a_{1n}x_n] \\ x_2 = \dfrac{1}{a_{22}}[y_2 - a_{21}x_1 - a_{23}x_3 - \cdots - a_{2n}x_n] \\ \cdots\cdots\cdots\cdots\cdots\cdots\cdots\cdots\cdots\cdots \\ x_n = \dfrac{1}{a_{nn}}[y_n - a_{n1}x_1 - a_{n2}x_2 - \cdots - a_{n,n-1}x_{n-1}] \end{array}\right\} \quad (12\text{-}38)$$

The procedure involves an initial assumed value for each of the independent variables ($x_1^{(0)}, x_2^{(0)}, x_3^{(0)}, \ldots, x_n^{(0)}$) where the superscript (0) indicates an initial approximation. The initial values are often taken as

$$x_1^{(0)} = y_1/a_{11},\ x_2^{(0)} = y_2/a_{22}, \ldots, x_n^{(0)} = y_n/a_{33} \quad (12\text{-}39)$$

These numerical values are then substituted into Eq. 12-38, yielding

$$\left.\begin{array}{l} x_1^{(1)} = \dfrac{1}{a_{11}}[y_1 - a_{12}x_2^{(0)} - a_{13}x_3^{(0)} - \cdots - a_{1n}x_n^{(0)}] \\ x_2^{(1)} = \dfrac{1}{a_{22}}[y_2 - a_{21}x_1^{(0)} - a_{23}x_3^{(0)} - \cdots - a_{2n}x_n^{(0)}] \\ \cdots\cdots\cdots\cdots\cdots\cdots\cdots\cdots\cdots\cdots \\ x_n^{(1)} = \dfrac{1}{a_{nn}}[y_n - a_{n1}x_1^{(0)} - a_{n2}x_2^{(0)} - \cdots - a_{n,n-1}x_{n-1}^{(0)}] \end{array}\right\} \quad (12\text{-}40)$$

This results in new corrected values ($x_1^{(1)}, x_2^{(1)}, \ldots, x_n^{(1)}$) which will be substituted into the next iteration. In general, for the kth iteration

$$\left.\begin{array}{l} x_1^{(k)} = \dfrac{1}{a_{11}}[y_1 - a_{12}x_2^{(k-1)} - a_{13}x_3^{(k-1)} - \cdots - a_{in}x_n^{(k-1)}] \\ x_2^{(k)} = \dfrac{1}{a_{22}}[y_2 - a_{21}x_1^{(k-1)} - a_{31}x_3^{(k-1)} - \cdots - a_{2n}x_n^{(k-1)}] \\ \cdots\cdots\cdots\cdots\cdots\cdots\cdots\cdots\cdots\cdots \\ x_n^{(k)} = \dfrac{1}{a_{nn}}[y_n - a_{n1}x_1^{(k-1)} - a_{n2}x_2^{(k-1)} - \cdots - a_{n,n-1}x_{n-1}^{(k-1)}] \end{array}\right\} \quad (12\text{-}41)$$

Power-system problems are well conditioned and will in general converge. Generally this convergence is better assured if the a_{ii} (diagonal) coefficients are larger than the a_{ij} coefficients, where $i \neq j$.

12-7. The Gauss-Seidel Iteration

This method is a modification of the Gauss iteration. This modification will hopefully reduce the number of iterations necessary to cause $x_i^{(k)} \doteq x_i^{(k-1)}$ for all values of i. Deviation from the straight Gauss iteration is as follows. Instead of substituting the $k - 1$ approximations into all of the equations in the kth iteration, the k approximations are used as soon as they are found. For example,

$$x_1^{(k)} = \frac{1}{a_{11}} [y_1 - a_{12}x_2^{(k-1)} - a_{13}x_3^{(k-1)} - \cdots - a_{1n}x_n^{(k-1)}]$$

But in the next equation of the kth iteration, let

$$x_2^{(k)} = \frac{1}{a_{22}} [y_2 - a_{21}\boxed{x_1^{(k)}} - a_{23}x_3^{(k-1)} - \cdots - a_{2n}x_n^{(k-1)}]$$

and

$$x_3^{(k)} = \frac{1}{a_{33}} [y_3 - a_{31}\boxed{x_1^{(k)}} - a_{32}\boxed{x_2^{(k)}} - \cdots - a_{3n}x_n^{(k-1)}]$$

$$\cdots\cdots\cdots\cdots\cdots\cdots\cdots\cdots \qquad (12\text{-}42)$$

$$x^{(k)} = \frac{1}{a_{nn}} [y_n - a_{n1}\boxed{x_1^{(k)}} - a_{n2}\boxed{x_2^{(k)}} - \cdots - a_{n,n-1}\boxed{x_{n-1}^{(k)}}]$$

The circled values indicate an early substitution of the $x^{(k)}$ values found in preceding steps of the kth iteration.

12-8. The Newton-Raphson Approach to Nonlinear Equations

This technique also involves certain approximations and iterations. Recall that any function of x can be written as the summation of a power series. In particular, Taylor's series is written as

$$f(x) = f(a) + f'(a)(x - a) + \frac{f''(a)}{2!}(x - a)^2 + \cdots + \frac{f^n(a)(x - a)^n}{n!} + \cdots \qquad (12\text{-}43)$$

This series is known to converge rapidly for values of x near a. As an approximation, assume convergence after the first two terms. Now if we replace a by x and x by $(x + \Delta x)$ the series becomes

$$y = f(x + \Delta x) \doteq f(x) + f'(x)\,\Delta x + \underbrace{\cdots}_{\text{assumed 0}} \qquad (12\text{-}44)$$

Equation 12-44 suffices for one unknown (x). While we are primarily concerned with many unknowns $(x_1, x_2, x_3, \ldots, x_n)$ it is well to note here (as a by-product) a common application of one equation and one unknown. Let $x^{(0)}$ be the first approximation and $x^{(1)}$ the second, where $x^{(1)} = x^{(0)} + \Delta x$. Substituting into Eq. 12-44,

$$f(x^{(1)}) = f(x^{(0)}) + f'(x^{(0)})(x^{(1)} - x^{(0)}) \tag{12-45}$$

Now if $x^{(1)}$ is to be a closer approximation than x^0, we could assume that the equations curve crosses the x-axis where $y^{(1)} = f(x^{(1)}) = 0$. Then

$$0 = f(x^{(0)}) + f'(x^{(0)})(x^{(1)} - x^{(0)}) \tag{12-46}$$

Solving Eq. 12-46 for the second approximation in terms of the first,

$$x^{(1)} = x^{(0)} - \frac{f(x^{(0)})}{f'(x^{(0)})} \tag{12-47}$$

Successive approximations follow and, in general, for the $(k + 1)$ approximation

$$x^{(k+1)} = x^k - \frac{f(x^{(k)})}{f'(x^{(k)})} \quad \text{special case of one equation, one unknown} \tag{12-48}$$

Equation 12-48 is often used in such applications as determining the first real root of a *cubic* equation in one unknown. The iteration process is merely continued until the original equation is satisfied.

Equation 12-47 may also be found in another manner other than by Taylor's expansion. This is done by assuming that the tangent line on the curve of Fig. 12-2 at the first approximation point $(x^{(0)}, y^{(0)})$ intersects the x-axis at a point $(x^{(1)}, y^{(1)} = 0)$ which is closer to the solution than was the first approximation. Writing the equation for this tangent line or slope at point $(x^{(0)}, y^{(0)})$,

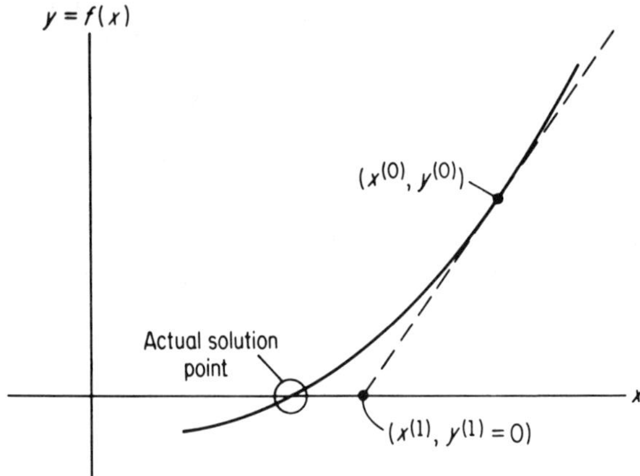

Fig. 12-2. Alternate concept for deriving Newton's approximation formula.

$$\left.\frac{dy}{dx}\right|_{x=x^0} = f'(x^0) = \frac{f(x^{(0)}) - y^{(1)}}{x^{(0)} - x^{(1)}}$$

Now setting $y^{(1)} = 0$,

$$f'(x^{(0)}) = \frac{f(x^{(0)})}{x^{(0)} - x^{(1)}}$$

Solving for the new approximation $x^{(1)}$,

$$x^{(1)} = x^{(0)} - \frac{f(x^{(0)})}{f'(x^{(0)})}$$

which is identical with Eq. 12-47.

Now if Eq. 12-44 is applied to a *whole set* of n nonlinear equations in n unknowns $(x_1, x_2, \ldots, x_n)$ the result is

$$\left.\begin{aligned} y_1 &\doteq f_1(x_1, x_2, \ldots, x_n) + \Delta x_1 \frac{\partial f_1}{\partial x_1} + \Delta x_2 \frac{\partial f_1}{\partial x_2} + \cdots + \Delta x_n \frac{\partial f_1}{\partial x_n} \\ &\cdots\cdots\cdots\cdots\cdots\cdots\cdots\cdots\cdots\cdots\cdots \\ y_n &\doteq f_n(x_1, x_2, \ldots, x_n) + \Delta x_1 \frac{\partial f_n}{\partial x_1} + \Delta x_2 \frac{\partial f_n}{\partial x_2} + \cdots + \Delta x_n \frac{\partial f_n}{\partial x_n} \end{aligned}\right\} \quad (12\text{-}49)$$

As a first approximation, let the unknowns be represented by $x_1^{(0)}, x_2^{(0)}, \ldots, x_n^{(0)}$. Then the first iteration for obtaining a second approximation is

$$\left.\begin{aligned} y_1 &= f_1(x_1^{(0)}, x_2^{(0)}, \ldots, x_n^{(0)}) + \Delta x_1^{(0)} \left.\frac{\partial f_1}{\partial x_1}\right|_0 + \cdots + \Delta x_n^{(0)} \left.\frac{\partial f_1}{\partial x_n}\right|_0 \\ &\cdots\cdots\cdots\cdots\cdots\cdots\cdots\cdots\cdots\cdots\cdots \\ y_n &= f_n(x_1^{(0)}, x_2^{(0)}, \ldots, x_n^{(0)}) + \Delta x_1^{(0)} \left.\frac{\partial f_n}{\partial x_1}\right|_0 + \cdots + \Delta x_n^{(0)} \left.\frac{\partial f_n}{\partial x_n}\right|_0 \end{aligned}\right\} \quad (12\text{-}50)$$

Hereafter we will refer to the term $f_k(x_1^{(0)}, x_2^{(0)}, \ldots, x_n^{(0)})$ as $f_k^{(0)}$. The values of $y_1, y_2, \ldots, y_n$ are normally known. The partials can be calculated as numerical quantities as can be $f_1^{(0)}, f_2^{(0)}, \ldots, f_n^{(0)}$. The unknowns are the $\Delta x^{(0)}$ quantities. Equation 12-50 may be rewritten as

$$\left.\begin{aligned} y_1 - f_1^{(0)} &= \left.\frac{\partial f_1}{\partial x_1}\right|_0 \Delta x_1^{(0)} + \left.\frac{\partial f_1}{\partial x_2}\right|_0 \Delta x_2^{(0)} + \cdots + \left.\frac{\partial f_1}{\partial x_n}\right|_0 \Delta x_n^{(0)} \\ y_2 - f_2^{(0)} &= \left.\frac{\partial f_2}{\partial x_1}\right|_0 \Delta x_1^{(0)} + \left.\frac{\partial f_2}{\partial x_2}\right|_0 \Delta x_2^{(0)} + \cdots + \left.\frac{\partial f_2}{\partial x_n}\right|_0 \Delta x_n^{(0)} \\ &\cdots\cdots\cdots\cdots\cdots\cdots\cdots\cdots\cdots\cdots\cdots \\ y_n - f_n^{(0)} &= \left.\frac{\partial f_n}{\partial x_1}\right|_0 \Delta x_1^{(0)} + \left.\frac{\partial f_n}{\partial x_2}\right|_0 \Delta x_2^{(0)} + \cdots + \left.\frac{\partial f_n}{\partial x_n}\right|_0 \Delta x_n^{(0)} \end{aligned}\right\} \quad (12\text{-}51)$$

The original nonlinear equations have been reduced to linear equations in $\Delta x_1^{(0)}, \Delta x_2^{(0)}, \ldots, \Delta x_n^{(0)}$ for the first iteration.

After having solved for these incremental unknowns, the new x approximations for the next iteration will be

$$x_1^{(1)} = x_1^{(0)} + \Delta x_1^{(0)}; \quad x_2^{(1)} = x_2^{(0)} + \Delta x_2^{(0)}, \ldots, x_n^{(1)} = x_n^{(0)} + \Delta x_n^{(0)} \qquad (12\text{-}52)$$

The iteration process of Eq. 12-51 is then repeated until the Δx values are small enough to give satisfaction. Equation 12-51 can also be conveniently expressed in matrix form as

$$\begin{bmatrix} y_1 - f_1^{(0)} \\ y_2 - f_2^{(0)} \\ \cdots\cdots \\ y_n - f_n^{(0)} \end{bmatrix} = \begin{bmatrix} \left.\dfrac{\partial f_1}{\partial x_1}\right|_0 & \cdots\cdots\cdots & \left.\dfrac{\partial f_1}{\partial x_n}\right|_0 \\ \vdots & & \vdots \\ \left.\dfrac{\partial f_n}{\partial x_1}\right|_0 & \cdots\cdots\cdots & \left.\dfrac{\partial f_n}{\partial x_n}\right|_0 \end{bmatrix} \begin{bmatrix} \Delta x_1^{(0)} \\ \Delta x_2^{(0)} \\ \vdots \\ \Delta x_n^{(0)} \end{bmatrix} \qquad (12\text{-}52a)$$

The matrix of partial derivitive coefficients is sometimes referred to as the *Jacobian* matrix. Of course, any determination of the unknown Δx array would involve some sort of a reduction or inversion process on the coefficient matrix.

Example 12-4. Given the two equations

$$f_1(x_1,x_2) = 2x_1^3 + 3x_1^2 x_2 - x_2^2 - 2 = 0$$

and

$$f_2(x_1,x_2) = x_1 x_2^2 + 2x_1^2 - 3x_2^2 + 16 = 0$$

find x_1 and x_2 by the Newton-Raphson method. The partial derivatives are:

$$\partial f_1/\partial x_1 = 6x_1^2 + 6x_1 x_2; \quad \partial f_2/\partial x_1 = x_2^2 + 4x_1$$
$$\partial f_1/\partial x_2 = 3x_1^2 - 2x_2; \quad \partial f_2/\partial x_2 = 2x_1 x_2 - 6x_2$$

As a first approximation let $x_1^{(0)} = 2$ and $x_2^{(0)} = 2$

$$f_1^{(0)} = 16 + 24 - 4 - 2 = 34$$
$$f_2^{(0)} = 8 + 8 - 12 + 16 = 20$$

$$\partial f_1/\partial x_1|_0 = 24 + 24 = 48; \qquad \partial f_2/\partial x_1|_0 = 4 + 8 = 12$$
$$\partial f_1/\partial x_2|_0 = 12 - 4 = 8; \qquad \partial f_2/\partial x_2|_0 = 8 - 12 = -4$$

From Eq. 12-51, $\Delta x_1^{(0)}$, and $\Delta x_2^{(0)}$ are found:

$$(1) \quad y_1 - f_1^{(0)} = \partial f_1/\partial x_1|_0 \Delta x_1^{(0)} + \partial f_1/\partial x_2|_0 \Delta x_2^{(0)}$$
$$(2) \quad y_2 - f_2^{(0)} = \partial f_2/\partial x_1|_0 \Delta x_1^{(0)} + \partial f_2/\partial x_2|_0 \Delta x_2^{(0)}$$

Substituting numerical values into Eqs. (1) and (2),

$$(1)\ 0 - 34 = 48\Delta x_1^{(0)} + 8\Delta x_2^{(0)}$$
$$(2)\ 0 - 20 = 12\Delta x_1^{(0)} + (-4)\Delta x_2^{(0)}$$

Multiply Eq. (2) by 2 and add to Eq. (1):

$$(1)\ -34 = 48\Delta x_1^{(0)} + 8\Delta x_2^{(0)}$$
$$(2)\ \underline{-40 = 24\Delta x_1^{(0)} - 8\Delta x_2^{0}}$$
$$-74 = 72\Delta x_1^{(0)}$$
$$\Delta x_1^{(0)} = -74/72 = \underline{\underline{-1.03}}$$

Substituting $\Delta x_1^{(0)}$ in Eq. (1),

$$-34 = 48(-1.03) + 8\Delta x_2^{(0)}$$

$$\Delta x_2^{(0)} = \frac{-34 + 49.4}{8} = \underline{\underline{1.9}}$$

The new approximations are

$$x_1^{(1)} = x_1^{(0)} + \Delta x_1^{(0)} = 2 - 1.03 = \underline{\underline{0.97}}$$
$$x_2^{(1)} = x_2^{(0)} + \Delta x_2^{(0)} = 2.0 + 1.9 = \underline{\underline{3.9}}$$

Equations of 12-51 for the next iteration are found to be

$$(1)\ 4.3 = 28.4\Delta x_1^{(1)} - 5.0\Delta x_2^{(1)}$$
$$(2)\ 13.0 = 19.1\Delta x_1^{(1)} - 15.8\Delta x_2^{(1)}$$

Solution of (1) and (2) yields $\Delta x_1^{(1)} = \underline{\underline{0.0085}}$ and $\Delta x_2^{(1)} = -\underline{\underline{0.812}}$. Then

$$x_1^{(2)} = x_1^{(1)} + \Delta x_1^{(1)} = 0.97 + 0.0085 = \underline{\underline{0.98}}$$

and

$$x_2^{(2)} = 3.9 + (-0.812) = \underline{\underline{3.09}}$$

It can be shown that the values of $x_1 = 1.0$ and $x_2 = 3.0$ will satisfy the original equations. It is obvious that the iterations are rapidly converging toward these results.

12-9. Defining the Load-Flow Problem

Load-flow studies of this chapter are normally concerned with balanced, three-phase systems, and with calculations on a per phase, per unit basis. As mentioned in Sec. 12-1, the d-c analyzer board does not qualify for load studies. The a-c analyzer, while satisfactory from the standpoint of accuracy, is giving way to computer methods. In either case, more detailed information is required than for the short-circuit study. Such data as complex impedances, transformer tap settings, and shunt-line capacitance are to be considered. The objectives of the load study may vary to some extent. Generally the voltage magnitudes and phase angles are important for each node. The phasor line currents and/or real and reactive power flow in the individual lines might also be required.

Normally three kinds of nodes or buses are identified in a power network. First of all there is the *swing bus*, or *slack bus* as it is sometimes called. This is the bus of the generator which is first to respond to a changing load condition. Voltage magnitude and phase angle are specified for this bus. Other *generator buses* will usually specify real power output and voltage magnitude, while the *load buses* will specify only the real and reactive power taken by the load (P, Q). If there is no load or generation at a bus, then it can be considered as a load bus where P_{load} and Q_{load} are both zero. To summarize the three node types:

1. Swing or slack generators bus—specifies $|V|$ and δ_v.
2. Other generator buses—specify $|V|$ and P.
3. Load buses—specify P and Q.

In each of the node types cited, four variables are present; P, Q, $|V|$ and δ_v. In each case two variables are known and two are unknown as demonstrated in Fig. 12-3. Specifying three variables on one bus would place an undue restriction upon the system. This is not to infer that a third restriction is not possible. For example, a generator bus with $|V|$ and P given may also require that the reactive vars Q be held within certain limits. The load-flow problem could be solved without the Q

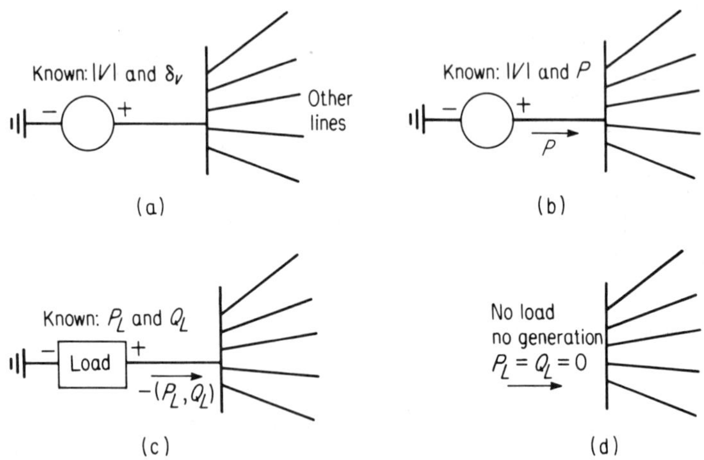

Fig. 12-3. Demonstrating various node types. Power flowing into bus is taken as positive (a) Swing generator bus. (b) Generator bus. (c) Load bus. (d) Load bus with $P_L = Q_L = 0$.

limitation, whereupon the solution would reveal the actual Q at the bus. If the actual Q is out of the desired range, then something must be altered to bring the generator Q back within limits (such as the addition of capacitor banks).

Some additional explanation should be given regarding the information normally specified on the three types of buses. First of all, the slack generator could be equipped with the most sensitive speed governor, thereby being the first machine to pick up new load or drop load, as the situation demands. Thus the P and Q is not specified as the unit adjusts to system change, taking up the slack in the system. At the same time, the shaft power input for the other generators could be preset to some constant value for the sake of efficiency. If all units have automatic voltage magnitude control, this accounts for the constant $|V|$ of all units. Some reference angle is needed in the system and angle of the slack generator δ_v is a likely choice. Notice that the reactive vars Q is not specified on generators. However, it should be realized that voltage regulation of these units will be accomplished by automatic adjustment of the d-c field excitation. Should the system demand the generator to deliver additional lagging amperes, the regulator will see a drop in terminal voltage and pass this information on so as to initiate an increase in field excitation. Recall that an overexcited synchronous generator delivers current at a lagging power factor. Therefore the Q of a generator need not be specified. The approximate loading requirement and corresponding load power factor will normally be known at the load buses, accounting for the given P and Q of these buses. This disussion is not an attempt to justify mathematically the fact that two specified variables per bus are necessary and sufficient. Rather, it is the purpose here to give a better physical concept to the system as a whole.

In setting up the load-flow problem, the system equations will be linear or nonlinear, depending upon our choice of methods. Three possibilities are suggested here.

1. Current equations can be written at the kth node in terms of the unknown voltages. These equations are of the form

$$I_k = Y_{k1}V_1 + Y_{k2}V_2 + \cdots + Y_{kn}V_n \tag{12-53}$$

This has been a very popular approach, yielding a set of linear equations. The known current source I_k can be written in terms of P_k, Q_k, and V_k if necessary. The linear equations may be written in matrix form utilizing the bus admittance matrix. This will be covered in more detail in Sec. 12-10. Furthermore the set of equations are often solved by an iterative process in preference to a direct approach.

2. Voltage equations can be written for the kth loop in terms of unknown currents and of the general form

$$E_k = Z_{k1}I_1 + Z_{k2}I_2 + \cdots + Z_{kn}I_n \tag{12-54}$$

Again this yields a set of linear algebraic equations. The kth loop can be identified with the kth bus if the loop current I_k is also the actual current flowing into (or from) bus k from (or into) the load or generator branch in question. In other words (as will be shown in Sec. 12-11), a bus impedance matrix with self and transfer impedances will identify the kth loop with the kth bus current. This matrix is similar to the s-c matrix of Chapter 10. However it does not yield as radical a simplification to the load-flow problem as was achieved for short-circuit applications. In fact an iterative solution to Eq. 12-54 is described in Sec. 12-11.

3. One might also choose to write the node equations in terms of the real and reactive power moving into or out of the individual nodes. In this case, the equation for power flowing to or from bus k takes the general nonlinear form of

$$\begin{aligned} P_k + jQ_k = {} & |Y_{k1}||E_1||E_k| \underline{/\delta_k - \delta_1 - \theta_{k1}} \\ & + |Y_{k2}||E_2||E_k| \underline{/\delta_k - \delta_2 - \theta_{k2}} + \cdots \\ & + |Y_{kn}||E_n||E_k| \underline{/\delta_k - \delta_n - \theta_{kn}} \end{aligned} \tag{12-55}$$

Equation 12-55 is easily derived by a substitution of Eq. 12-53 into a general form of Eq. 3-7. Of the four variables at the kth node (P_k, Q_k, $|E_k|$, and δ_k) two of the variables are known and two unknown. The Newton-Raphson method has been proven to be an effective method for solving the set of equations represented by Eq. 12-55.

12-10. Bus Admittance Concept for Load Studies with the Gauss-Seidel Iteration

It is recommended that Sec. 5-4 be reviewed before continuing on with this section, with particular attention to a thorough understanding of the nodal Eqs. 5-18, 5-19 and 5-20.

To begin, we will examine the per phase passive network of Fig. 12-4,

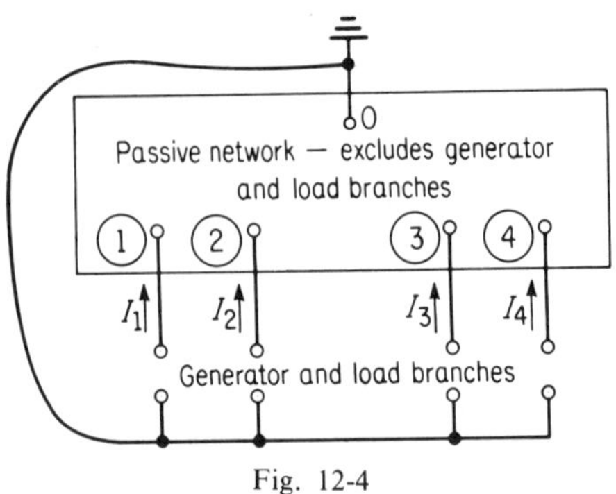

Fig. 12-4

which excludes only the generator and load branches. These branches are considered external to the box. Current equations are written at these generator and load nodes according to Eq. 12-53.

$$\left.\begin{array}{l} I_1 = Y_{11}V_1 + Y_{12}V_2 + Y_{13}V_3 + Y_{14}V_4 \\ \cdots\cdots\cdots\cdots\cdots\cdots\cdots\cdots\cdots\cdots \\ I_4 = Y_{41}V_1 + Y_{42}V_2 + Y_{43}V_3 + Y_{44}V_4 \end{array}\right\} \tag{12-56}$$

Consider for the present that I_1, I_2, I_3 and I_4 are the known current sources. The matrix equation is

$$\begin{bmatrix} I_1 \\ I_2 \\ I_3 \\ I_4 \end{bmatrix} = \begin{bmatrix} Y_{11} & Y_{12} & Y_{13} & Y_{14} \\ Y_{21} & Y_{22} & Y_{23} & Y_{24} \\ Y_{31} & Y_{32} & Y_{33} & Y_{34} \\ Y_{41} & Y_{42} & Y_{43} & Y_{44} \end{bmatrix} \begin{bmatrix} V_{01} \\ V_{02} \\ V_{03} \\ V_{04} \end{bmatrix} \tag{12-57}$$

In the admittance matrix of Eq. 12-57 the diagonal elements Y_{kk} are the summation of all admittance surrounding node k, while $Y_{kl} = Y_{lk}$ is the negative of the branch admittance joining node k with node 1. We again refer to this matrix as the *bus admittance matrix* or Y_{bus}.

Next refer to Fig. 12-5 in which we have attached typical generator

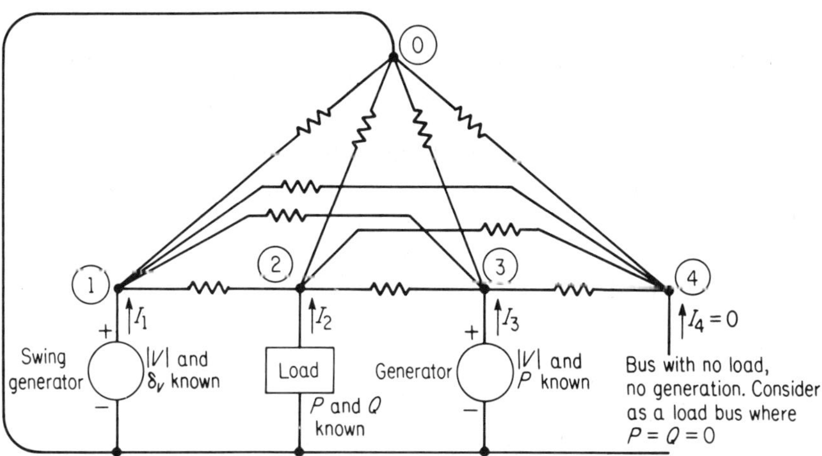

Fig. 12-5. Bus admittance (or mesh) equivalent for passive network of Fig. 12-4. Example demonstrates one of each type of node possible.

and load branches. Now it is generally true in power applications that the known quantities at the buses will be something other than the magnitudes and phase angles of the currents. In fact, Sec. 12-9 stated that the individual node information in the load-flow problem normally takes one of three forms. One of each type is demonstrated in Fig. 12-5. For the swing or slack generator of bus 1, $|V|$ and δ_v are known. For other generators $|V|$ and P are known and for load buses P and Q are known. Buses with neither generator or load may be considered as load buses where $P = Q = 0$. If such buses are of no particular interest, a node elimination could be performed as explained in Sec. 5-5. Otherwise the node will be retained for a voltage calculation.

Although currents entering the nodes from generator and load branches are not known, they may be written in terms of P, Q, and V. For example, from Eq. 3-8 we may write the current I_2 as

$$I_2 = \frac{P_2 - jQ_2}{V_2^*} \tag{12-58}$$

where V_2^* is the conjugate of the phasor V_2. Since currents entering the nodes are considered positive, then power flow into the node is also considered positive, and power dissipated in a resistive load would be entered as a negative numerical value. For example a load branch drawing power at 5,000 watts and inductive vars of 8,000 would be entered as $I_2 = [-5{,}000 - j(-8{,}000)]/V_2^* = (-5{,}000 + j8{,}000)/V_2^*$. The nodal equation for bus 2 becomes

$$\frac{P_2 - jQ_2}{V_2^*} = Y_{21}V_1 + Y_{22}V_2 + Y_{23}V_3 + Y_{24}V_4 \tag{12-59}$$

At this point we will move from the specific case of Fig. 12-5 to the more general network with n nodes (plus reference node 0). The set of equations represented by Eq. 12-56 may be solved in a number of ways for $V_1, V_2, \ldots, V_n$. The *Gauss-Seidel* iteration technique will be stressed here. The node 2 Eq. 12-59 as applied to the n bus network can be solved for V_2 as

$$V_2 = \frac{1}{Y_{22}}\left[\frac{P_2 - jQ_2}{V_2^*} - Y_{21}V_1 - Y_{23}V_3 - \cdots - Y_{2n}V_n\right]$$

likewise

$$V_3 = \frac{1}{Y_{33}}\left[\frac{P_3 - jQ_3}{V_3^*} - Y_{31}V_1 - Y_{32}V_2 - Y_{34}V_4 - \cdots - Y_{3n}V_n\right] \tag{12-60}$$

....

In general, for the kth bus,

$$V_k = \frac{1}{Y_{kk}}\left[\frac{P_k - jQ_k}{V_k^*} - \sum_{i=1}^{i=n} Y_{ki}V_i\right]; \quad \text{for } i \neq k \tag{12-61}$$

Notice that V_k has been written in terms of itself and the other voltages. Also note that the first equation involving swing bus 1 is omitted, since V_1 has already been specified with respect to both magnitude and phase angle. The iteration procedure is as follows:

1. Assume initial phasor values of load bus voltages and angles for generator bus voltages (except for the swing bus). Assumed values are $V_2^{(0)}, V_3^{(0)}, \ldots, V_n^{(0)}$.
2. Calculate $V_2^{(1)}$ in terms of the initial assumed voltages, or

$$V_2^{(1)} = \frac{1}{Y_{22}}\left[\frac{P_2 - jQ_2}{V_2^{(0)*}} - Y_{21}V_1^{(0)} - Y_{23}V_3^{(0)} - \cdots - Y_{2n}V_n^{(0)}\right] \tag{12-62}$$

3. Substitute the corrected values of $V_2^{(1)}$ back into the right-hand side of Eq. 12-62 for a new corrected value of V_2. Continue this process for a specified number of iterations. The last iteration here will not, in general, yield the correct answer for V_2, since the correction also depends upon other assumed voltages.
4. Go to bus 3, again performing several iterations, while making use of the corrected value voltage of V_2 from step 3 as well as other assumed values. Continue this procedure on bus 4, etc., until all buses have been considered. Always use the latest corrected voltage values already found for preceding buses.
5. Repeat the iteration process (of steps 1 through 4) for the network until the voltage correction necessary for each bus is less than some predetermined precision index.

The above procedure has (for the sake of simplicity) been geared primarily to the load bus where P and Q are known and V and δ_v are both assumed. A modification must be made in the procedure for a generator bus k, where P_k and $|V_k|$ are known and Q_k unknown. In this case an approximate value of Q is inserted into Eq. 12-61. The equation for calculating this approximate Q is found below:

$$I_{\text{gen}} = \frac{P_k - Q_k}{V_k^*} = Y_{k1}V_1 + Y_{k2}V_2 + \cdots + Y_{kn}V_n$$

$$P_k - jQ_k = V_k^*\left(\sum_{i=1}^{i=n} Y_{ki}V_i\right)$$

$$Q_k = -\text{Im}\left\{V_k^*\left(\sum_{i=1}^{i=n} Y_{ki}V_i\right)\right\} \tag{12-63}$$

This is the value of Q_k to be substituted into Eq. 12-61 for a generator bus where the symbol "Im" means "the imaginary part of." Values of V_k^* and V_i are taken from the best (latest) iterations for use in Eq. 12-63. Recall that the magnitude $|V_k|$ is known, so that when the calculation for $|V'_k| \,\underline{/\delta_k}$ is made from Eq. 12-61, the angle δ_k will be maintained while $|V'_k|$ will be corrected back to V_k by multiplying by the ratio of $|V_k|/|V'_k|$. Most likely, $|V'_k| \,\underline{/\delta_k}$ will be in rectangular form in which case both real and imaginary voltage components are multiplied by $|V_k|/|V'_k|$, again maintaining the new angle δ_k.

Example 12-5. Given the three-phase load-flow problem represented by the phase a diagram of Fig. 12-6, all given information is included on the drawing.

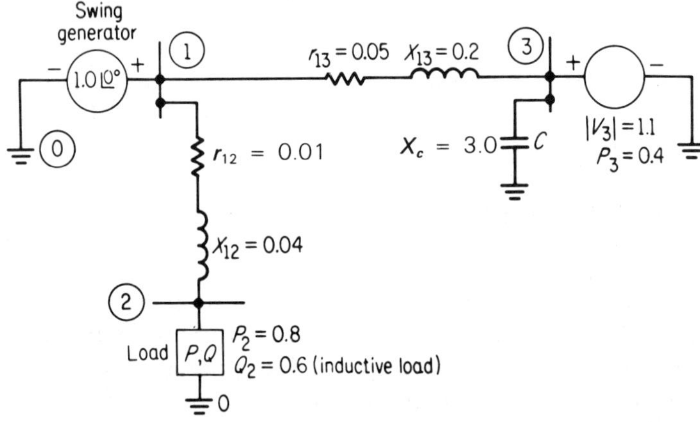

Fig. 12-6. Phase a diagram for Example 12-5. All values are given in per unit.

Find the first iteration of voltage (both magnitude and angle) for bus 2 and the voltage angle of bus 3, using the Gauss-Seidel iteration method.

This problem is of such a simple nature that it is possible that substitution could be made directly into the iteration Eqs. 12-61 and 12-63. However the concept of the bus admittance matrix will also be brought out in this example. For this reason the circuit is redrawn in Fig. 12-7. Small y values are used to represent individual branch admittances. The problem here is solved to slide-rule accuracy.

$$y_{12} = 1/z_{12} = \frac{1}{0.01 + j0.04} = 5.88 - j23.5 \text{ per unit mho}$$

$$y_{13} = 1/z_{13} = \frac{1}{0.05 + j0.2} = 1.175 - j4.71$$

Elements for the bus admittance matrix are found as

$$Y_{12} = Y_{21} = -y_{12} = -5.88 + j23.5$$

$$Y_{13} = Y_{31} = -y_{13} = -1.175 + j4.71$$

$$Y_{11} = \overset{0}{\cancel{y_{01}}} + y_{21} + y_{31} = 7.05 - j28.2$$

$$Y_{22} = \overset{0}{\cancel{y_{02}}} + y_{12} + \overset{0}{\cancel{y_{32}}} = 5.88 - j23.5$$

$$Y_{33} = y_{03} + y_{13} + \overset{0}{\cancel{y_{23}}} = 1.175 - j4.38$$

The bus admittance matrix is

$$Y_{bus} = \begin{bmatrix} Y_{11} & Y_{12} & Y_{13} \\ Y_{21} & Y_{22} & Y_{23} \\ Y_{31} & Y_{32} & Y_{33} \end{bmatrix}$$

$$= \begin{bmatrix} (7.05 - j28.2) & (-5.88 + j23.5) & (-1.175 + j4.71) \\ (-5.88 + j23.5) & (5.88 - j23.5) & 0 \\ (-1.175 + j4.71) & 0 & (1.175 - j4.38) \end{bmatrix}$$

Let the initial assumed values of V_2 and δ_3 be

$$V_2^{(0)} = 1.0 + j0; \quad \delta_3^{(0)} = 0°$$

For the first iteration, apply Eq. 12-62:

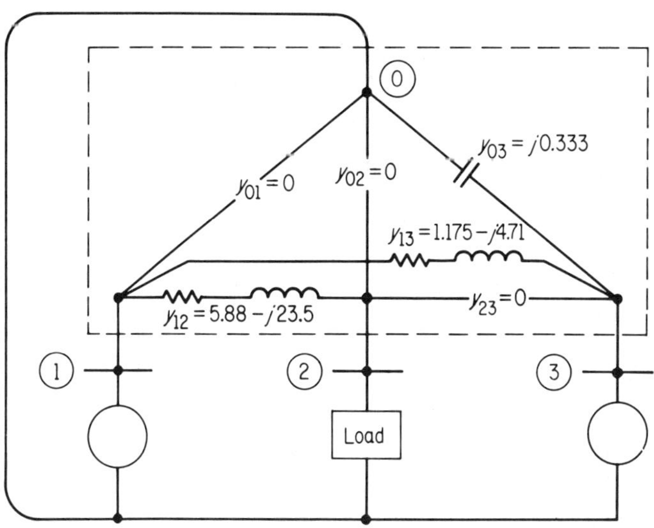

Fig. 12-7. Circuit redrawn in terms of admittances.

$$V_2^{(1)} = \frac{1}{Y_{22}}\left[\frac{P_2 - jQ_2}{V_2^{(0)*}} - Y_{21}V_1^{(0)} - Y_{23}V_3^{(0)}\right]$$

$$= \frac{1}{5.88 - j23.5}\left[\frac{-(0.8 - j0.6)}{1.0 + j0} - (-5.88 + j23.5)(1.0\,\underline{/0^\circ}) - (0)(1.1\,\underline{/0^\circ})\right]$$

$$= (0.01 + j0.04)(5.08 - j22.9)$$

$$= 0.0508 + 0.916 + j0.2032 - j0.229$$

$$= \underline{\underline{0.9668 - j0.0258}}$$

$$V_2^{(1)*} = 0.967 + j0.0258$$

The corrected value of $V_2^{(1)*}$ will be substituted back into the Eq. 12-62 in place of $V_2^{(0)*}$. Only two iterations will be made before going on to bus 3:

$$V_2^{(2)} = \frac{1}{Y_{22}}\left[\frac{P_2 - jQ_2}{V_2^{(1)*}} - Y_{21}V_1^{(0)} - Y_{23}V_3^{(0)}\right]$$

$$= (0.01 + j0.04)\left[\frac{-(0.8 - j6)}{0.967 + j0.0258} - (-5.88 + j23.5)1.0\,\underline{/0^\circ} - 0\right]$$

$$= (0.01 + j0.04)[(-0.8 + j0.6)(1.02 - j0.0272) + 5.88 - j23.5]$$

$$= (0.01 + j0.04)(5.08 - j22.9) = \underline{\underline{0.967 - j0.026}}$$

Since bus 3 is a generator bus, Q_3 must be approximated according to Eq. 12-63.

$$Q_3^{(0)} = -\text{Im}\,\{V_3^{(0)*}[Y_{31}V_1 + Y_{32}V_2^{(2)} + Y_{33}V_3^{(0)}]\}$$

$$= -\text{Im}\,\{1.1\,\underline{/0^\circ}[(-1.175 + j4.71)(1.0\,\underline{/0^\circ}) + (1.175 - j4.38)(1.1\,\underline{/0^\circ})]\}$$

$$= -\text{Im}\,\{1.1\,(0.118 - j0.100)\} = -\text{Im}\,(0.13 - j0.11) = \underline{\underline{+0.11}}$$

Substituting $Q_3^{(0)}$ into Eq. 12-61,

$$V_3^{(1)} = \frac{1}{Y_{33}}\left[\frac{P_3 - jQ_3^{(0)}}{V_3^{(0)*}} - Y_{31}V_1 - Y_{32}V_2^{(2)}\right]$$

$$= \frac{1}{1.175 - j4.38}\left[\frac{0.4 - j0.11}{1.1\,\underline{/0^\circ}} - (-1.175 + j4.71)(1.0\,\underline{/0^\circ}) - 0\right]$$

$$= \frac{0.363 - j0.100 + 1.175 - j4.71}{1.175 - j4.38} = \frac{1.538 - j4.81}{1.175 - j4.38}$$

$$= \frac{5.05\,\underline{/-72.3^\circ}}{4.53\,\underline{/-75^\circ}} = \underline{\underline{1.114\,/2.7^\circ}}$$

Since the magnitude $|V_3|$ is given as 1.1, correct $V_3^{(1)}$ by the ratio of $1.1/1.114$, which yields $V_3 = 1.1\ \underline{/2.7}$.

The complete iteration process could be repeated using the latest values of V_2 and V_3 to obtain a closer approximation. The foregoing calculation should, however, be sufficient to demonstrate the method.

Sometimes the number of iterations needed for arriving at the predetermined precision index may be reduced through the use of *acceleration factors*. In the choice of a factor, there is nothing to guarantee rapid convergence. At the same time, numerous studies have been made in an attempt to arrive at the most appropriate factor. To demonstrate the use of acceleration factors, suppose on the first round of iterations, the bus 4 voltage $V_4^{(1)}$ calculates $1.1 - j0.2$ whereas $V_4^{(0)}$ was initially assumed as $1.0 + j0$. Then

$$\Delta V_4^{(0)} = (1.1 - j0.2) - (1.0 + j0) = 0.1 - j0.2$$

When we speak of a predetermined precision index we mean that ΔV should not exceed some predetermined incremental value. This test is easily made in the computer program. However, if ΔV_4 is not small enough, we might speed up convergence by multiplying ΔV_4 by some acceleration factor. If the popular factor of 1.6 were applied here, our V_4 for the next iteration would be

$$\begin{aligned} V_4 &= V_4^{(0)} + \Delta V_4^{(0)} \times 1.6 \\ &= (1.0 + j0) + (0.1 - j0.2)(1.6) = \underline{\underline{1.16 - j0.32}} \end{aligned}$$

Given a load branch for which P_L and Q_L are both specified. See Fig. 12-8a. If the branch is supplied with an automatic tap-changing transformer for holding load voltage constant, extra information is specified in the form of $|V_L|$. Suppose $|V_L|$ is specified within certain limits. The load flow problem could be solved completely without specifying $|V_L|$, using one transformer turns ratio (N_1/N_2) or tap setting. Then if the calculated $|V_L|$ does not fall within the specified limits, a new tap setting could be assumed and the complete iteration procedure repeated.

Another approach in the handling of the load-tap-changing trans-

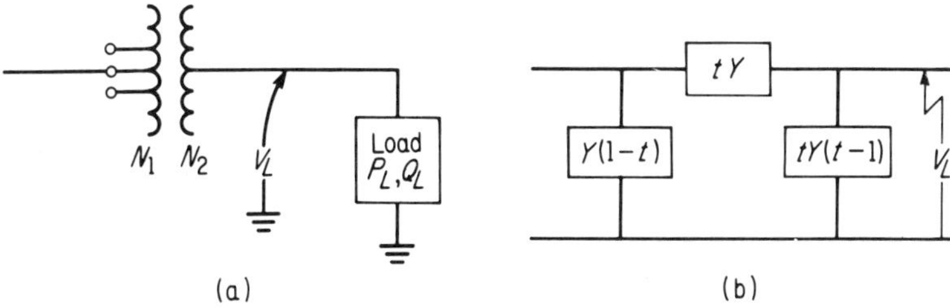

Fig. 12-8. (a) Load bus with automatic tap-changing transformer. (b) π model of LTC transformer.

former is to model the transformer using the π equivalent of Fig. 11-21. Now $|V_L|$ of Fig. 12-8b can be specified and the variable t can be exchanged for $|V_L|$ in the load-flow equations.

12-11. Bus Impedance Matrix Concept for Load Studies

Section 12-9 indicated that the bus impedance matrix and its corresponding equivalent circuit is similar in many respects to the short circuit matrix of Chapter 10. In that chapter, we neglected loads and considered generator voltages equal. The common terminal of generator excitation (behind generator reactances) became the reference bus for the positive sequence network. The neutral bus became reference for the negative and zero sequence networks. Theoretically, any bus could be chosen as reference for the load-flow problem.

If we were to choose the neutral bus as reference for the bus impedance matrix as was done in Fig. 12-4 for the bus admittance matrix, certain of the self-impedances (diagonal elements of the matrix) would approach infinity. This is because many buses have little or no path back to neutral (if generators and loads are considered as external to the network equivalent). What path does exist would be in the form of line capacitance to ground, transformer magnetizing impedance, static capacitors, etc.

In applying the bus impedance matrix and its equivalent circuit to the load-flow problem the *swing generator bus* will be chosen as the *reference.* Loads and generation will be considered external to the rake equivalent of the passive network. In addition, admittance paths to neutral (line capacitance, etc.) from the various buses will also be external to the equivalent. (Refer to Fig. 12-9.) The swing bus has been numbered

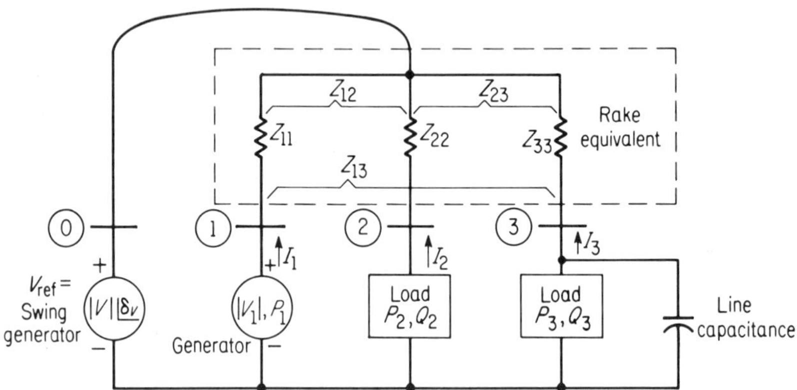

Fig. 12-9. Example of a rake equivalent concept for the bus impedance matrix with swing bus as reference. Generation, loads, and all other paths to neutral are shown external to the equivalent.

0 (rather than 1 as was done in Sec. 12-10) in order to more closely identify the self and transfer impedances with their corresponding buses. In other words, Z_{11} corresponds to bus 1, Z_{22} to bus 2, etc. Also shown in Fig. 12-9 are typical buses; generator bus 1 where $|V_1|$ and P_1 are known, load bus 2 where P_2 and Q_2 are known, and bus 3 which has both a load and a line capacitance to neutral.

When the voltage equation is written for the kth loop, the equations are of the general form of Eq. 12-54, or

$$E_k = Z_{k1}I_1 + Z_{k2}I_2 + \cdots + Z_{kn}I_n \qquad [12\text{-}54]$$

Loop current I_k for the rake equivalent is identical with the current entering the kth bus. To be more specific, the loop equations for Fig. 12-9 are

$$\left.\begin{aligned} V_1 - V_{\text{ref}} &= I_1Z_{11} + I_2Z_{12} + I_3Z_{13} \\ V_2 - V_{\text{ref}} &= I_1Z_{21} + I_2Z_{22} + I_3Z_{23} \\ V_3 - V_{\text{ref}} &= I_1Z_{31} + I_2Z_{32} + I_3Z_{33} \end{aligned}\right\} \qquad (12\text{-}64)$$

The general form of Eq. 12-64, for the kth bus of an n bus system is

$$V_k - V_{\text{ref}} = Z_{k1}I_1 + Z_{k2}I_2 + \cdots + Z_{kn}I_n \qquad (12\text{-}65)$$

The neutral bus and the reference bus 0 are not included among the n buses of Eq. 12-65.

Once again, Eq. 12-65 is not entirely in the form of our given data. In order to write the general expression for the current of bus k, we will assume current entering the equivalent from bus k is positive. Refer to

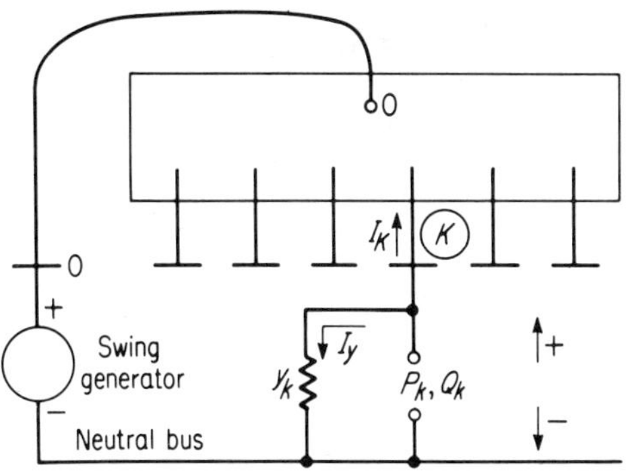

Fig. 12-10. Assumptions for Eq. 12-65.

Fig. 12-10. Let y_k be the admittance to ground of bus k (line capacitance, etc.). Then

$$I_k = I_{\text{branch}} - I_y$$

or

$$I_k = \frac{P_k - jQ_k}{V_k^*} - y_k V_k \tag{12-66}$$

When the kth branch delivers (generates) real and inductive reactive power to the bus, enter P and Q as positive values. If the kth branch draws power in the form of a real and inductive reactive load, then enter P and Q as negative values. Assume for example that bus 3 of Fig. 12-9 draws a per unit power of 0.4 and per unit inductive vars of 0.5. If per unit $Z_c = -j300$, or $Y_c = +j0.00333$, then Eq. 12-66 becomes

$$I_3 = \frac{(-0.4) - j(-0.5)}{V_3^*} - (j0.00333)V_3$$

V_3 is the voltage rise from neutral to bus k. *In this text a voltage rise is considered positive whether called by the symbol V or E.*

The general procedure from this point is very similar to that employed in Sec. 12-10 in which we used the Gauss-Seidel iteration approach. The current estimates of $I_1, I_2, \ldots, I_n$ from Eq. 12-66 may be substituted into Eq. 12-65.

$$\begin{aligned} V_k - V_{\text{ref}} &= Z_{k1}I_1 + \cdots + Z_{kn}I_n \\ &= Z_{k1}\left(\frac{P_1 - jQ_1}{V_1^*} - y_1 V_1\right) + \cdots + Z_{kn}\left(\frac{P_n - jQ_n}{V_n^*} - y_n V_n\right) \\ &= \sum_{i=1}^{i=n} Z_{ki}\left(\frac{P_i - jQ_i}{V_i^*} - y_i V_i\right) \end{aligned}$$

or

$$V_k = V_{\text{ref}} + \sum_{i=1}^{i=n} Z_{ki}\left(\frac{P_i - jQ_i}{V_i^*} - y_i V_i\right) \tag{12-67}$$

The iteration procedure is outlined here. An example problem of this procedure can be found incorporated as a part of the dispatch example Prob. 13-2. Refer specifically to Fig. 13-13. The procedure is

1. Assume initial voltages ($V_1^{(0)}$ through $V_n^{(0)}$) for all n buses.
2. Calculate $V_1^{(1)}$ from Eq. 12-67 in terms of the initial assumed voltages ($V_1^{(0)}$), or

$$V_1^{(1)} = \sum_{i=1}^{n} Z_{ki}\left(\frac{P_i - jQ_i}{V_i^{(0)*}} - y_i V_i^{(0)}\right) + V_{\text{ref}} \tag{12-68}$$

3. Substitute the corrected value of $V_i^{(1)}$ back into the foregoing equation for a new corrected value $V_1^{(2)}$. Continue this process for a specified number of iterations before proceeding to the bus 2 iterations.

4. Go to bus 2 for a corrected value of $V_2^{(1)}$, making use of the last corrected V_1 value of step 3. Again, a specified number of iterations can be made at this bus. Proceed in like manner to bus 3, 4, etc. until all buses have been included.
5. Repeat steps 2 through 4, always using the latest values of corrected voltages. These steps are repeated until the bus voltages corrections are less than some predetermined precision index.

The foregoing procedure is geared to load buses where P_i and Q_i of Eq. 12-67 are known. At the generator bus k where Q_k is not known, an approximation for Q_k is necessary. This approximation is derived from Eq. 12-67 as follows:

$$V_k = V_{\text{ref}} + \sum_{i=1}^{n} Z_{ki}\left(\frac{P_i - jQ_i}{V_i^*} - y_i V_i\right) \qquad [12\text{-}67]$$

Rearranging terms yields

$$Z_{kk}\left(\frac{P_k - jQ_k}{V_k^*} - y_k V_k\right) = V_k - V_{\text{ref}} - \sum_{\substack{i=1 \\ i \neq k}}^{n} Z_{ki}\left(\frac{P_i - jQ_i}{V_i^*} - y_i V_i\right)$$

Solving for $P_k - jQ_k$;

$$P_k - jQ_k = y_k V_k V_k^* + \frac{V_k^*}{Z_{kk}}\left[V_k - V_{\text{ref}} - \sum_{\substack{i=1 \\ i \neq k}}^{n} Z_{ki}\left(\frac{P_i - jQ_i}{V_i^*} - y_i V_i\right)\right]$$

$$= \frac{V_k^*}{Z_{kk}}\left[V_k(1 + Z_{kk} y_k) - V_{\text{ref}} - \sum_{\substack{i=1 \\ i \neq k}}^{n} Z_{ki}\left(\frac{P_i - jQ_i}{V_i^*} - y_i V_i\right)\right]$$

Solving for Q_k;

$$Q_k = -\text{Im}\,\frac{V_k^*}{Z_{kk}}\left[V_k(1 + Z_{kk} y_k) - V_{\text{ref}} - \sum_{\substack{i=1 \\ i \neq k}}^{n} Z_{ki}\left(\frac{P_i - jQ_i}{V_i^*} - y_i V_i\right)\right] \qquad (12\text{-}69)$$

In cases where the effect of admittance to ground (y_k) of Fig. 12-10 lumped together with the load current I_k, the terms involving y_i and y_k drop out and a simplified form of the Q_k approximation would be

$$Q_k = -\text{Im}\,\frac{V_k^*}{Z_{kk}}\left[V_k - V_{\text{ref}} - \sum_{\substack{i=1 \\ i \neq k}}^{n} Z_{ki}\,\frac{P_i - jQ_i}{V_i^*}\right] \qquad (12\text{-}70)$$

This approximation for Q_k is used in Example 13-2 in the load flow iterations which made up a part of the total dispatch problem.

Some observations should be made with regard to the *formation of the bus impedance matrix*, from which the rake equivalent of Fig. 12-9 was taken. The method of formation is essentially the same as that given for the forming of a short-circuit impedance matrix of Secs. 10-3 and 10-8. Of course, we have chosen the swing generator bus as reference here. Paths to the neutral bus are considered outside of the equivalent for this load-flow application.

12-12. The Relaxation Method Applied to Eq. 12-60

The relaxation method is an iteration approach similar to the Gauss methods, but some space is devoted here to its special application to the load-flow problem. Refer back to Sec. 12-10, in which the concept of the bus admittance equivalent (Fig. 12-5) was given. From this concept we obtained Eq. 12-60. Rewriting these equations in the form of Eq. 12-56, only with all terms on the same side of the equal sign, we obtain

$$\left.\begin{array}{l} Y_{21}V_1 + Y_{22}V_2 + \cdots + Y_{2n}V_n - I_2 = 0 \\ \cdots\cdots\cdots\cdots\cdots\cdots\cdots\cdots\cdots\cdots \\ Y_{n1}V_1 + Y_{n2}V_2 + \cdots + Y_{nn}V_n - I_n = 0 \end{array}\right\} \tag{12-71}$$

The swing-bus equation has again been omitted since its voltage V_1 is already completely specified.

To commence the iteration process, a first assumption is made for the voltages $(V_2^{(0)}, V_3^{(0)}, \ldots, V_n^{(0)})$. Upon substituting these initial values in Eq. 12-71, the result is a current residual R_2, R_3, etc., which indicates errors caused by the assumed voltages. The residuals approach zero only as our successive voltage approximations approach their true values. To demonstrate, assume a four bus system, similar to Fig. 12-5. The first equations are

$$\left.\begin{array}{l} Y_{21}V_1 + Y_{22}V_2^{(0)} + Y_{23}V_3^{(0)} + Y_{24}V_4^{(0)} - (P_2 - jQ_2)/V_2^{(0)*} = R_2 \\ Y_{31}V_1 + Y_{32}V_2^{(0)} + Y_{33}V_3^{(0)} + Y_{34}V_4^{(0)} - (P_3 - jQ_3)/V_3^{(0)*} = R_3 \\ Y_{41}V_1 + Y_{42}V_2^{(0)} + Y_{43}V_3^{(0)} + Y_{44}\textcircled{$V_4^{(0)}$} - (P_4 - jQ_4)/V_4^{(0)*} = \textcircled{R_4} \end{array}\right\} \tag{12-72}$$

The relaxation procedure is straightforward. First we compare the current residuals R_2, R_3, and R_4 to determine which is largest. Suppose R_4 is largest. Attention is focused upon the circled values of R_4 and $V_4^{(0)}$ of Eq. 12-72. The object would be to make a correction in our voltage approximation which will force R_4 toward zero and, hopefully, cause a reduction in the R_2 and R_3 residuals as well. The correction to be made will be on the voltage of node 4 for which the highest residual was found.

One might reason that this large current error or residual is more likely to be caused by an erroneous voltage at node 4 then by voltage further away from the point of error. In an attempt to force R_4 to zero, we then proceed to correct the term $Y_{44} V_4^{(0)}$ to $Y_{44}(V_4^{(0)} + \Delta V_4^{(0)})$, where

$$Y_{44} \Delta V_4^{(0)} = -R_4$$

or

$$\Delta V_4^{(0)} = -R_4 / Y_{44} \tag{12-73}$$

The corrected voltage is then

$$V_4^{(1)} = V_4^{(0)} + \Delta V_4^{(0)} \tag{12-74}$$

The correction on V_4 may be inserted into bus 2 and bus 3 of Eq. 12-72 and new residuals can be found at these buses as

$$\left. \begin{aligned} R_2\,(\text{new}) &= R_2\,(\text{old}) + Y_{24} \Delta V_4 \\ R_3\,(\text{new}) &= R_3\,(\text{old}) + Y_{34} \Delta V_4 \end{aligned} \right\} \tag{12-75}$$

Actually the residual R_4 (new) is not zero if we correct the V_4^* of I_4, where

$$I_4^{(1)} = (P_4 - jQ_4)/V_4^{(1)*}$$

$$R_4\,(\text{new}) = I_4^{(1)} - I_4^{(0)} \tag{12-76}$$

Again we compare the new values of R_2, R_3, and R_4 to determine the largest residual, whereupon the voltage correction process is repeated. It is possible that convergence could be speeded up by the application of an acceleration factor to the ΔV correction.

12-13. Load-Flow and the Newton-Raphson Method

Given the three types of buses in Fig. 12-11. Swing bus 1 specifies V_1 and δ_1; load bus 2 specifies P_2 and Q_2; generator bus 3 specifies P_3 and $|V_3|$. The bus admittance equivalent is used with the neutral bus 0 as reference. As was pointed out in Sec. 12-9, load equations may take several forms. This section will write the system equations in terms of the real and reactive power feeding the buses. Rectangular coordinates will be used for the voltages. It should be understood that while additional buses add to the size of the problem, this three-bus example should be sufficient to demonstrate the method and related concepts. Again, the swing bus equation is omitted as the voltage is completely specified. Equations are

$$\left. \begin{aligned} P_2 - jQ_2 &= V_2^* \,[Y_{21} V_1 + Y_{22} V_2 + Y_{23} V_3] \\ P_3 - jQ_3 &= V_2^* \,[Y_{31} V_1 + Y_{32} V_2 + Y_{33} V_3] \end{aligned} \right\} \tag{12-77}$$

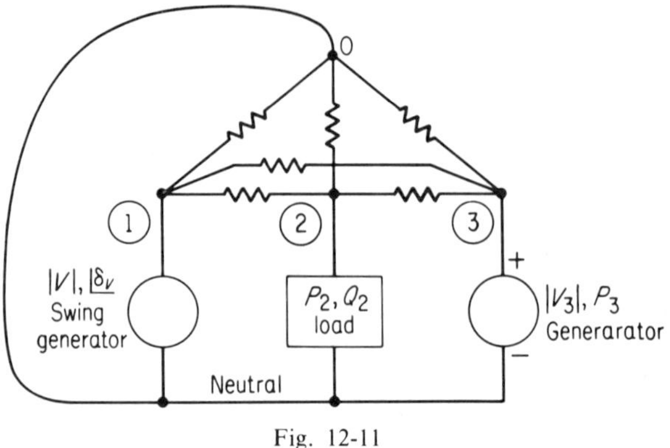

Fig. 12-11

A general form for bus k in an n bus system is

$$P_k - jQ_k = V_k^* \sum_{i=1}^{n} Y_{ik} V_i \tag{12-78}$$

Equation 12-78 expressed in rectangular form is

$$P_k - jQ_k = (e_k - jh_k) \sum_{i=1}^{n} (G_{ik} - jB_{ik})(e_i + jh_i) \tag{12-79}$$

Equation 12-79 can be broken up into real and imaginary parts:

$$\begin{aligned} P_k - jQ_k &= (e_k - jh_k) \sum_{i=1}^{n} [(G_{ik}e_i + B_{ik}h_i) - j(B_{ik}e_i - G_{ik}h_i)] \\ &= \sum_{i=1}^{n} [e_k(G_{ik}e_i + B_{ik}h_i) + h_k(G_{ik}h_i - B_{ik}e_i)] \\ &\qquad - j \sum_{i=1}^{n} [h_k(G_{ik}e_i + B_{ik}h_i) + e_k(B_{ik}e_i - G_{ik}h_i)] \end{aligned}$$

or

$$P_k = \sum_{i=1}^{n} [e_k(G_{ik}e_i + B_{ik}h_i) + h_k(G_{ik}h_i - B_{ik}e_i)] \tag{12-80}$$

$$Q_k = \sum_{i=1}^{n} [h_k(G_{ik}e_i + B_{ik}h_i) + e_k(B_{ik}e_i - G_{ik}h_i)] \tag{12-81}$$

where P_k and Q_k are functions of e_i, e_k, h_i, and h_k. For load buses, P_k and Q_k are known. However, after having substituted for assumed values of e and h, we obtain values of $P_k^{(0)}$ and $Q_k^{(0)}$ which differ from the known values by ΔP_k and ΔQ_k:

$$\left.\begin{aligned} \Delta P_k^{(0)} &= P_k - P_k^{(0)} \\ \Delta Q_k^{(0)} &= Q_k - Q_k^{(0)} \end{aligned}\right\} \tag{12-82}$$

So far our discussion applies to the load bus, where P_k and Q_k are given. For the generator bus m, where $|V_m|$ and P_m are given the Q equation of 12-81 is replaced by a $|V_m|$ equation, where

$$|V_m|^2 = e_m^2 + h_m^2 \tag{12-83}$$

and

$$\Delta |V_m^{(0)}|^2 = |V_m|^2 - |V_m^{(0)}|^2 \tag{12-84}$$

Now we will apply Eq. 12-52 to the simplified network of Fig. 12-11. Refer to the column matrix on the left-hand side of the equal sign of Eq. 12-52. The $(y - f^{(0)})$ elements of this column correspond to $P - P^{(0)}$, $Q - Q_0$ and $|V|^2 - |V^{(0)}|^2$ of our network application. From Eq. 12-82 and 12-84 we can also say that the $(y - f^{(0)})$ elements correspond to ΔP, ΔQ, and $\Delta |V|^2$. Keep in mind that the unknowns of our matrix equation are the Δe and Δh corrections, corresponding to the Δx elements of Eq. 12-52. Our matrix equation for Fig. 12-11 becomes

$$\begin{bmatrix} \Delta P_2^{(0)} \\ \Delta P_3^{(0)} \\ \Delta Q_2^{(0)} \\ \Delta |V_3^{(0)}|^2 \end{bmatrix} = \begin{bmatrix} \left.\frac{\partial P_2}{\partial e_2}\right|_0 & \left.\frac{\partial P_2}{\partial e_3}\right|_0 & \left.\frac{\partial P_2}{\partial h_2}\right|_0 & \left.\frac{\partial P_2}{\partial h_3}\right|_0 \\ \left.\frac{\partial P_3}{\partial e_2}\right|_0 & \left.\frac{\partial P_3}{\partial e_3}\right|_0 & \left.\frac{\partial P_3}{\partial h_2}\right|_0 & \left.\frac{\partial P_3}{\partial h_3}\right|_0 \\ \left.\frac{\partial Q_2}{\partial e_2}\right|_0 & \left.\frac{\partial Q_2}{\partial e_3}\right|_0 & \left.\frac{\partial Q_2}{\partial h_2}\right|_0 & \left.\frac{\partial Q_2}{\partial h_3}\right|_0 \\ \left.\frac{\partial |V_3|^2}{\partial e_2}\right|_0 & \left.\frac{\partial |V_3|^2}{\partial e_3}\right|_0 & \left.\frac{\partial |V_3|^2}{\partial h_2}\right|_0 & \left.\frac{\partial |V_3|^2}{\partial h_3}\right|_0 \end{bmatrix} \begin{bmatrix} \Delta e_2^{(0)} \\ \Delta e_3^{(0)} \\ \Delta h_2^{(0)} \\ \Delta h_3^{(0)} \end{bmatrix} \tag{12-85}$$

Notice in this rectangular form that two equations are required for each bus (except the swing bus and reference bus) in order to accommodate both real and imaginary terms. The partial coefficients are written in terms of P_2, P_3, Q_2, and $|V_3|^2$, all of which are functions of e_2, e_3, h_2, and h_3. These coefficients can be calculated numerically by substituting initial assumed values into the partial derivitive equations. The unknown column array of Δe and Δh terms may be found for the first iteration by any one of the matrix methods previously outlined, involving inversion reduction, etc.

The approximations for e_2, e_3, h_2, and h_3 are obtained for the next iteration by

$$e_2^{(1)} = e_2^{(0)} + \Delta e_2^{(0)}; \qquad h_2^{(1)} = h_2^{(0)} + \Delta h_2^{(0)}, \text{etc.} \tag{12-86}$$

Iterations are continued until the ΔP, ΔQ, and $\Delta |V|^2$ increments are small enough. Refer back to Example 12-4 for a numerical solution employing the Newton-Raphson method.

12-14. Summary

Section 12-9 pointed up three ways in which the load flow equations might be written.
They are:

I. Bus current equations, using the bus admittance concept.
II. Bus voltage equations, using the bus impedance concept.
III. Bus power equations.

The first two approaches yield sets of linear equations, while the last approach results in a set of nonlinear equations.

Sections 12-2 through 12-5 are devoted to certain direct numerical methods which apply to sets of linear equations. They include pivotal condensation, Gauss reduction, Crout reduction and performing elementary row operations upon an augmented matrix.

Sections 12-6, 12-7, 12-10, 12-11 and 12-12 cover several indirect, iterative methods to solve n linear equations in n unknowns. They are the Gauss iterative, the Gauss-Seidel and the relaxation methods.

The Newton-Raphson method of Sec. 12-8 is useful in the solution of nonlinear equations. Application of this method to load studies is outlined in Sec. 12-13.

The concept of the short-circuit impedance matrix of Chapter 10 and its rake equivalent was also useful in Sec. 12-11, where the bus impedance matrix was emphasized. However, we chose the swing generator bus as reference bus when applying the bus impedance matrix to the load-flow problem. The neutral bus is the reference bus for negative and zero sequence networks in short-circuit studies of Chapter 10 and for the sections of Chapter 12 which utilize the bus-admittance matrix. The common excitation bus (behind generator reactance) was reference for the positive-sequence network of Chapter 10.

While the study of numerical methods does of itself encompass a wide field, certain selected methods have a most appropriate place in this chapter. It is felt that the load-flow problem can (in this manner) be treated so as to require a minimum of reference work and a maximum of continuity.

Problems

12-1. Given the following equations I_1, I_2 and I_3:

$$\begin{aligned} 7I_1 - 5I_2 &= 100 \\ 5I_1 - 14I_2 + 6I_3 &= 25 \\ 6I_2 - 10I_3 &= 50 \end{aligned}$$

Solve for these currents using determinants and Cramer's rule as outlined in Appendix C.

12-2. Repeat Prob. 12-1 using matrices and the conventional method of matrix inversion, also covered in Appendix C. Keep in mind that this method is most cumbersome for large system studies.

12-3. Solve for I_1 of Prob. 12-1 using pivotal condensation as applied to determinants.

12-4. Solve for I_3 of Prob. 12-1 using the Gauss reduction method of Sec. 12-3.

12-5. Solve for all currents of Prob. 12-1 using the Gauss-Jordan reduction variation of Sec. 12-3.

12-6. Solve for all currents of Prob. 12-1 using the Crout reduction method.

12-7. Find the inverse matrix and the unknown currents of Prob. 12-1 using elementary row operations as explained in Sec. 12-5.

12-8. Given the following set of equations:

$$\begin{aligned} 3I_1 - 2I_2 &= 1.5 \\ -2I_1 + 6I_2 &= 1.0 \end{aligned}$$

Use the Gauss iterative technique longhand for solving these two equations. Let the first approximations for both I_1 and I_2 be 1.0. Continue the iterations longhand until the change in both $I_1^{(k)}$ or $I_2^{(k)} \leqq 2$ percent.

12-9. Repeat Prob. 12-8 using the Gauss-Seidel modifications according to Sec. 12-7. Note the more rapid convergence than was attainable in Prob. 12-8.

12-10. Revise the given equations in example Prob. 12-4 by changing the constant term in f_1 from -2 to -48 and the constant term in f_2 from $+16$ to $+8$. Solve using the Newton-Raphson method. As a start, let $X_1^{(0)} = X_2^{(0)} = 3$. Carry through four iterations.

12-11. In Example 12-5, what is the total P and Q delivered from the swing generator? What current flows in the line between nodes 1 and 3?

12-12. In Example 12-5, interchange the given generator information for the two machines and rework the problem.

12-13. In Example 12-5, move the capacitor from node 3 to node 1 and rework the problem, again by the Gauss-Seidel technique utilizing the nodal admittance matrix.

12-14. Work Example 12-5, using the bus impedance matrix method of Sec. 12-11. Use the swing generator as reference.

12-15. Apply the relaxation technique in the solution of the equations of Example 12-1. Assume $I_1^{(0)} = I_2^{(0)} = I_3^{(0)} = 5.0$.

12-16. Work Example 12-5 using the Newton-Raphson method.

chapter 13

ECONOMIC DISPATCH AND UNIT COMMITMENT

13-1. Introduction

The subject of economic load dispatch is too broad to be covered in great detail within the chapter allotted. Rather, a brief treatment of some of the more important aspects of the problem will be given. Many developments have been witnessed in recent years with regard to:

1. More accurate load forecasting—both short-term and long-term.
2. Economic scheduling of generators (unit commitment).
3. Economic loading of the units, once on the line.

The above three subjects are closely related. By short-term forecasting we mean the hour-by-hour prediction of system load. By unit commitment we refer to the determination and use of an optimum combination of generators to provide the hourly load demand. The third factor (economic load dispatch), which has received more attention in the past, refers to the optimizing of load levels for each of the individual units with a given combination of generators running.

Without a reasonable load prediction, it is not possible to be assured of an economic solution to the commitment problem. The scheduling program must look ahead into the future hours to weigh in such factors as fuel costs, maintenance costs, line losses, start-up costs, and shutdown costs. A poor load forecast could obviously result in the running of a poor combination of units from an economic standpoint.

13-2. Operating Constraints

A number of operating restrictions must be taken into account. These would include:

(a) *Capacity restrictions of individual generators.* These constraints are determined in the main by thermal restrictions and boiler capabilities. Also involved is the start-up time of a unit, as well as the rate at which the unit can pick up load after start-up. The amount of power transferred from the generator bus to the system could also be limited by stability considerations.

There is a limit (from the standpoint of I^2R heating) on the amount of real and reactive power which a generator can deliver. These limits are imposed by both the a-c armature winding and the d-c field winding. Refer to Fig. 13-1 which depicts a capability curve for a generator with

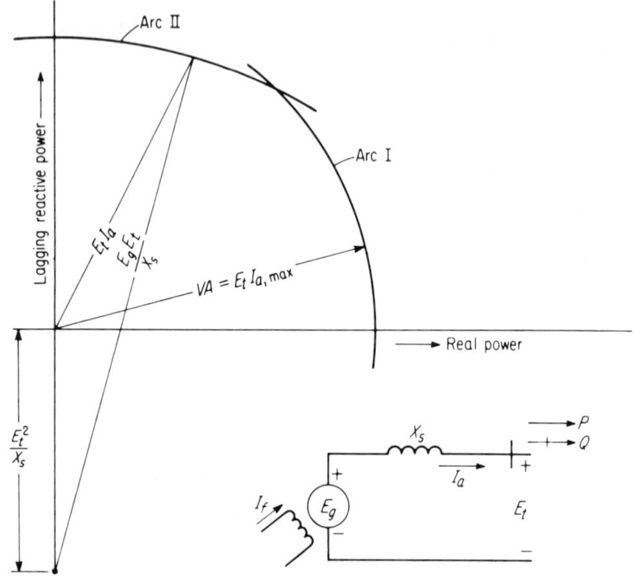

Fig. 13-1. Capability curves of a synchronous generator, demonstrating output restrictions imposed by I^2R heating.

constant excitation voltage (E_g) delivering power to a system at constant terminal voltage (E_t), through the synchronous reactance (X_s). The capability curve is bounded by arcs of two circle diagrams. Arc I represents the I^2R restriction imposed by the maximum allowable current in the a-c winding ($I_{a,\max}$). This arc centers about the origin with a constant volt-ampere radius. Real power (P) is plotted on the horizontal axis with reactive power (Q) on the vertical. The machine may safely operate within (to the left of) this arc insofar as armature heating is concerned. Arc II represents the heating restriction imposed by the allowable field current and its corresponding maximum excitation voltage (E_g). This second arc has its center at $(-|E_t|^2)/X_s$ on the negative Q axis and its radius equal to $(|E_g||E_t|)/X_s$. These facts relating to arc II are proven in App. B, and were specifically demonstrated by Fig. B-10b. Now the two arcs of Fig. 13-1 form the permissible boundaries of $P + jQ$. However, if the machine should go leading by underexcitation of the field winding, the kvar capacity will be reduced considerably, as additional heating occurs in the end iron. This fact is not readily obvious, but the leading kvar permissible may be reduced as much as one-half that of the lagging output capacity. Stability will place additional boundaries upon the machine as well, but this consideration will be treated in Chapter 14.

(b) *Reserve requirements for system security.* This will account for such emergencies as outages on lines, generators, or transformers. Excess spinning reserve is required as a margin to account for such forced outages in addition to possible forecast errors. Some regard must also be given to geographic scheduling of reserve in cases where transmission line outages would create the partial isolation of an area.

13-3. Short-Term Load Forecasting

Load forecasting is often subdivided as "long-term," where seasonal load peaks, etc., are predicted on a long-range basis and "short-term," where hour-by-hour predictions are made for the particular day at hand. In a typical arrangement of the past, the production department might be making the short-term forecast, relying upon the traditional methods of experience, judgment, and intuition. Based upon this forecast, units would be scheduled (or committed) according to some priority list. The operation groups would work hand-in-hand with the production group to guarantee that system security was maintained. A 2 or 3 percent forecast error (with the prediction on the low side) might necessitate the use of inefficient, oil-fired turbines generators or "peaker units," all of which proves to be most costly. On the other hand, a high forecast error would hold excessive generation in hot reserve. Accuracy on the order of 1 percent has been taken to be a desirable objective, although not always attainable. Studies at Detroit Edison Co. have shown that a temperature variance alone of 3°F can vary the total load by 1 percent. This points up the importance of reliable weather information to a good load forecast.

The short-term forecast problem is not a simple one, as many factors enter into the expression for total load. Among these factors are the effects of *light*, *weather*, *daily and seasonal patterns*, and *industrial demand.* Even if it were possible to accurately predict such factors, there are always random factors which can upset the predictions, such as unexpected storms, strikes, etc.

One technique used for obtaining a short-term load forecast is termed *regression analysis*. For example, suppose the total load P could be expressed as the summation of several linear expressions in W_1, W_2, $\ldots$, W_n, where these independent W values are weather variables, related to temperature, humidity, wind velocity, visibility, etc. The load-forecast expression would then be

$$P = C_0 + C_1 W_1 + C_2 W_2 + \cdots + C_n W_n \tag{13-1}$$

Net System Load
De form SL 26 6-66
Date Friday, June 27, 1969
Output (mwhr) 102,595
60 minute max. load (mw) 5,221
3,953 (Same day last year)
Load factor % 81.9
Primary sales (est) (mwhr) 44,853
Temperature max. °F 91 min. °F 74
Average light meter 30
Total system load - megawatts
Interchange
In
Out
Light meter
Eastern standard time (75th meridian)
The Detroit Edison Company
Control Department
Statistical Division
By
Approved

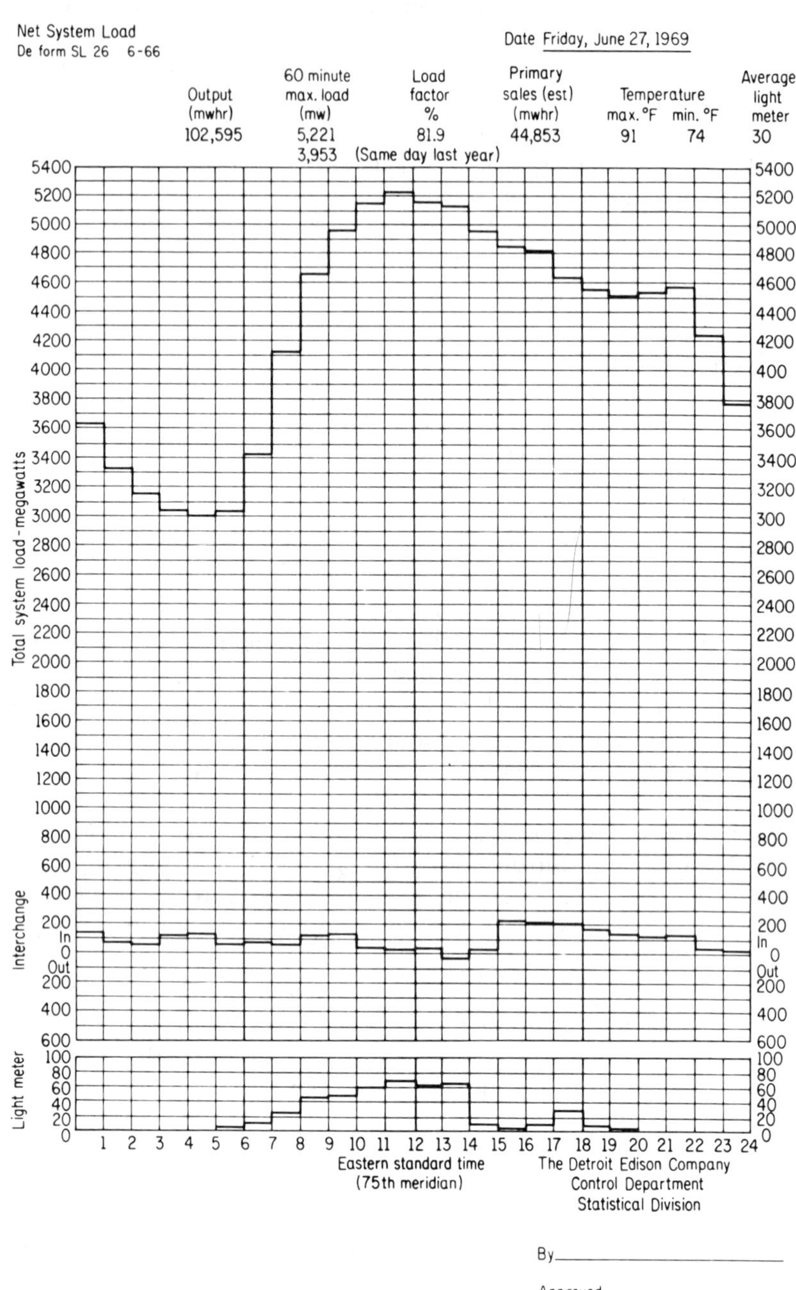

Fig. 13-2

From a number of separate observations, where the P and W values are known, the C coefficients could be determined by a so-called "regression program," and the aid of the least-squares approximation method. Space will not be taken for coverage of the details of this technique. It is, rather, the author's purpose to make the student aware of the general problems of short-term load forecasting and its importance to unit commitment and economic load dispatch.

Refer to Fig. 13-2 for a typical load curve on a weekday in June.

13-4. Economic Load Distribution Between Generators Within a Station

Several factors will be named as important in the production of electrical energy at minimum costs. They are (1) operating efficiencies of the generators, (2) fuel costs, and (3) transmission-line losses. Obviously, the most efficient generator in the system does not guarantee the least cost per megawatt-hour as it may be far removed from the load and/or it may be located in an area where fuel cost is high.

This section assumes that units considered are either within the same station or that they are close enough together that the effect of transmission-line losses may be neglected. Consideration of line losses for the benefit of economy will be treated in the next section. The objective at this stage is to distribute plant load between units in such a way as to minimize the station fuel costs.

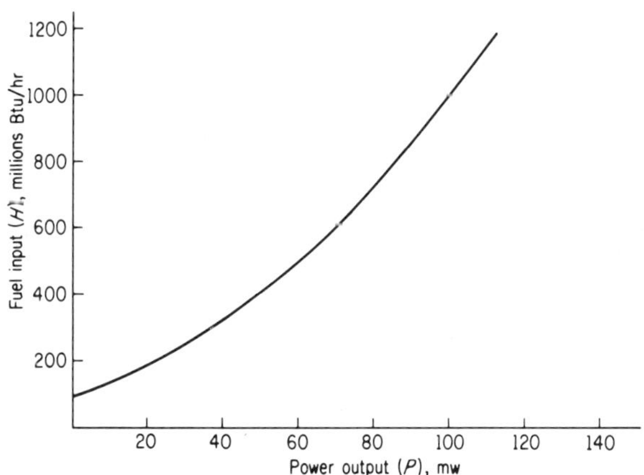

Fig. 13-3. Typical input-output curve for a generating unit.

The input-output curves of a thermal plant are important in describing the plant's efficiency. Such a curve is a plot of fuel input (Btu/hr) vs. electrical power output (kw or mw) as shown in Fig. 13-3.

Machine (or plant) heat rate may be expressed as the Btu input per mw hr output, and this value, of course, varies with the output. A heat rate curve should not be confused with an incremental fuel rate curve (to be treated shortly) even though the coordinate units are the same. Points for a heat rate curve can be obtained from an input-output curve merely by dividing the ordinates of various points on the input-output curves by corresponding abscissa values. However, an incremental fuel-rate curve is a plot of the slope of the input-output curve vs. output. The heat rates for generating units using fossil fuels have improved considerably over the years. For example, in 1940 the average heat rate was more than 16,000 Btu/kw-hr. Twenty years later, values on the order of 10,000 Btu/kw-hr or less were not uncommon. Next, refer to Fig. 13-4,

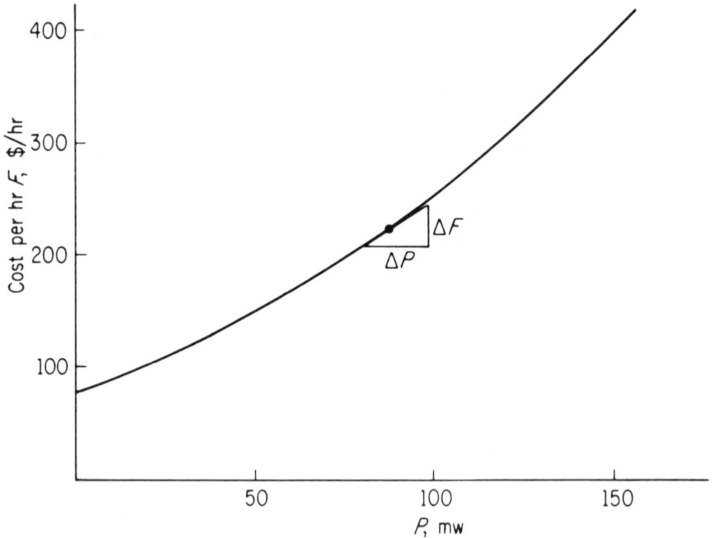

Fig. 13-4. Example of an input-output curve with ordinate converted to $/hr.

which has converted the ordinate scale from Btu/hr (H) to $/hr ($F$). Also of importance is the slope ($\Delta F/\Delta P$) of Fig. 13-4 for various outputs. A plot of this slope (λ) vs. P will yield an incremental cost curve of the type shown in Fig. 13-5. The incremental cost curve is a measure of how costly it will be to produce the next increment of power. Notice that there are two curves in Fig. 13-5. The dotted curve represents the actual incremental fuel cost curve, while the solid curve has included what is often referred to as *incremental maintenance costs*, also called OTF costs (other than fuel costs). These OTF costs take in such costs as incremental labor costs and incremental supplies, etc. It is customary to lump these costs together in the incremental fuel cost curve. One company may find, for

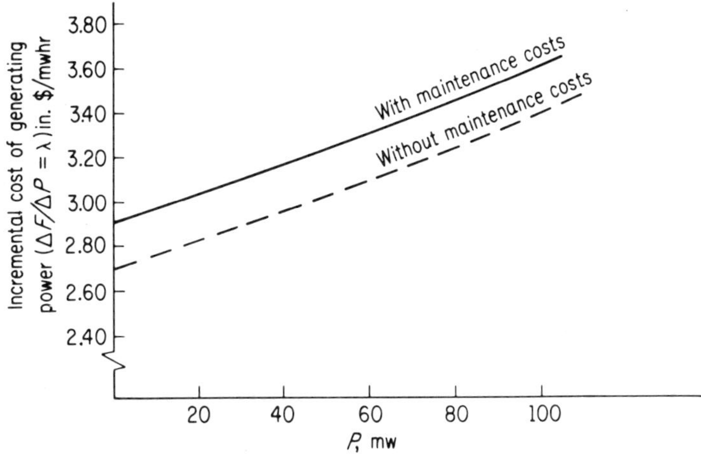

Fig. 13-5. Incremental cost curve for a generator. A plot of the slope of curve of Fig. 13-4.

example, that maintenance costs will cause the total incremental fuel cost curve to ride approximately 10 percent above the curve which considers fuel cost alone. The determination of the true effect of the maintenance cost contribution is often difficult and is somewhat of an approximation. Because the incremental cost curve approaches a linear relationship, it may be represented mathematically by one or more straight-line equations.

Now let us assume that two or more generators within the same plant are to be operated together in the most economical manner. The basic criteria for such an operation is that each unit operate at the same value of incremental fuel cost (λ). Stated mathematically,

$$\lambda = dF_1/dP_1 = dF_2/dP_2 = \cdots = dF_n/dP_n \tag{13-2}$$

This principle is used in Fig. 13-6 for units A and B.

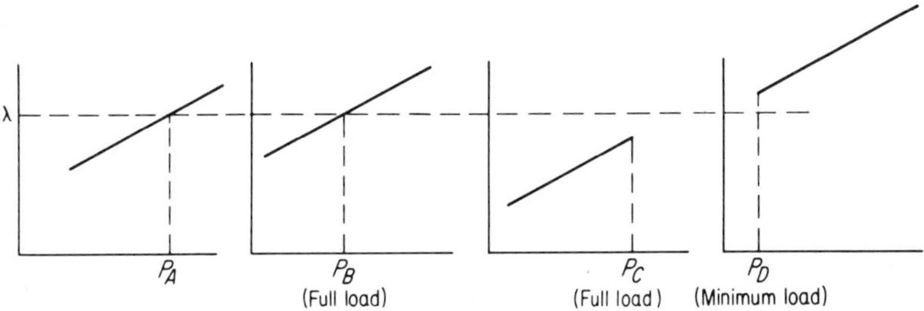

Fig. 13-6. Demonstrating the economic distribution of load between four generator units. $P_{\text{total}} = P_A + P_B + P_C + P_D$.

Of course a unit with an entire range of incremental costs which is below that of the other units (see curve for unit C) should be fully loaded before the others are brought above their minimum load level, at least from the economic point of view. Similarly, unit D, with a higher range of incremental costs, might be used as reserve, being loaded to the minimum level. Notice the value of λ will be such that the total plant output is satisfied, or

$$P_{\text{total}} = P_A + P_B + P_C + P_D$$

If the plant output is to be increased, λ must increase accordingly.

There are several ways of demonstrating the principle of the *equal* λ *criterion* for economic operation. The method chosen here is the graphical proof, although the method of Lagrangian multipliers is also often used. With reference to the graphical approach, we will assume first that only two units are involved and that there are no slope reversals on either of the input-output curves. Refer to Fig. 13-7. The curve for unit A is plotted normally while the axis of the unit B curve is rotated by 180°. The origin OB of unit B is placed at a distance from O_A equal to the total

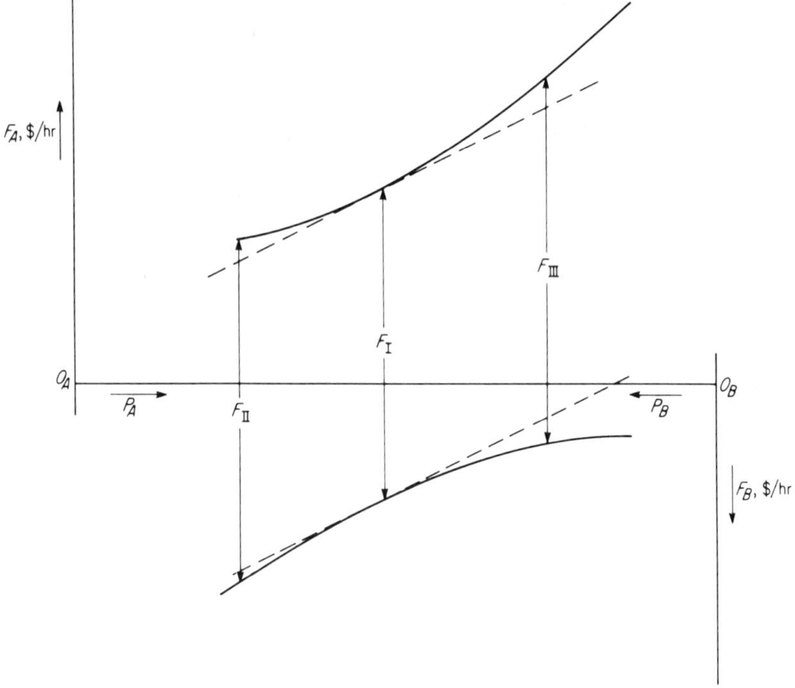

Fig. 13-7. Demonstrating the equal λ criterion for economic distribution of load between two generators.

load, or $P_A + P_B$. A vertical line intersecting the curves will therefore represent a given operating condition. The question to be answered graphically concerns the optimum (most economic) point of operation. First, let the vertical line F_I represent the point of operation where the slopes of both curves are equal. F_I represents the total cost per hour of operating both units. To determine whether there could be any cost advantage in operating on either side of the line of equal slopes, we need but to test the cost lines corresponding to conditions II and III. Dotted lines are drawn tangent to the curves at the line of equal slopes. Any vertical distance between these parallel lines is equal to the cost (F_{I}) and it is obvious that this distance is less than either F_{II} or F_{III}.

The equal-slope criterion has thus been demonstrated for two units and will hold as well for more than two units. In fact, the incremental cost curves of two (or more) units can easily be combined as one by considering equal λ operation as shown in Fig. 13-8. This is done by adding

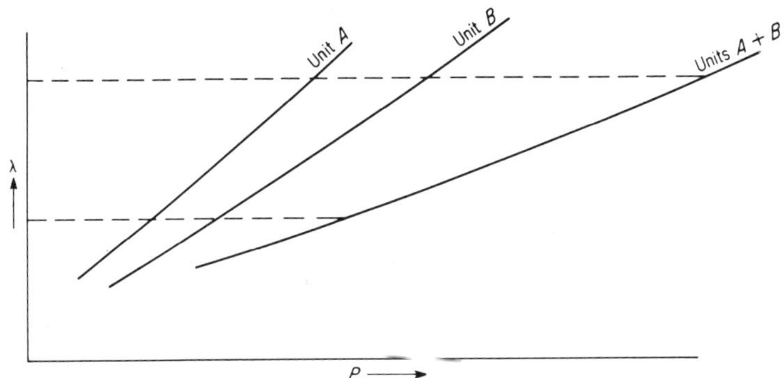

Fig. 13-8. Combining incremental cost curves for equal λ operation.

the abscissa P values of the two incremental cost curves together for equal λ points.

The incremental cost curves are often considered linear, at least over a given range. For this reason, the mathematical expression for such curves will take the form of $\lambda = mP + b$, where m is the slope of the curve and b is the y intercept. The following example should serve to demonstrate the use of the equal λ criterion of generators within a plant.

Example 13-1. The total output of a two-generator station is to be 315 mw. Determine how this load should be shared to give the most economic distribution. The incremental cost curves can be represented mathematically as straight-line equations. The equations are:

$$\lambda_A = dF_A/dP_A = 0.004P_A + 2.2 \qquad (\$/\text{mw-hr})$$

$$\lambda_B = dF_B/dP_B = 0.007P_B + 2.0$$

Setting the incremental cost equal,

$$(1)\ 0.004P_A + 2.2 = 0.007P_B + 2.0$$

$$(2)\ P_A + P_B = 315$$

Solving Eqs. (1) and (2) for P_A & P_B, from Eq. (2),

$$P_A = 315 - P_B$$

Substitute into Eq. (1),

$$0.004(315 - P_B) + 2.2 = 0.007P_B + 2.0$$

$$1.26 - 0.004P_B + 2.2 = 0.007P_B + 2.0$$

$$0.011P_B = 3.46 - 2.0 = 1.46$$

$$P_B = \underline{\underline{133}}\ \text{mw}$$

$$P_A = 315 - 133 = \underline{\underline{182}}\ \text{mw}$$

The plant λ corresponding to this operation is:

$$\lambda = 0.004P_A + 2.2$$

$$= 0.004 \times 182 + 2.2$$

$$= \underline{\underline{2.93}}\ (\$/\text{mw-hr incremental cost})$$

13-5. The Penalty-Factor Technique for Considering Line Losses

The previous section emphasized the equal λ criterion for economic distribution of loads between generators which are either in the same plant or close enough electrically to neglect line losses. A popular extension of this method uses the concept of "penalty factors" for correcting the plant λ's to include the effects of transmission losses. A different correction factor might apply to each plant, and in every case, consideration would be given to the plant's distance (electrically) to the loads it serves. The consideration is somewhat complex as would be expected, since the distribution of currents into the lines (as currents move from generation to loads) must enter into the picture.

A concept is presented here of the penalty factor L_k with the aid of a relatively simple analysis. Recall that the incremental fuel cost λ_k for a generator can be expressed as the slope of the input-output curve for generator k, or

$$\lambda_k = \frac{\Delta F_k}{\Delta P_k} \tag{13-3}$$

It would seem reasonable that, if λ_k is to be corrected for transmission losses by some penalty factor, then the new λ_k' would be related to the rate of change of production cost with respect to the change in power which actually reaches the load, or

$$\lambda_k' = \frac{\Delta F_k}{\Delta P_{\text{load}}} \tag{13-4}$$

also

$$\lambda_k' = L_k \lambda_k \tag{13-5}$$

Substituting Eqs. 13-3 and 13-4 into Eq. 13-5,

$$\frac{\Delta F_k}{\Delta P_{\text{load}}} = L_k \frac{\Delta F_k}{\Delta P_k}$$

Solving for L_k,

$$L_k = \frac{\Delta P_k}{\Delta P_{\text{load}}} \tag{13-6}$$

However, ΔP_{load} can be expressed in terms of ΔP_k and line losses as

$$\Delta P_{\text{load}} = \Delta P_k - \Delta P_{\text{losses}} \tag{13-7}$$

Substituting Eq. 13-7 into Eq. 13-6 yields

$$L_k = \frac{\Delta P_k}{\Delta P_k - \Delta P_{\text{losses}}}$$

or

$$L_k = \frac{1}{1 - \Delta P_{\text{losses}}/\Delta P_k} \tag{13-8}$$

Generally the form of Eq. 13-8 is expressed in terms of a partial derivative ($\delta P_{\text{loss}}/\delta P_k$) indicating we are changing the power of unit k only. This defines the penalty factor L_n in its more common terms as

$$L_k = \frac{1}{1 - \delta P_{\text{losses}}/\delta P_k} \tag{13-9}$$

While units within the same plant are to be loaded to equal values of incremental cost with respect to the power *generated*, an economy of dispatch which considers transmission losses will then require units to be loaded to equal values of incremental costs with respect to power delivered. In other words,

$$\lambda_{\text{system}}' = \lambda_1' = \lambda_2' = \cdots = \lambda_n' \tag{13-10}$$

13-6. Determination of the Loss Equation

One major step in the solution of the dispatch problem is in the determination of the penalty factor L_k for correcting the λ_k of generator k to λ_k'. This will call, in particular, for a solution of the incremental loss term $(\partial P_{loss}/\partial P_k)$ which appears in Eq. 13-9. In order to obtain this partial derivative, on must first be able to express the system losses (P_{loss}) mathematically in terms of the generator outputs ($P_1, P_2, \ldots, P_n$), or

$$P_L = f(P_1, P_2, \ldots, P_n) \tag{13-11}$$

There are various ways of arriving at a loss equation. Two variations are pointed out, the first of which has been used for some time. The second approach will be emphasized here in more detail.

1. The first approach expresses P_{loss} in terms of *loss coefficients* for the network. Here the system transmission losses are written as a function of these so-called "B coefficients" and the generator outputs according to the equation,

$$P_{loss} = \sum_{i=1}^{n} \sum_{k=1}^{n} P_i B_{ik} P_k \tag{13-12}$$

where P_i and P_k are the power outputs of generators (or plants) i and k respectively. In matrix form, Eq. 13-12 can be rewritten as

$$P_{loss} = [P]_{tr}[B][P] \tag{13-13}$$

where for n sources,

$$[P]_{tr} = [P_1 P_2 \cdots P_n] \tag{13-14}$$

and

$$[B] = \begin{bmatrix} B_{11} & B_{12} & \cdots & B_{1n} \\ B_{21} & B_{22} & \cdots & B_{2n} \\ \vdots & & & \vdots \\ B_{n1} & B_{n2} & \cdots & B_{nn} \end{bmatrix} \tag{13-15}$$

and

$$B_{ik} = B_{ki}$$

As an example of a three-generator system, the form of the loss equation is

$$P_{loss} = B_{11}P_1^2 + B_{22}P_2^2 + B_{33}P_3^2 + 2B_{12}P_1P_2 + 2B_{13}P_1P_3 + 2B_{23}P_2P_3 \tag{13-16}$$

The method for obtain these B coefficients will not be covered here. It should be pointed out that a number of simplifying assumptions are necessary in making use of this method. In some cases the effect of these assumptions may be great enough as to require that Eq. 13-12 be modified.

2. The second approach for obtaining the loss equation utilizes the *bus impedance matrix* for expressing the loss formula. Basic principles of method 2 will be given. Again (as in the short circuit and load flow applications of Chapters 10 and 12) a special emphasis is placed upon the physical concept and equivalent circuit of the matrix model. As before, the network equivalent which represents the Z-bus matrix will be referred to as the "rake" equivalent.

Refer to Fig. 13-9 which represents a small system of three generators serving one load center. There is good reason for choosing the load bus as reference for this application. In this way the system losses can be expressed in terms of generator currents (or powers) while excluding load powers as variables. Equation 13-9 requires that we take the derivative of P_{loss} with respect to P_k, holding all other generator powers constant. While P_k is changing by some increment, this change is primarily absorbed by the load (plus a slight loss change). Another observation is made with regard to Fig. 13-9, that being the presence of only one load bus. While this is not the normal situation, it is always possible, with proper matrix transformations, to eliminate all load buses in favor of one composite load bus. More will be said of this in Sec. 13-7. The self impedance elements on the rake equivalent of Fig. 13-9 are identical with the diagonal elements of the bus impedance matrix, while the mutual or transfer impedances of the equivalent are the off-diagonal elements of the matrix. For the system shown,

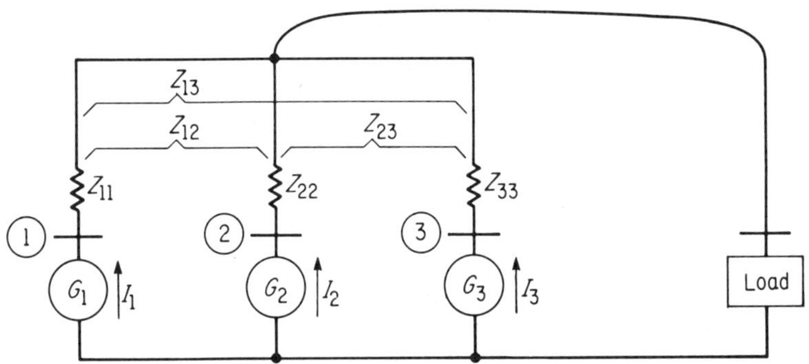

Fig. 13-9. Rake equivalent of a bus impedance matrix for use in obtaining the loss equation.

$$\left[Z_{\text{bus}}\right] = \begin{bmatrix} Z_{11} & Z_{12} & Z_{13} \\ Z_{21} & Z_{22} & Z_{23} \\ Z_{31} & Z_{32} & Z_{33} \end{bmatrix} \tag{13-17}$$

For details in obtaining the Z-bus matrix by computer, refer to Chapter 10. Obtaining Z_{bus} with the network analyzer is treated in Chapter 4. For the system of Fig. 13-9, the real and reactive loss in branch 3, for example, of the rake equivalent is equal to the product of the conjugate of current in branch 3 times the branch voltage, or

$$(P_3 + jQ_3)_{\text{loss}} = I_3^* \cdot V_{03}$$

The total loss in the network is the sum total of branch losses, or

$$(P + jQ)_{\text{loss}} = I_1^* \cdot V_{01} + I_2^* \cdot V_{02} + I_3^* \cdot V_{03} \tag{13-18}$$

In matrix form,

$$(P + jQ)_{\text{loss}} = [I_1^* I_2^* I_3^*] \begin{bmatrix} V_{01} \\ V_{02} \\ V_{03} \end{bmatrix} \tag{13-19}$$

$$(P + jQ)_{\text{loss}} = [I^*]_{\text{transpose}} \cdot [V_{\text{branch}}] \tag{13-20}$$

The voltage elements of V_{branch} can be written as

$$\left.\begin{aligned} V_{01} &= Z_{11}I_1 + Z_{12}I_2 + Z_{13}I_3 \\ V_{02} &= Z_{21}I_1 + Z_{22}I_2 + Z_{23}I_3 \\ V_{03} &= Z_{31}I_1 + Z_{32}I_2 + Z_{33}I_3 \end{aligned}\right\} \tag{13-21}$$

In matrix form,

$$[V_{\text{branch}}] = [Z_{\text{bus}}][I] \tag{13-22}$$

Substituting Eq. 13-22 into Eq. 13-20 yields

$$(P + jQ)_{\text{loss}} = [I^*]_t [Z_{\text{bus}}][I] \tag{13-23}$$

The loss Eq. 13-23 can also be expressed with the use of index notation. For a more general n bus (plus reference bus) system, the system loss equation corresponding to Eq. 13-19 becomes

$$(P + jQ)_{\text{loss}} = \sum_{k=1}^{n} I_k^* V_{ok} \tag{13-24}$$

Equation 13-21 can also be expressed as

$$V_{ok} = \sum_{i=1}^{n} Z_{ki} I_i \tag{13-25}$$

Substituting Eq. 13-25 into Eq. 13-24 yields

$$(P + jQ)_{\text{loss}} = \sum_{k=1}^{n} \sum_{i=1}^{n} I_k^* Z_{ki} I_i \tag{13-26}$$

While either Eq. 13-23 or Eq. 13-26 will serve as the basic loss equation for the Z-bus method, this is not to say that they are in the most usable form. Refer to Sec. 12-9 regarding the information to be specified at any particular bus. As in the case of load-flow studies, there is some choice as to the two quantities to be stipulated at a given bus. Recall that it is often practical to stipulate P and Q for a load bus, $|V|$ and P for a generator and $|V|$ and θ_v for the swing generator. This means that the bus currents appearing in Eq. 13-26 might be better expressed in terms of P, Q, etc. First, in order to separate P_{loss} and Q_{loss} of Eq. 13-26, the equation will be broken up into its real and imaginary components:

$$\sum_{k=1}^{n} \sum_{i=1}^{n} I_k^* Z_{ki} I_i = \sum_{k=1}^{n} \sum_{i=1}^{n} (I_{kRE} - jI_{kIM})(R_{ki} + jX_{ki})(I_{iRE} + jI_{iIM}) \quad (13\text{-}27)$$

The real part of Eq. 13-27 for computing P_{loss} is

$$P_{\text{loss}} = \sum_{k=1}^{n} \sum_{i=1}^{n} (I_{kRE} R_{ki} I_{iRE} - I_{kRE} X_{ki} I_{iIM} + I_{kIM} X_{ki} I_{iRE} + I_{kIM} R_{ki} I_{iIM}) \quad (13\text{-}28)$$

The second and third terms of Eq. 13-28 cancel out as they have like terms when expanded over the total summation. Then

$$P_{\text{loss}} = \sum_{k=1}^{n} \sum_{i=1}^{n} I_{kRE} R_{ki} I_{iRE} + I_{kIM} R_{ki} I_{iIM} \quad (13\text{-}29)$$

but

$$I_k = \frac{P_k - jQ_k}{V_k^*} = \frac{P_k - jQ_k}{|V_k|} (\cos\theta_k + j\sin\theta_k)$$

where

θ_k = angle on bus voltage V_k

$$I_k = \underbrace{\frac{P_k \cos\theta_k + Q_k \sin\theta_k}{|V_k|}}_{I_{kRE}} + \underbrace{\frac{j(P_k \sin\theta_k - Q_k \cos\theta_k)}{|V_k|}}_{I_{kIM}} \quad (13\text{-}30)$$

Equation 13-30 also holds for bus current I_i, where the subscript i replaces k throughout the equation. Substituting the real and imaginary components of buses i and k into Eq. 13-29,

$$P_{\text{loss}} = \sum_{k=1}^{n} \sum_{i=1}^{n} R_{ki} \left[\frac{(P_k \cos\theta_k + Q_k \sin\theta_k)(P_i \cos\theta_i + Q_i \sin\theta_i)}{|V_k|\,|V_i|} + \frac{(P_k \sin\theta_k - Q_k \cos\theta_k)(P_i \sin\theta_i - Q_i \cos\theta_i)}{|V_k|\,|V_i|} \right] \quad (13\text{-}31)$$

$$P_{\text{loss}} = \sum_{k=1}^{n} \sum_{i=1}^{n} \frac{R_{ki}}{|V_k|\,|V_i|} [P_k P_i(\cos\theta_k \cos\theta_i + \sin\theta_k \sin\theta_i) + P_k Q_i(\sin\theta_i \cos\theta_k - \sin\theta_k \cos\theta_i) + Q_k P_i(\sin\theta_k \cos\theta_i - \cos\theta_k \sin\theta_i) + Q_k Q_i(\sin\theta_k \sin\theta_i + \cos\theta_k \cos\theta_i)] \quad (13\text{-}32)$$

Applying trigonometric identities to Eq. 13-32 yields

$$P_{\text{loss}} = \sum_{k=1}^{n} \sum_{i=1}^{n} \frac{R_{ki}}{|V_k|\,|V_i|} [P_k P_i \cos(\theta_i - \theta_k) + P_k Q_i \sin(\theta_i - \theta_k) - Q_k P_i \sin(\theta_i - \theta_k) + Q_k Q_i \cos(\theta_i - \theta_k)] \quad (13\text{-}33)$$

Collecting terms;

$$P_{\text{loss}} = \sum_{k=1}^{n} \sum_{i=1}^{n} \left[\frac{R_{ki}\cos(\theta_i - \theta_k)}{|V_k|\,|V_i|}(P_k P_i + Q_k Q_i) + \frac{R_{ki}\sin(\theta_i - \theta_k)}{|V_k|\,|V_i|}(P_k Q_i - Q_k P_i) \right] \quad (13\text{-}34)$$

As an additional observation, note that the last of the two products of Eq. 13-34 is, in general, small with respect to the first product. This is especially true where the voltage angles θ_i and θ_k are nearly equal, in which case the value of $\sin(\theta_i - \theta_k)$ approaches zero. Also the factor $(P_k Q_i - P_i Q_k)$ will generally be smaller than the sum of $P_k P_i + Q_k Q_i$. As an approximation, it may then be appropriate in some problems to simplify Eq. 13-34 as

$$P_{\text{loss}} \doteq \sum_{k=1}^{n} \sum_{i=1}^{n} \frac{R_{ki}\cos(\theta_i - \theta_k)}{|V_k|\,|V_i|}(P_k P_i + Q_k Q_i) \quad (13\text{-}35)$$

where $(\theta_i - \theta_k)$ is small.

Equations 13-34 and 13-35 can also be placed in matrix form. Let

$$C_{ki} = \frac{R_{ki}\cos(\theta_i - \theta_k)}{|V_k|\,|V_i|} \quad (13\text{-}36)$$

and

$$D_{ki} = \frac{R_{ki}\sin(\theta_i - \theta_k)}{|V_k|\,|V_i|} \quad (13\text{-}37)$$

Rewriting Eq. 13-34 in terms of C_{ki} and D_{ki},

$$P_{\text{loss}} = \sum_{k=1}^{n} \sum_{i=1}^{n} (P_k C_{ki} P_i + Q_k C_{ki} Q_i + P_k D_{ki} Q_i - Q_k D_{ki} P_i) \qquad (13\text{-}38)$$

Where the approximate form of the loss equation may be justified, the D coefficients go to 0 and

$$P_{\text{loss}} = \sum_{k=1}^{n} \sum_{i=1}^{n} (P_k C_{ki} P_i + Q_k C_{ki} Q_i) \qquad (13\text{-}39)$$

The matrix form for Eq. 13-38 is

$$P_{\text{loss}} = \Big[\underbrace{P_1 P_2 \cdots P_n}_{P_{\text{tr}}} \quad \underbrace{Q_1 Q_2 \cdots Q_n}_{Q_{\text{tr}}}\Big]$$

$$\times \left[\begin{array}{cccc|cccc} \multicolumn{4}{c|}{\overbrace{\qquad\qquad\qquad}^{\text{Ⓒ}}} & \multicolumn{4}{c}{\overbrace{\qquad\qquad\qquad}^{\text{Ⓓ}}} \\ C_{11} & C_{12} & \cdots & C_{1n} & D_{11} & D_{12} & \cdots & D_{1n} \\ C_{21} & C_{22} & \cdots & C_{2n} & D_{21} & D_{22} & \cdots & D_{2n} \\ \vdots & & & & \vdots & & & \\ C_{n1} & C_{n2} & \cdots & C_{nn} & D_{n1} & D_{n2} & \cdots & D_{nn} \\ \hline -D_{11} & -D_{12} & \cdots & -D_{1n} & C_{11} & C_{12} & \cdots & C_{1n} \\ -D_{21} & -D_{22} & \cdots & -D_{2n} & C_{21} & C_{22} & \cdots & C_{2n} \\ \vdots & & & & \vdots & & & \\ -D_{n1} & -D_{n2} & \cdots & -D_{nn} & C_{n1} & C_{n2} & \cdots & C_{nn} \end{array}\right] \left[\begin{array}{c} P_1 \\ P_2 \\ \vdots \\ P_n \\ \hline Q_1 \\ Q_2 \\ \vdots \\ Q_n \end{array}\right] \qquad (13\text{-}40)$$

Matrix Eq. 13-40 is written below in abbreviated form as

$$P_{\text{loss}} = \begin{bmatrix} \overline{P}_{\text{tr}} & \overline{Q}_{\text{tr}} \end{bmatrix} \begin{bmatrix} \overline{C} & \overline{D} \\ -\overline{D} & \overline{C} \end{bmatrix} \begin{bmatrix} \overline{P} \\ \overline{Q} \end{bmatrix} \qquad (13\text{-}41)$$

$\overline{P}$ and $\overline{Q}$ are $n \times 1$ column vectors, while $\overline{C}$ and $\overline{D}$ are $n \times n$ matrices. Again, where the approximate equation is justified, merely set elements of the $\overline{D}$ matrix to 0.

Recall from Eqs. 13-5 and 13-9 that, before one can determine the penalty factor and the corrected incremental cost of a generator, the partial derivative of the P_{loss} equation must be taken with respect to P_k. All generator P and Q variables are considered constant except for P_k. As was pointed out in Chapter 11, a change in power output (and power angle δ) of generator k has a lesser effect upon altering the Q supplied to the circuit. The distribution of vars between generators was in general, shown to be more of a function of relative voltage magnitudes. Taking the derivative of Eq. 13-38 will result in the following:

$$\frac{\partial P_{\text{loss}}}{\partial P_k} = 2\sum_{i=1}^{n} (C_{ki}P_i + D_{ki}Q_i) \qquad (13\text{-}42)$$

Whenever the D coefficients are to be neglected, the equation becomes

$$\frac{\partial P_{\text{loss}}}{\partial P_k} \doteq 2\sum_{i=1}^{n} C_{ki}P_i \qquad (13\text{-}43)$$

13-7. Solution Procedure for the Dispatch Problem

This section assumes that the reference frame used (for the bus impedance matrix and its rake equivalent) is of the form of Fig. 13-9, where all loads have been represented as one composite load, the load bus being reference. If such is not the case, refer to Sec. 13-8 in which the necessary transformations are demonstrated.

A procedure summary for solving the dispatch problem is as follows:

1. Obtain the bus impedance matrix for the network of the form of Fig. 13-9. A computer method is described for this in chapter ten and the network analyzer procedure is given in Chapter 4. Example 13-2 will use a longhand approach to a simple network. If transformations are necessary in arriving at the composite load bus as reference, refer to Sec. 13-8.

2. Find the incremental-loss equation for each unit $\left(\frac{\partial P_L}{\partial P_1}, \frac{\partial P_L}{\partial P_2}, \text{etc.}\right)$ in terms of the generator powers according to either Eq. 13-42

$$\frac{\partial P_L}{\partial P_k} = 2\sum_{i=1}^{n} (C_{ki}P_i + D_{ki}Q_i) \qquad [13\text{-}42]$$

or the more approximate form

$$\frac{\partial P_L}{\partial P_k} \doteq 2\sum_{i=1}^{n} C_{ki}P_i \qquad [13\text{-}43]$$

Recall that the C and D coefficients are functions of self- and transfer resistances (R_{ki}) in the bus impedance matrix. They are also functions of bus voltage magnitudes and phase angles according to Eqs. 13-36 and 13-37.

$$C_{ki} = \frac{R_{ki}\cos(\theta_i - \theta_k)}{|V_k|\,|V_i|}; \qquad D_{ki} = \frac{R_{ki}\sin(\theta_i - \theta_k)}{|V_k|\,|V_i|}$$

Of course, generator voltage phase angles are as yet unknown since the economic dispatch problem is yet unsolved—the final powers to be supplied by the generators are as yet undetermined. Therefore, for a first approximation for the C and D coefficients, one could use either (1) the voltage values of an approximate load flow or (2) assume for a first approximation of C and D that $|V_k| = |V_i| = 1.0$ and $\theta_i = \theta_k$, which sets $C_{ki} = R_{ki}$ and $D_{ki} = 0$.

3. Given the load power received, add to this the best estimate of system losses in order to arrive at the total system demand.

4. Suppose the incremental cost equations are given for the individual plants, possibly as linear functions of the plant outputs, or $\lambda_k = m_k P + b_k$. Substituting this λ_k into the corrected incremental cost equations in λ'_k yields,

$$\lambda'_k = \lambda_k L_k$$

$$\lambda'_k = (m_k P_k + b_k) \frac{1}{1 - \partial P_L / \partial P_k} \tag{13-44}$$

or

$$\lambda'_k = f_k(P_1, P_2, \ldots, P_n)$$

Now assume a numerical value for λ'_k and, realizing that the λ' values must be set equal for each unit, a set of n equations (for n generators) is obtained in $P_1, P_2, \ldots, P_n$, or

$$\left.\begin{aligned} \lambda'_1 &= f_1(P_1, P_2, \ldots, P_n) \\ \lambda'_2 &= f_2(P_1, P_2, \ldots, P_n) \\ &\vdots \\ \lambda'_n &= f_n(P_1, P_2, \ldots, P_n) \end{aligned}\right\} \tag{13-45}$$

The above equations are solved for $P_1, P_2, \ldots, P_n$, then tested against the demand, or

$$\sum_{i=1}^{n} P_i \overset{?}{=} \text{demand} \tag{13-46}$$

If the summation is greater than the demand, reduce the assumed value of λ' and resolve Eq. 13-45.

5. Place the power information of step 4 into a load flow, possibly specifying P and $|V|$ for each generator bus, P and Q for each load bus, and $|V|$ and θ_v for the swing generator. From this load flow,

(a) Calculate θ_k and Q_k for each generator bus. Any of several load flow techniques can be utilized here, as outlined in Chapter 12.

These values of θ_k and $|V_k|$ will be used to revise the C and D coefficients of step 2.

(b) Calculate the power of the swing generator. If it differs from the power calculated in step 4, correct the estimated losses of part 3 by the difference. This new P_{loss} will replace the estimated P_{loss} of step 3 which is again used in determining the total system demand.

6. Repeat steps 4 and 5 until no significant change is noted in the system losses.

Example 13-2 follows for the purpose of following through (longhand) the economic dispatch problem using the bus impedance matrix. It is necessary to gain a familiarity with the longhand procedure even though it is understood that the method lends itself to computer application. The problem is similar to that of Example 13-1, with the exception that the transmission losses are now to be taken into account. The complete procedure will be given, including determination of Z-bus as well as the load flow problem.

Example 13-2. Given the same two generators of Example 13-1, where the incremental cost equations are again,

$$\lambda_A = \frac{dF_A}{dP_A} = 0.004\ P_A + 2.2$$

$$\lambda_B = \frac{dF_B}{dP_B} = 0.007\ P_A + 2.0$$

These generators are no longer at the same plant but are feeding a load through the network of Fig. 13-10. A single-phase representation is given on a per unit basis, where 300 mva is taken as the base mva. The load is given as 300 mw (1.0 pu) and 200 mvar (0.667 pu). Both generator buses are held to a voltage magnitude of 1.0 pu volts.

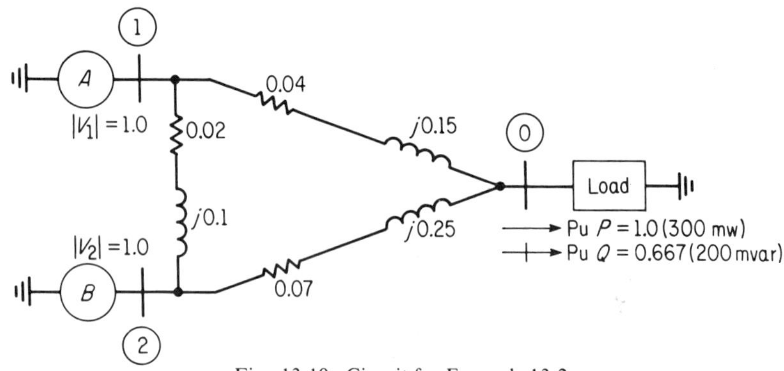

Fig. 13-10. Circuit for Example 13-2.

Find the most economic distribution of loads for the generators, taking both fuel costs and transmission losses into account.

Solution. As a first step the Z-bus matrix will be found and its "rake equivalent" demonstrated, using the load bus 0 as reference. Here, for example, Z_{11} will be the driving-point impedances (also referred to as self-impedance or "through" impedance) from bus 1 to bus 0 with generator B open. Z_{12} is the transfer or mutual impedance between the two legs of the rake equivalent. In calculating these values by hand (as opposed to the techniques of Chapters 4 and 10), first consider 1.0 pu ampere flowing from the reference bus and out bus 1 as per Fig. 13-11. The through im-

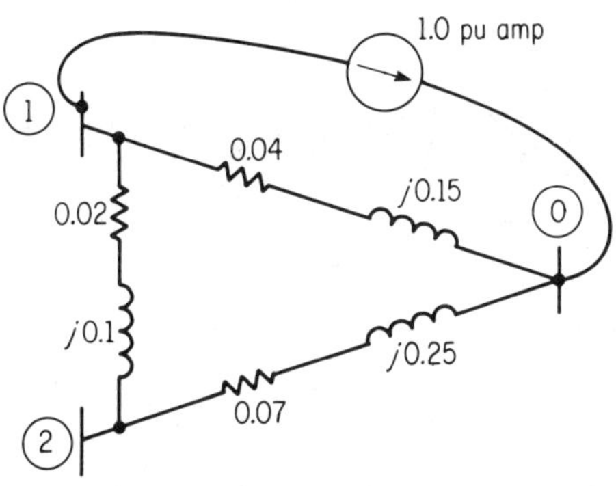

Fig. 13-11

pedance from 0 to 1 becomes equal to the voltage from 0 to 1 since $1.0 \times Z_{11} = V_{01}$. Also the transfer impedance (Z_{12}) equals the voltage induced between buses 0 and 2 with a current of 1.0 flowing out of bus 1, or $Z_{12} = V_{02}$. Then

$$Z_{11} = \frac{(0.04 + j0.15)(0.09 + j0.35)}{(0.13 + j0.5)} = \frac{(0.155 \underline{/75.07^\circ})(0.361 \underline{/75.6^\circ})}{0.516 \underline{/75.45^\circ}}$$

$$= 0.1084 \underline{/75.22^\circ} = \underline{\underline{0.0277 + j0.1047}}$$

$$Z_{12} = \frac{(0.07 + j0.25) \times Z_{11}}{(0.09 + j0.35)} = \frac{0.26 \underline{/74.36^\circ}}{0.361 \underline{/75.6^\circ}} (0.1084 \underline{/75.22^\circ})$$

$$= 0.078 \underline{/74^\circ} = \underline{\underline{0.0215 + j0.075}}$$

The through impedance Z_{22} is

$$Z_{22} = \frac{(0.07 + j0.25)(0.06 + j0.25)}{0.13 + j0.5} = \frac{(0.26 \underline{/74.36^\circ})(0.257 \underline{/76.5^\circ})}{0.516 \underline{/75.44^\circ}}$$

$$= 0.129 \underline{/74.42^\circ} = \underline{\underline{0.0346 + j0.124}}$$

The Z-bus matrix is then,

$$\overline{Z}_{\text{bus}} = \begin{bmatrix} Z_{11} & Z_{12} \\ Z_{21} & Z_{22} \end{bmatrix} = \begin{bmatrix} 0.1084 \angle 75.22^\circ & 0.078 \angle 74^\circ \\ 0.078 \angle 74^\circ & 0.129 \angle 74.42^\circ \end{bmatrix}$$

Figure 13-12 shows the rake equivalent corresponding to these Z-bus values.

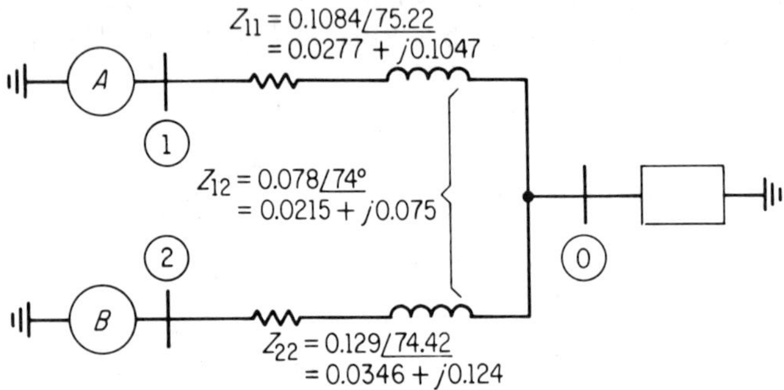

Fig. 13-12. Passive network replaced with rake equivalent.

Proceeding to step 2, the first approximations for the C and D coefficients will be made and used to determine the first incremental-loss equations for each generator. From Eqs. 13-36 and 13-37,

$$C_{ki} = \frac{R_{ki} \cos(\theta_i - \theta_k)}{|V_i|\,|V_k|} \qquad [13\text{-}36]$$

$$D_{ki} = \frac{R_{ki} \sin(\theta_i - \theta_k)}{|V_i|\,|V_k|} \qquad [13\text{-}37]$$

Normally, θ and $|V|$ values from the latest load flow could be used in the equations for C and D. However, for our first approximation we will assume $\theta_1 = \theta_2$ while $|V_1|$ and $|V_2|$ are given as 1.0. Then,

$$C_{ki}^{(1)} = R_{ki} \quad \text{and} \quad D_{ki}^{(1)} = 0$$

$$C_{11}^{(1)} = R_{11} = \underline{\underline{0.0277}}, \quad C_{12}^{(1)} = R_{12} = \underline{\underline{0.0215}}, \quad C_{22}^{(1)} = R_{22} = \underline{\underline{0.0346}}$$

Substituting these numerical coefficients into incremental-loss Eq. 13-43,

$$\begin{aligned} \partial P_L / \partial P_1 &= 2 \sum_{i=1}^{2} C_{1i} P_i = 2(C_{11} P_1 + C_{12} P_2) \\ &= 2(0.0277\, P_1 + 0.0215\, P_2) \\ &= \underline{\underline{0.0554\, P_1 + 0.043\, P_2}} \end{aligned}$$

$$\begin{aligned} \partial P_L/\partial P_2 &= 2\sum_{i=1}^{2} C_{2i}P_i = 2(C_{21}P_1C_{22}P_2) \\ &= 2(0.0215\,P_1 + 0.0346\,P_2) \\ &= \underline{\underline{0.043\,P_1 + 0.0692\,P_2}} \end{aligned}$$

On to step 3, we estimate the line losses and add them to the load requirement to arrive at a first approximation of total demand. With an estimate of 5 percent losses,

$$\begin{aligned} \text{Demand} &= P_{\text{received}} + P_L = 1.0 + 0.05 \\ &= \underline{\underline{1.05 \text{ pu mw}}} \end{aligned}$$

For step 4 we will assume a first value of system λ' of \$3.0/mw-hr. Consistent with the equal λ criteria, we set both λ_1' and λ_2' at this same value and solve for the most economic P_1 and P_2 values (first try). Check $(P_1 + P_2)$ against the total demand of 1.05. Our λ_1' equation from Eq. 13-44 is

$$\lambda_1' = \lambda_1 \frac{1}{1 - \partial P_L/\partial P_1} \tag{13-44}$$

or

$$\lambda_1'(1 - \partial P_L/\partial P_1) = \lambda_1 \tag{13-47}$$

where the incremental fuel cost for plant 1 is given as

$$\lambda_1 = 0.004 \times 300(\text{pu } P_1) + 2.2$$

The factor of 300 on P_1 is for the purpose of changing P_1 of the given equation from mw to pu mw. Substituting, Eq. 13-47 becomes

$$3.0[1 - (0.0554\,P_1 + 0.043\,P_2)] = 1.2\,P_1 + 2.2$$

$$3.0 - 0.166\,P_1 - 0.129\,P_2 = 1.2\,P_1 + 2.2$$

$$(1)\quad 1.37\,P_1 + 0.129\,P_2 = 0.8$$

Likewise,

$$\lambda_2'\left(1 - \frac{\partial P_L}{\partial P_2}\right) = \lambda_2$$

or

$$3.0[1 - (0.043P_1 + 0.0692P_2)] = 0.007 \times 300P_2 + 2.0$$

$$3.0 - 0.129P_1 - 0.2076P_2 = 2.1\,P_2 + 2.0$$

$$(2)\quad 0.129P_1 + 2.31\,P_2 = 1.0$$

Solving Eqs. (1) and (2) for P_1 and P_2, from Eq. (2),

$$P_1 = \frac{1.0 - 2.31P_2}{0.129}$$

Substituting in Eq. (1),

$$1.37\left(\frac{1.0 - 2.31P_2}{0.129}\right) + 0.129P_2 = 0.8$$

$$1.37 - 3.17P_2 + 0.0167P_2 = 0.1032$$

$$315P_2 = 1.27$$

$$\text{pu } P_2 = \underline{\underline{0.404}} \quad (121 \text{ mw})$$

Substituting this P_2 back into Eq. (2) yields

$$\text{pu } P_1 = \underline{\underline{0.528}}$$

Now check $P_1 + P_2$ against the first estimate of demand:

$$0.528 + 0.404 = 0.932 \neq 1.05$$

λ^1 was then assumed too small. Letting $\lambda' = 3.1$ and repeating the foregoing procedure yields

(1) $1.37P_1 + 0.132P_2 = 0.9$

(2) $0.133P_1 + 2.315P_2 = 1.1$

Solving the new set of equations in P_1 & P_2 yields

$$P_1 = \underline{\underline{0.601}}\ (180.3 \text{ mw}), \qquad P_2 = \underline{\underline{0.441}}\ (132.3 \text{ mw})$$

Again checking against demand,

$$P_1 + P_2 = 0.601 + 0.441 = 1.042\ (313 \text{ mw})$$

This total of 1.042 will be accepted for the first load flow as our first demand approximation was based purely upon an estimate.

Step 5 requires *load-flow* information for the purpose of correcting the C and D coefficients. Refer to Fig. 13-13, where a more typical load-flow rake equivalent is shown, using one generator bus as a reference. The steps for obtaining the equivalent of Fig. 13-13 from Fig. 13-10 will not be repeated again. However, results of the Z-bus (load flow) are entered on the figure, where

$$Z_{00} = 0.0277 + j0.105 = 0.1088\ \underline{/75.25^\circ}$$

$$Z_{22} = 0.0172 + j0.0801 = 0.082\ \underline{/77.88^\circ}$$

$$Z_{02} = 0.0062 + j0.03 = 0.0307\ \underline{/78.3^\circ}$$

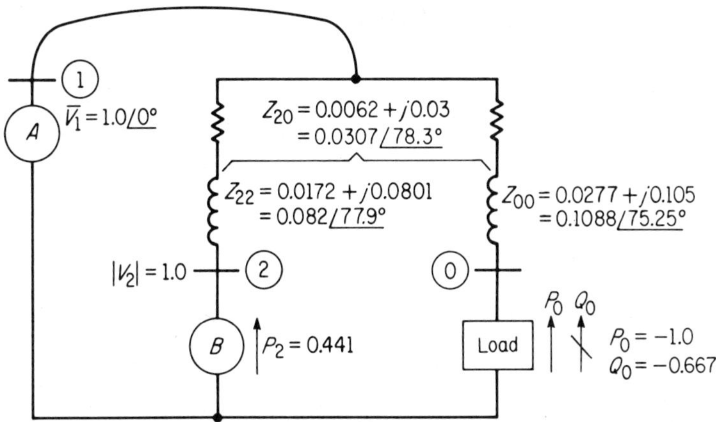

Fig. 13-13. Equivalent circuit for load-flow calculation.

Recall from Chapter 12 that two of the four quantities (P, Q, $|V|$, θ_v) are stipulated for each bus. The following quantities are given for the first load flow:

$$V_{ref} = V_1 = 1.0\underline{/0^\circ}, \qquad P_2 = 0.441,$$
$$|V_2| = 1.0, \qquad P_0 = -1.0, \qquad Q_0 = -0.667$$

Load-flow equations for Fig. 13-13 are:

$$(1)\ V_0 = \frac{Z_{00}(P_0 - jQ_0)}{V_0^*} + \frac{Z_{02}(P_2 - jQ_2)}{V_2^*} + V_1$$

$$(2)\ V_2 = \frac{Z_{20}(P_0 - jQ_0)}{V_0^*} + \frac{Z_{22}(P_2 - jQ_2)}{V_2^*} + V_1$$

Unknowns in the above equations are θ_2, Q_2, $|V_0|$, and θ_0. For a first approximation, let $\theta_2 = 0^\circ$, $V_0 = 1.0\underline{/0^\circ}$, and Q_2 will be approximated from Eq. 12-70 as

$$Q_2^{(0)} = -\operatorname{Im}\frac{V_2^{*(0)}}{Z_{22}}\left[V_2^{(0)} - V_{ref} - \frac{(P_0 - jQ_0)}{V_0^{*(0)}}Z_{20}\right]$$

$$= -\operatorname{Im}\frac{1.0\underline{/0^\circ}}{0.082\underline{/77.9^\circ}}$$

$$\times\left[1.0\underline{/0^\circ} - 1.0\underline{/0^\circ} - \frac{(-1.0 + j0.667)(0.0307\underline{/78.3^\circ})}{1.0\underline{/0^\circ}}\right]$$

$$= -\text{Im } 12.2\,\underline{/-77.9^\circ}\,[(1.2\,\underline{/-33.7^\circ})(0.0307\,\underline{/78.3^\circ})]$$

$$= -\text{Im } 0.450\,\underline{/-33.3^\circ} = -\text{Im}(0.376 - j0.247)$$

$$= \underline{\underline{0.247}} \text{ pu mvar}$$

The first iteration for V_0 in Eq. (1) becomes

$$V_0^{(1)} = \frac{Z_{00}(P_0 - jQ_0)}{V_0^{*(0)}} + \frac{Z_{02}(P_2 - jQ_2^{(0)})}{|V_2|\,\underline{/-\theta_2^{(0)}}} + V_1$$

$$= \frac{0.1088\,\underline{/75.25^\circ}(-1.0 + j0.667)}{1.0\,\underline{/0^\circ}} + \frac{0.0307\,\underline{/78.3^\circ}(0.441 - j0.247)}{1.0\,\underline{/0^\circ}} + 1.0\,\underline{/0^\circ}$$

$$= -0.1306\,\underline{/41.6^\circ} + (0.0307\,\underline{/78.3^\circ})(0.505\,\underline{/-29.2^\circ}) + 1.0\,\underline{/0^\circ}$$

$$= -0.0976 - j0.0866 + 0.01015 + j0.0117 + 1.0$$

$$= 0.9126 - j0.0749 = \underline{\underline{0.913\,/-4.71^\circ}}$$

Before proceeding to the $V_2^{(1)}$ iteration, a second iteration is made on V_0, using the new $V_0^{(1)}$ value. The result is:

$$V_0^{(2)} = 0.906\,\underline{/-4.69^\circ}$$

The first iteration on V_2 is

$$V_2^{(1)} = \frac{Z_{20}(P_0 - jQ_0)}{V_0^{*(2)}} + \frac{Z_{22}(P_2 - jQ_2^{(0)})}{|V_2|\,\underline{/-\theta_2^{(0)}}} + V_1$$

$$= \frac{0.0307\,\underline{/78.3^\circ}(-1.2\,\underline{/-33.7^\circ})}{0.906\,\underline{/+4.69^\circ}} + \frac{(0.082\,\underline{/77.9^\circ})\,0.505\,\underline{/-29.25^\circ}}{1.0\,\underline{/0^\circ}} + 1.0\,\underline{/0^\circ}$$

$$= -0.0406\,\underline{/40.3^\circ} + 0.0414\,\underline{/48.6^\circ} + 1.0\,\underline{/0^\circ}$$

$$= -0.0310 - j0.0263 + 0.0274 + j0.031 + 1.0$$

$$= 0.9964 + j0.0048 = 0.996\,\underline{/0.3^\circ}$$

Since $|V_2|$ was specified as 1.0, use $V_2^{(1)}$ as $1.0\,\underline{/0.3^\circ}$.

Now, using the latest values of V_0 and V_2, we will reiterate for $Q_2^{(1)}$, again using Eq. 12-70. The result is:

$$Q_2^{(1)} = \underline{\underline{+0.306}}$$

Again back to V_0 yields $V_0^{(3)} = \underline{\underline{0.897\,/-4.8^\circ}}$.

For our purposes, the following load-flow values will be considered acceptable for use in the dispatch problem,

$$V_0 = 0.897\,\underline{/-4.8^\circ}, \quad V_2 = 1.0\,\underline{/0.3^\circ}, \quad Q_2 = 0.306$$

Using this same load-flow equivalent, the P_1 and Q_1 of the swing generator are found:

$$I_1 = -(I_2 + I_{\text{load}})$$

or

$$\frac{P_1 - jQ_1}{V_1^*} = \frac{-(P_2 - jQ_2)}{V_2^*} + \frac{-(P_0 - jQ_0)}{V_0^*}$$

$$P_1 - jQ_1 = -V_1^*\left(\frac{P_2 - jQ_2}{V_2^*}\right) - V_1^*\left(\frac{P_0 - jQ_0}{V_0^*}\right)$$

$$= \frac{-1.0\,\underline{/0^\circ}\,(0.441 - j0.306)}{1.0\,\underline{/-0.3^\circ}} - \frac{1.0\,\underline{/0^\circ}\,(-1.2\,\underline{/-33.7^\circ})}{0.897\,\underline{/4.8^\circ}}$$

$$= -0.536\,\underline{/-34.5^\circ} + 1.34\,\underline{/-38.5^\circ}$$

$$= -0.442 + j0.303 + 1.05 - j0.835$$

$$= 0.608 - j0.532$$

$$P_1 = 0.608, \qquad Q_1 = 0.532$$

Now as directed in step 5(b), we compare this value of $P_1 = 0.608$ with that of 0.601 (calculated using an estimated loss of 0.042). The first approximation of loss was then off by the difference of $0.608 - 0.601 = 0.007$ pu mw.

Viewed more directly, losses obtained from the first load flow are

$$P_{\text{loss}} = P_1 + P_2 - P_{\text{load}}$$

$$= 0.608 + 0.441 - 1.0 = \underline{\underline{0.049}} \text{ pu mw}$$

$$\text{The corrected demand} = \underline{\underline{1.049}} \text{ pu mw}$$

One complete sequence of steps has now been completed in our dispatch procedure. Latest values obtained for voltages, demand, etc., will be used as we return to step 2 to repeat the procedure. The C and D coefficients show no significant change in this particular problem. Likewise, the incremental-loss equations are essentially unchanged. The same can be said for the new equations. However, the new demand has increased from 1.042 to 1.049, so we will increase the numerical value of λ' above the last one assumed (was \$3.1/mw-hr). Interpolation can be used to advantage here, using old results below,

	λ'	Calculated Demand	
0.1	3.0	0.932	0.11
	3.1	1.042	
x	$3.1 + x$	1.049	0.007

From the table above, let

$$x = (0.1)\,\frac{(0.007)}{(0.11)} = 0.00635$$

The new λ' will be taken as $3.1 + 0.006 = 3.106$.

Solving for P_1 and P_2 as before,

$$P_1 = 0.607\ (182.1\ \text{mw}), \qquad P_2 = 0.442\ (132.6\ \text{mw})$$

The load has not changed enough in this iteration to justify another long-hand iteration, so the values of 182.1 mw and 132.6 mw will be accepted as the most economic split for generation, corresponding to a total demand of 314.7 mw and losses of 14.7 mw.

At this point, it is well to compare the results with those obtained in Example 13-1, in which losses were not considered. Demand for Example 13-1 was purposely given as 315 mw for comparison purposes. Results are very similar in the two problems, with plant B being penalized only slightly more than plant A because of transmission losses.

13-8. Transforming an *m*-Load System into a System with One Composite Load Bus

The transformation of a bus impedance matrix from the form represented in Fig. 13-14 to that of Fig. 13-15 may be necessary in order to solve the dispatch problem of Sec. 13-7. This transformation is possible

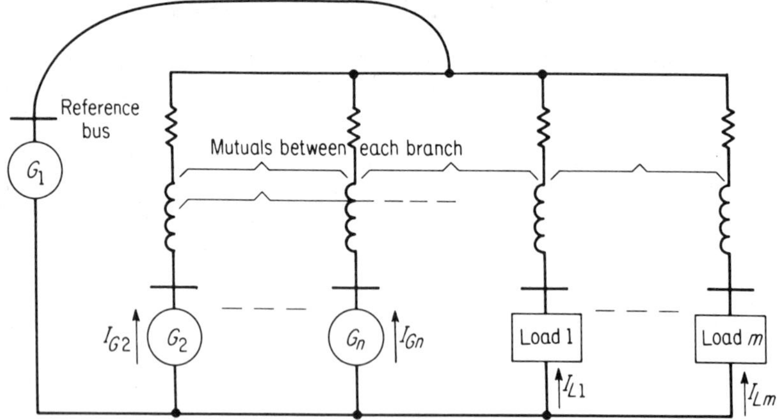

Fig. 13-14. Z-bus (rake) equivalent for system using G_1 bus as reference. (Reference frame 1.)

by applying methods of Gabriel Kron. Such transformation methods were also used in Chapter 6 of this text in which we changed from one reference frame to another by means of a connection matrix. Figure

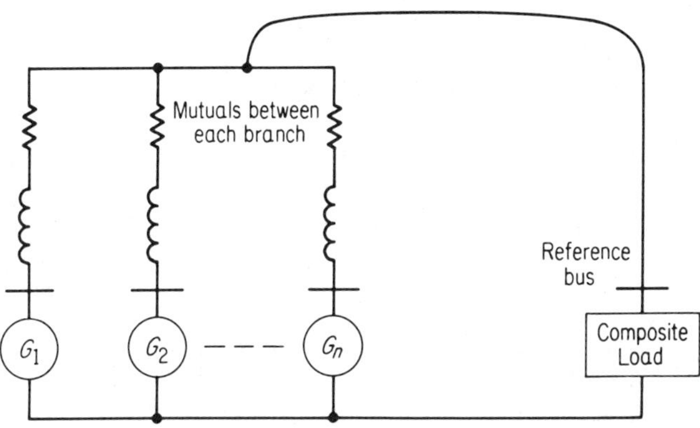

Fig. 13-15. Rake equivalent in the final form as used in Sec. 13-7, with one composite load bus as reference. (Reference frame 3.)

13-14 has been referred to as reference frame 1, while Fig. 13-15 is reference frame 3. An intermediate step (reference frame 2) is necessary in getting from Figs. 13-14 to 13-15.

Three basic equations are necessary in performing such a transformation. They were developed in Chapter 6 for the specific purpose of transforming from branch to loop systems. However, these three equations also apply when any system (old) is to be transformed to a new system, the total network real and reactive power being invarient (not changing with the reference frame). The three matrix equations are:

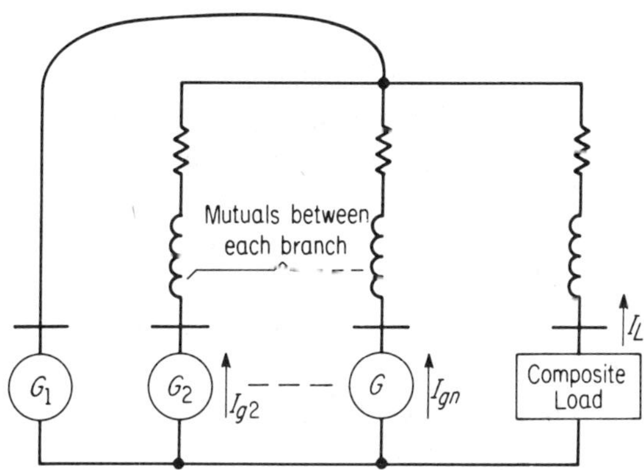

Fig. 13-16. Z-bus (rake) equivalent. (Reference frame 2.)

$$\bar{I}_{\text{old}} = \bar{C}\,\bar{I}_{\text{new}} \tag{13-48}$$

$$\bar{E}_{\text{new}} = \bar{C}_t^*\,\bar{E}_{\text{old}} \tag{13-49}$$

$$\bar{Z}_{\text{new}} = \bar{C}_t^*\,\bar{Z}_{\text{old}}\,\bar{C} \tag{13-50}$$

As a first step, the Z-bus form of Fig. 13-14 (reference frame 1) will be transformed into that of Fig. 13-16 (reference frame 2). This second form still regards the generator bus 1 as a reference bus, but all loads have been combined into one total or composite load. Refer to Fig. 13-16. The connection matrix $C_{1\text{-}2}$ is found for the first transformation by using Eq. 13-48, relating the old currents to the new. The individual load currents are assumed to be always related to the total load current by the constants $K_1, K_2, \ldots, K_m$. Or $I_{L1} = K_1 I_L$, $I_{L2} = K_2 I_L, \ldots$, etc. In matrix form;

$$\begin{bmatrix} I_{g1} \\ I_{g2} \\ \vdots \\ I_{gn} \\ \hline I_{L1} \\ I_{L2} \\ \vdots \\ I_{Lm} \end{bmatrix} = \begin{bmatrix} 1 & 0 & & 0 & 0 \\ 0 & 1 & & 0 & 0 \\ \vdots & & & \vdots & \vdots \\ & 0 & \cdots & 1 & 0 \\ \hline 0 & 0 & \cdots & 0 & K_1 \\ 0 & 0 & & 0 & K_2 \\ \vdots & & & \vdots & \vdots \\ 0 & 0 & \cdots & 0 & K_m \end{bmatrix} \begin{bmatrix} I_{g1} \\ I_{g2} \\ \vdots \\ I_{gn} \\ \hline I_L \end{bmatrix} \qquad (13\text{-}51)$$

In shorthand form,

$$\bar{I}_{\text{old}} = \bar{C}_{1\text{-}2}\,\bar{I}_{\text{new}}$$

Once the $C_{1\text{-}2}$ matrix is established as in Eq. 13-51, this matrix can be applied directly to Eq. 13-50 in order to determine the Z-bus matrix for reference frame 2. It is

$$[Z\text{-bus}_2] = \bar{C}^*_{1-2(t)}\,[Z\text{-bus}_1]\,\bar{C}_{1-2} \qquad (13\text{-}52)$$

For the sake of brevity, these techniques are indicated here without benefit of example. For more detailed coverage of this transformation, refer to Leon K. Kirchmeyer, *Economic Operation of Power Systems* (New York: John Wiley & Sons, Inc., 1958).

Having obtained Z-bus$_2$, it is desirable to move the reference bus from generator 1 to the load bus, resulting in the form of Fig. 13-15. The total load current is related to the generator currents by

$$I_{g1} + I_{g2} + \cdots + I_{gn} = -I_{\text{load}} \qquad (13\text{-}53)$$

Again, using Eq. 13-48, the connection matrix $\bar{C}_{2-3}$ (for transforming from reference frame 2 to reference frame 3 of Fig. 13-15) is determined as:

$$\begin{bmatrix} I_{g1} \\ I_{g2} \\ \vdots \\ I_{gn} \\ \hline -I_L \end{bmatrix} = \underbrace{\begin{bmatrix} 1 & 0 & \cdots & 0 \\ 0 & 1 & & 0 \\ \vdots & & \ddots & \vdots \\ 0 & 0 & & 1 \\ \hline 1 & 1 & \cdots & 1 \end{bmatrix}}_{\bar{C}_{2-3}} \begin{bmatrix} I_{g1} \\ I_{g2} \\ \vdots \\ I_{gn} \end{bmatrix} \tag{13-54}$$

Now using $\bar{C}_{2-3}$ along with Eq. 13-50, the final Z-bus matrix for Fig. 13-16 is:

$$[Z\text{-bus}_3] = \bar{C}^*_{2-3(t)}\,[Z\text{-bus}_2]\,\bar{C}_{2-3} \tag{13-55}$$

It should be mentioned that Z-bus matrices of Eqs. 13-52 and 13-55 will not (in general) be symmetrical, or element Z_{ik} will not (in general) equal Z_{ki}.

13-9. Unit Commitment (Scheduling) of Generators

It may seem at first glance that the subject of scheduling should precede that of economic dispatch since units must first be committed, but once committed, load must be dispatched constantly so as to minimize costs. However, ecnomic scheduling must look at all cost factors, including the fuel costs of the dispatch problem. Precalculations are made of relative fuel costs for the various possible combinations of units. For this reason, the rather involved fuel-cost procedures of economic dispatch were considered first in this chapter. Many items must be weighed in a unit commitment program. These items include:

1. A short-term load forecast.
2. System reserve requirements.
3. System security.
4. Start-up costs for all units.
5. Shutdown costs for all units.
6. Minimum level fuel costs for all units.
7. Incremental fuel costs of units.
8. Maintenance costs.
9. Costs due to transmission-line loss.
10. Cost of purchasing interchange power.

A brief treatment of item 1 was given in Sec. 13-3, while items 2 and 3 were covered in Sec. 13-2. Items 7 and 8 were considered in Sec. 13-4 where fuel costs and maintenance costs were considered without regard to line losses. Where line losses are included, the dispatch problem of Secs. 13-5 through 13-8 included items 7, 8, and 9.

Costs of *interchange power* are agreed upon by neighboring utilities, being largely a function of the incremental cost of producing the last increment of power ($/mw-hr). If one utility is heavily loaded including costly peaker operation, this last incremental cost could come high. In lighter load periods when more efficient units are still available to take on load, the incremental cost would naturally be lower.

Start-up costs include costs of fuel, labor, and maintenance. If a boiler has not been banked, it cools exponentially with time. A banked boiler is one which is left operating during generator shutdown but is disconnected from the turbines. Again, for an *unbanked* boiler, one equation for expressing the start-up cost is

$$\text{Start-up cost} = C_0(1 - e^{-at}) + K \tag{13-56}$$

where C_0 = cost of starting the boiler cold
a = boiler cooling constant
t = hours the unit has been down
K = constant costs of starting the turbine plus labor and maintenance costs

If the boiler is banked, then the start-up cost is given by the equation:

$$\text{Start-up cost} = Bt + K \tag{13-57}$$

Where B is the cost per hour of banking the boiler (supplying fuel, etc). Figure 13-17 demonstrates the shape of the boiler start-up curves for both the banked and unbanked boilers.

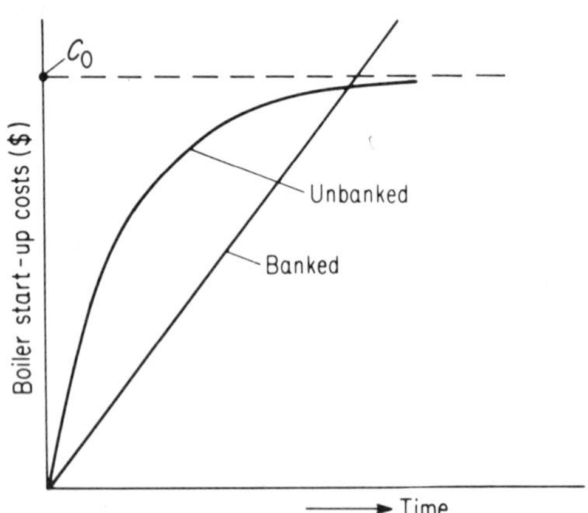

Fig. 13-17

13-10. Methods for Scheduling Units

A number of techniques have been used in attempting to schedule the best possible combination of units. Three of these methods are covered briefly.

A. *Priority list scheduling*. This method essentially ranks the unit in some order of preference. The units could, for example, be ranked in ascending order according to their fuel costs per mw-hr. The order of preference could be modified to correct for area security, system losses, etc. In the actual use of this list, the dispatcher will take the sum of the hourly load demand plus reserve requirements and schedule enough units (in order of the priority) with sufficient capacity to match this sum. For the shutting down of units, the priority list would be used in reverse order. However, before shutdown of a unit, costs of continuing the operation of the unit should be weighed against shutdown and start-up costs. A rule for shutdown of each unit could be related to the number of hours before the unit is again needed. Once units have been committed to service, the equal λ criterion for loading would be automatically applied, assuming the system has an economic dispatch computer in service.

Often the priority list method of scheduling is subject to errors of human judgment. However, a considerable degree of sophistication can be incorporated by including additional cost factors and forecast information to modify the priority list.

B. *Dynamic programming*. This technique is another attempt to optimize the process of scheduling over a given horizon or period of time. The horizon is broken up into time segments or stages, usually of one hour each. Decisions are to be made at the beginning of each hour as to whether to place units into service or withdraw them from service. The load demand and reserve requirements are assumed known over the total horizon, along with other operating constraints. Total costs over the horizon, or period are to be minimized. The procedure is as follows:

(a) List all possible combinations for each future hour which will meet the total demand of the load forecast, system reserve, and all operating constraints. As an example, suppose all possible combinations for a two-hour period are as shown in Fig. 13-18, where the letter symbols (A, B, C, etc.) represent generator units running in the various combinations.

(b) The total operating costs are calculated in going from hour 0 to each of the possible combinations in hour 1. These costs would include the operating costs of hour 0 plus production costs of hour 1, plus any start-up costs incurred in going from hour 0 to hour 1.

(c) Next, for each of the possible combinations of hour 2, deter-

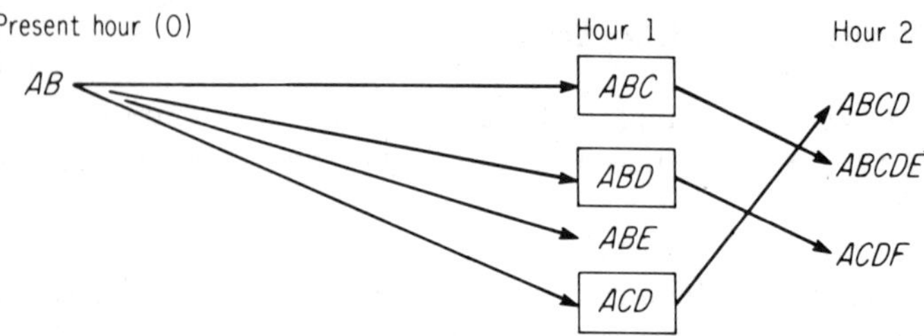

Fig. 13-18. Demonstrating all possible combinations and three least-cost paths through hour 2.

mine the least-cost path from hour 0 through hour 2. For example, in Fig. 13-18, the least-cost path for combination $ABCD$ is by way of ACD in hour 1. The three optimum paths are shown to the end of hour 2.

(d) Proceed to hour 3 and repeat the general procedure of step (c) with regard to obtaining the least-cost path through hour 3. Continue with this procedure for each hour of the horizon. If there are 24 hours considered in the horizon and 7 possible combinations in hour 24, then the 7 best paths are found. While a dollar figure is associated with each path, the lowest figure does not necessarily represent the best path since the problem is theoretically a never-ending one with hours extending toward infinity. The error introduced by choosing the one least-cost path through hour 24 is called "end-effect error." This error is generally held to a minimum by choosing the peak-load hour as the last hour of the horizon.

Nothing has been said concerning line-loss differences in the various possible combinations. While line losses may not be greatly significant in the scheduling of units, they could be taken into account by the assigning of penalty factors to the units.

C. *Integer programming*. This latter method of scheduling units will only be mentioned in brief. The objective is again the same, being that of minimizing costs while meeting load demand plus various constraints. Methods of *linear programming* are used in optimizing the cost expression, where all variables are integer valued (with suitable scaling). The subject of linear programming is beyond the scope of this text.

Problems

13-1. The expression for the input-output curves of a turbine generator, where F is fuel input in Btu/hr and P is output in mw, is

$$F = 2.5 \times 10^4 P^2 + 1.0 \times 10^7 P + 1.0 \times 10^4$$

(a) Plot the input-output curve (Btu/hr vs. mw).
(b) Plot the heat-rate curve (Btu)/mw-hr vs. mw).
(c) Rewrite the expression in F, changing units on F from Btu/hr to $/hr, considering a fuel cost of 28¢ per million Btu.
(d) From the cost equation of part (c), find the incremental cost equation and plot λ in $/mw-hr vs. mw.

13-2. Refer to Example 13-1. Calculate the new λ and economic load split for a total output of 250 mw. Again, the effect of line losses is neglected here.

13-3. Assume the load of 315 mw in Example 13-1 is shared equally between the two machines. What loss in $/hr does this arrangement create as compared with the economic split?

13-4. In Example 13-2, show all required work for arriving at the bus impedance matrix used in the load-flow portion of the economic dispatch problem. Resulting values from this matrix are shown in Fig. 13-13.

13-5. Given information of Example 13-2. On step 3, change the first estimate of 5 percent losses to an estimate of 10 percent losses and rework steps 3, 4, and 5 through one iteration.

13-6. In Example 13-2, change the impedance from bus 2 to bus 0 from $0.07 + j0.25$ to $0.15 + j0.5$ and rework the problem through two complete iterations.

chapter 14

SYSTEM STABILITY

14-1. Introduction

When synchronous machines are electrically tied in parallel, they must operate at the same average frequency. Recall the equation for speed n where

$$n = \frac{120f}{\text{number of poles}} \tag{14-1}$$

The speed n and frequency f are proportional, and although minor speed fluctuations will occur, the same average frequency otherwise must exist throughout the system.

Refer to Fig. 14-1, in which two machines are shown paralleled through a total reactance (X). Any sudden disturbance (fault, load

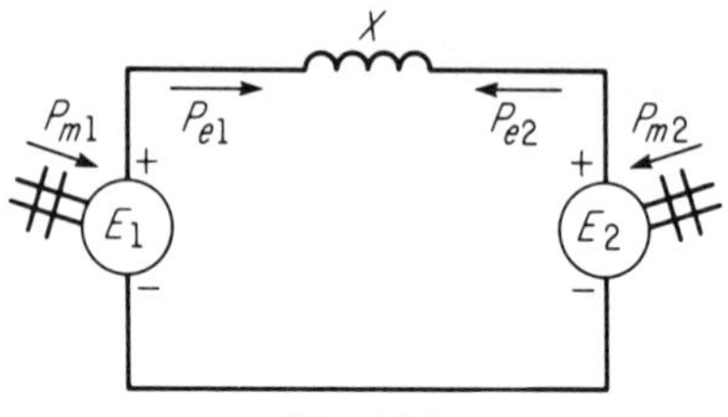

Fig. 14-1

change, line opening) would suddenly change the electrical output of the machines before the mechanical inputs have time to react. Neglecting losses, if $P_{\text{mech 1}} > P_{\text{elect 1}}$, the extra input would appear as accelerating power, attempting to increase the speed (and power angles) of the generator. As will be shown in Sec. 14-2, the electrical power output will increase with this angle δ until P_{e1} again equals $P_{\text{mech 1}}$ (assuming a stable situation where the rotor does not advance too far). Paralleled synchronous machines do have some natural tendency to maintain stability.

A system that is able to develop restoring forces sufficient to overcome the disturbing forces and restore equalibrium is said to remain stable. For example, any sudden advance in speed and rotor position of machine 1 would increase the generated output, with a corresponding slowing effect. At the same time, the additional electrical power transferred to machine 2 tends to speed up the rotor of machine 2, thus bringing the machine angles (and frequencies) back toward one another. Here again, mechanical inputs are assumed constant.

Power-system stability problems are commonly classified into two basic types—steady-state and transient. The study of steady-state stability is basically concerned with the effect of gradual, infinitesimal power changes. Transient stability problems deal with the effects of large, sudden system disturbances such as line faults, the sudden switching of lines, or the sudden application or removal of loads. Sections 14-2 through 14-7 deal primarily with the steady-state problems, while the remainder of the chapter is devoted to the transient problem.

14-2. The Electrical and Physical Significance of δ

Consider the generator of Fig. 14-2a which is tied to a very large system (infinite bus). Resistance has been neglected and X will represent all of the reactance from source to infinite bus, including generator syn-

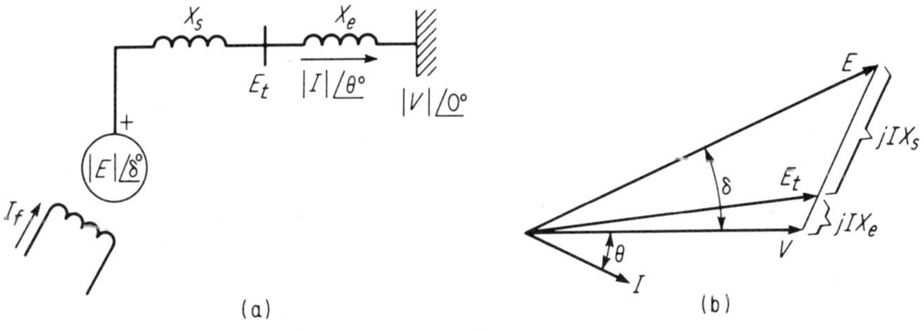

Fig. 14-2

chronous reactance X_s plus equivalent system reactance X_e to the infinite bus. The higher-speed turboalternators are usually cylindrical rotor. The reluctance of magnetic path for both quadrature and direct axes is much the same with the quadrature axis reluctance slightly greater than that of the direct axis. We will assume a cylindrical rotor in this case, where X_s is taken to be equal to $X_d = X_q$, as explained in Sec. 4-2. It is clearly seen in Fig. 14-2b that the power angle δ is the angle between the excitation voltage E and the voltage V. The physical significance, however, can be visualized as follows.

Recall from Chapter 4 that the excitation voltage E behind synchronous reactance is a function of the field flux alone. The synchronous reactance includes effects of both the armature current leakage flux and the more substantial flux of armature reaction. In general, as seen from the vector diagram, lagging current opposes the main field flux (in establishing V_t) while leading current adds to the main field flux. The resulting terminal voltage E_t is due to the combination of field and armature flux, and is nearly proportional to this total flux, or

$$E_t \doteq 4.44 f N \Phi_{\substack{\max \\ \text{total}}} \tag{14-2}$$

With no load current flowing in the stator winding, the reference voltage (V) and the excitation voltage E would be equal and in phase. Refer to Fig. 14-3a in which the no-load rotor position is depicted as being cen-

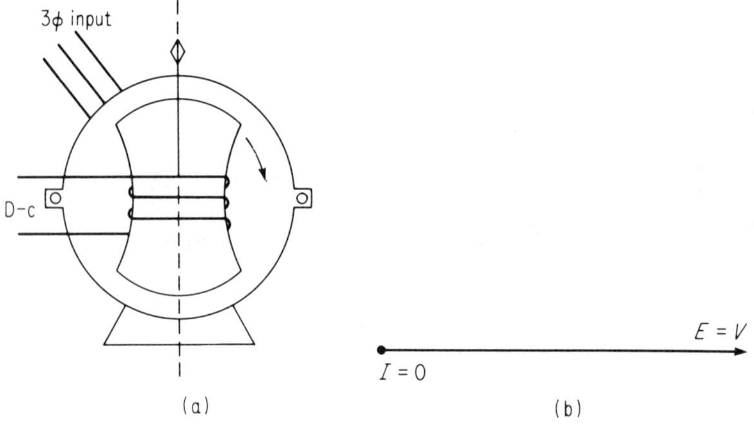

Fig. 14-3. (a) No-load rotor position centered on phase a. (b) No-load vector diagram.

tered between the stator turn of phase a. The revolving flux corresponding to V for the no-load case is along the physical reference axis (dotted)—corresponding to $t = 0$ on the vector diagram of Fig. 14-3b.

Next, consider the case of increased shaft power input for a generator. The rotor of Fig. 14-4a has advanced by an angle approximately equal to the electrical angle δ. This can be seen by realizing that a voltage E induced into the stator coils by the field flux will reach the stator coils δ degrees before the flux of the no-load case. In summary it can be said that δ is not only the electrical angle between the E and V of Fig. 14-2b but it is also the approximate angle between the poles center and the physical reference axis for the open-circuit condition. Poles are advanced (in direction of rotation) for generator action and retarded for motor action.

It may be well to look briefly at the cases of so-called "over and under excitation" of the d-c fields. When excessive magnetizing ampere

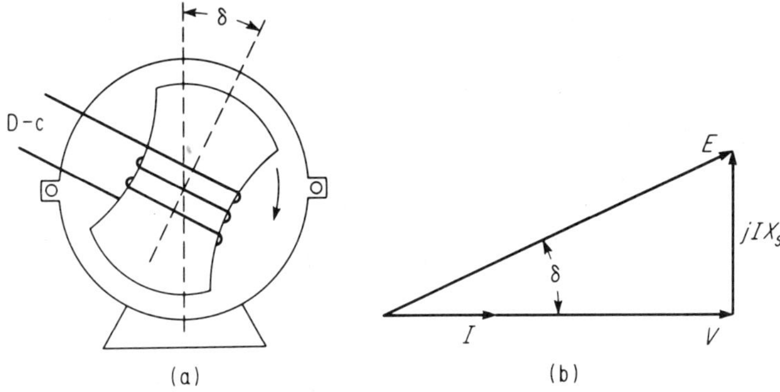

Fig. 14-4

turns are provided from the d-c field (more than called for by the terminal voltage) this excess mmf appears at the stator terminals as lagging amperes *delivered*. On the other hand, an underexcited machine will call upon the a-c system to make up the deficiency of magnetizing amperes by *drawing* lagging amperes from the system. Refer to Figs. 14-5a and b.

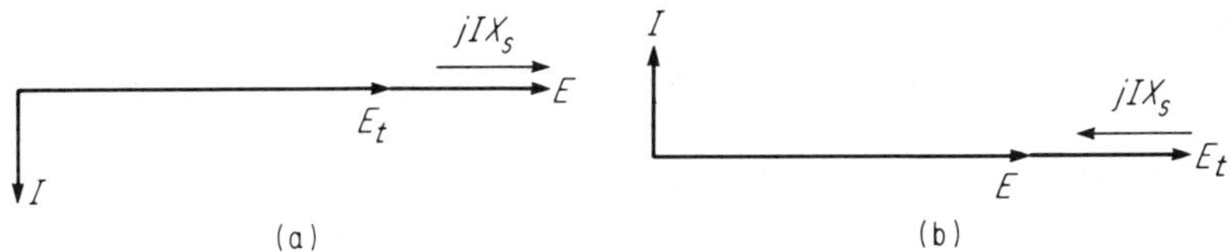

Fig. 14-5. (a) Over excited case where power = 0. (b) Unexcited case where power = 0.

14-3. The Power-Transfer Equation

The amount of power transferred from a machine to a large system was given in Eq. 4-1, where the infinite bus was considered as connected to the machine through an equivalent system reactance (X_e). Refer to Fig. 14-6.

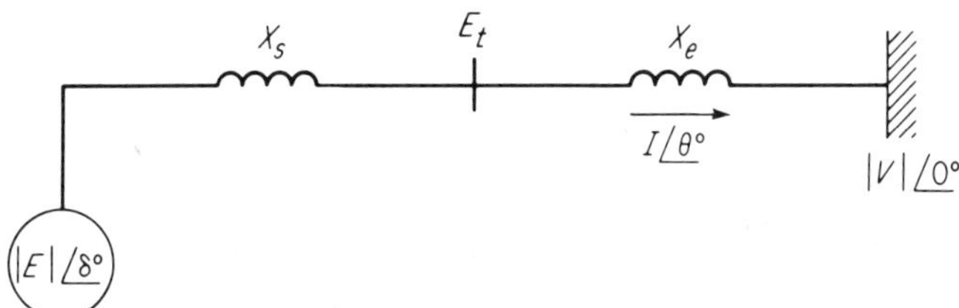

Fig. 14-6

Derivation of the power-transfer equation for a cylindrical rotor machine is given as follows. \

$$P_e = |E|\ |I| \cos(\delta - \theta_I) \tag{14-3}$$

The current is expressed as

$$I = \frac{E - V}{jX} = \frac{E}{jX} - \frac{V}{jX}$$

$$= \underbrace{\frac{|E| \angle \delta - 90^\circ}{X}}_{I_1} + \underbrace{\frac{|V| \angle +90^\circ}{X}}_{I_2}$$

The two components of currents will be treated separately and substituted into Eq. 14-3. Again, keep in mind that $X = X_s + X_e = X_d + X_e$.

$$P_e = |E|\ |I_1| \cos(\delta - \theta_1) + |E|\ |I_2| \cos(\delta - \theta_2)$$

$$= |E| \frac{|E|}{X} \cos[\delta - (\delta - 90^\circ)] + |E| \frac{|V|}{X} \cos(\delta - 90^\circ)$$

$$P_e = \frac{|E|\ |V|}{X_d + X_e} \sin \delta \tag{14-4}$$

The relationship of transferred electrical power P_e versus δ (from Eq. 14-4) is plotted in Fig. 14-7. This plot assumes $|E|$ and V constant. As pointed out in Sec. 14-2, the positive power angle corresponds to generator action and a rotor advanced in the direction of rotation.

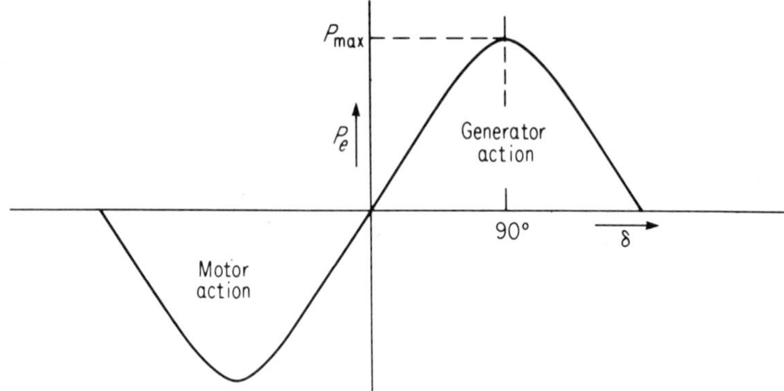

Fig. 14-7. Power transfer curve.

14-4. Steady-State Stability Limit

In treating the subject of steady-state stability, we are concerned with the maximum power transfer permissible from a synchronous machine—while maintaining stability during a gradual power increase. Again, refer to Figs. 14-6 and 14-7. Notice the maximum power that can be delivered from the machine occurs when δ has been advanced to the 90° point, or

$$P_{max} = \frac{EV}{X} \tag{14-5}$$

Obviously, if an attempt were made to advance δ further by further increasing the shaft input, the electrical power output will decrease from the P_{max} point. Therefore the angle δ of excitation voltage E increases further as the machine accelerates, thus driving the machine and system apart electrically. The value of P_{max} if often referred to as the *pull-out power.* Another term given this value is the "steady-state stability limit."

14-5. Short-Circuit Ratio (SCR) and Its Effect upon Stability

The term "short-circuit ratio" of a synchronous machine is defined as the ratio of field current required to give rated stator voltage at no load and rated speed to that field current required to produce rated armature current with a sustained three-phase short circuit. The direct axis synchronous reactance X_d for steady-state studies is often taken as the reciprocal of the SCR as will be demonstrated. Machines of high SCR are inherently more stable from both the steady-state and transient viewpoint. Short-circuit ratio is, thus somewhat of a measure of stability.

It follows from Eq. 14-4 that steady-state stability is improved with higher SCR machines, since

$$P_{max} \; \alpha \; \frac{1}{X_d + X_e}, \qquad \text{and} \qquad X_d \doteq \frac{1}{\text{SCR}}$$

The effect of higher SCR values in improving transient stability is related to the higher mass of the rotor. Any given accelerating power ($P_a = P_{mech} - P_{elect}$) associated with sudden system disturbances will not increase the rotor angle so rapidly. The accelerating power is proportional to the product of rotor mass and acceleration. Therefore, the time for the angles of E and V to move apart will be extended.

The question should be raised as to why the higher SCR (lower X_d) corresponds to a heavier rotor (other things being equal). The following reasoning is offered. Suppose a particular rotor is increased in size and weight. If the iron is to be worked to the same density, this increase in iron corresponds to an increase in total rotor core flux (Φ_t). See Fig. 14-8. Less armature turns (N_a) are now required on the stator at no load to

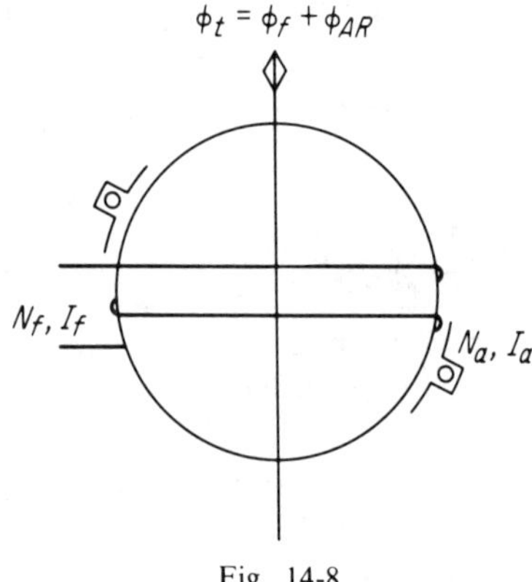

Fig. 14-8

satisfy the stator coil voltage. This is seen by examining the voltage equation for any coil, where

$$V = 4.44\, f \overset{\downarrow}{N_a} \quad \overset{\uparrow}{\Phi}$$

This also means that the effect of any armature current I_a upon the field has been lessened since the relative mmf ($N_a I_a$) of armature reaction has now been reduced. But synchronous reactance is a direct function of the effect of armature reaction upon the field as demonstrated in the vector diagrams of Figs. 4-2a and 14-5. Actually, synchronous reactance X_s of a cylindrical rotor or direct axis reactance X_d of a salient-pole rotor combine the effects of leakage reactance and armature reaction, the latter effect being the greater. This was pointed out in Chapter 4. Now, if the relative effect of armature reaction upon the total field has been lessened for the heavier rotor, then X_s (or X_d) has been reduced and the SCR increased.

To illustrate the meaning and measurement of SCR and its relationship to X_d, consider the shorted machine of Fig. 14-9. Sufficient field current is applied to permit rated armature current to flow. An unsatu-

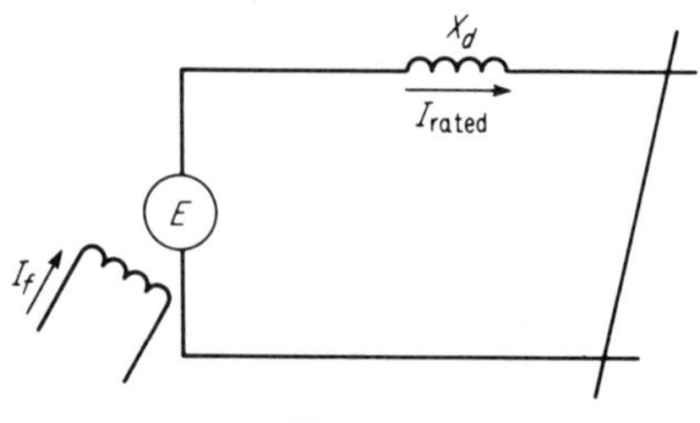

Fig. 14-9

rated synchronous reactance (X_{du}) would be determined as

$$\text{pu } X_{du} = \frac{\text{pu } E_u}{\text{pu } I} = \frac{\text{pu } E_u}{1.0} \tag{14-6}$$

For determining the unsaturated E_u, refer to the short-circuit character-

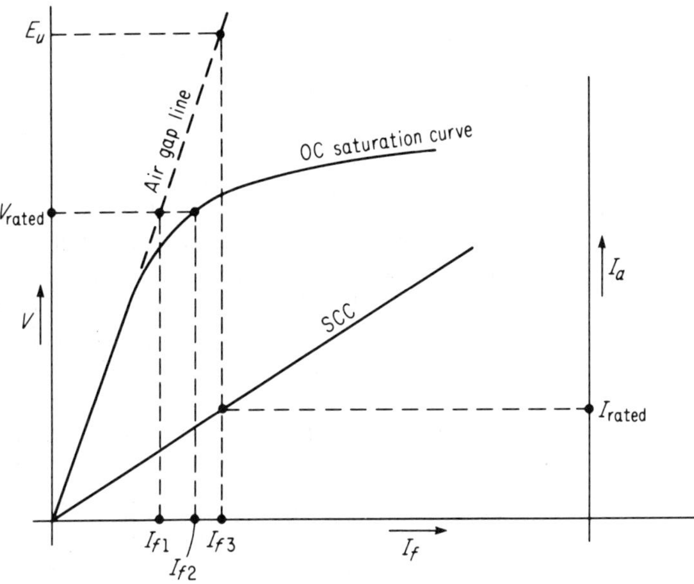

Fig. 14-10. Open-circuit saturation and short-circuit characteristic curves.

istic curve and o-c saturation curve (with air-gap line) of Fig. 14-10. Similar triangles show that

$$\begin{aligned} \text{pu } E_u &= V_{\text{rated}} \times \frac{I_{f3}}{I_{f1}} \\ &= 1.0 \times \frac{I_{f3}}{I_{f1}} \end{aligned} \tag{14-7}$$

Substituting into Eq. 14-6 gives the unsaturated X_{du} as

$$\text{pu } X_{du} = I_{f3}/I_{f1} \tag{14-8}$$

A more realistic synchronous reactance (X_d) which includes the effect of no-load saturation could be written in terms of the saturation factor k, where

$$k = \frac{I_{f2}}{I_{f1}} > 1 \tag{14-9}$$

Then

$$X_d = \frac{X_{du}}{k} = \frac{I_{f3}}{I_{f1}} \times \frac{I_{f1}}{I_{f2}}$$

or

$$X_d = \frac{I_{f3}}{I_{f2}} \tag{14-10}$$

The definition of SCR has already been defined. In equation form it is

$$\text{SCR} = I_{f2}/I_{f3} = \frac{1}{X_d} \tag{14-11}$$

In spite of the advantages to be gained from the heavier, higher SCR machines, the trend in design today is toward the lighter rotors (lb/mva). The designer has moved in this direction in an attempt to hold the rotor size to practical limits in the large 2-pole, 3,600-rpm machines. Better methods of cooling have contributed greatly to the smaller size per kva in the later designs. Typical values of SCR in 1955 were on the order of 1.0. By 1968 a figure of 0.5 was feasible. If stability suffers as a result of the lower SCR, other means for improving the stability have made up the difference. Two such improvements are *higher exciter responses* and *fast steam valving* on the input, both of which will extend the critical time before breakers are required to clear the fault.

14-6. Prevention of Pullout Under Steady-State Conditions

There must be some practical guarantee, from an operational standpoint, that a given machine will not go into an unstable region of operation. This means that the P and Q delivered from the generator terminals to the system must not go beyond certain predetermined limits. Refer to the machine of Fig. 14-11 where terminal voltage (E_t) is to be held constant. Suppose the system voltage rises. In order to maintain E_t constant the field current and excitation voltage backs off and the output vars go leading. If E drops too low, the steady-state stability may

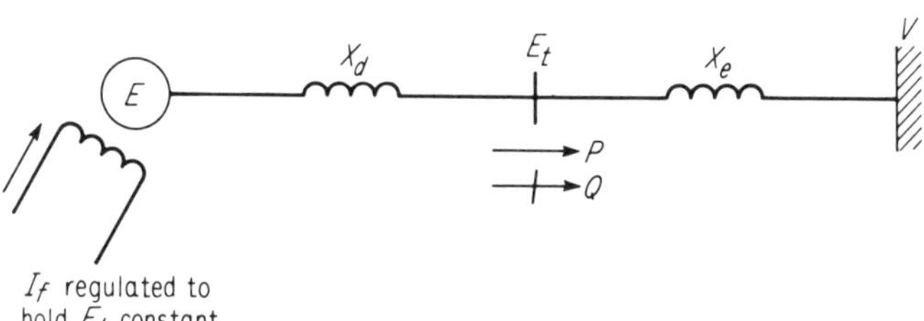

Fig. 14-11

be endangered as δ increases toward the 90° pullout point (P mechanical being constant). This is demonstrated by again referring to the power-transfer equation, where

$$P = \frac{|E|\ |V| \sin \delta}{X_d + X_e} \qquad [14\text{-}4]$$

For each given value of P (E_t being constant) there is some value of Q beyond which the machine cannot go. It is the purpose of this section to determine the minimum excitation limit (MEL) permissible for a machine and to discuss briefly the controls needed for holding to this limit.

There is always the possibility that a line close to the generator be out of service due either to an emergency or a prearranged shutdown. The power-transfer equation may be revised using an increased value of X_e to allow for this possibility. Also, in establishing minimum excitation limits, a reduced value of E_t may be used to represent the minimum terminal voltage condition. This might, for example, be taken as 95 percent if the generator kv rating is specified at nominal voltage ±5 percent.

The vector diagram of Fig. 14-12 corresponds to the circuit of Fig. 14-11. *Given* the terminal voltage E_t as constant. *Find* the relationship

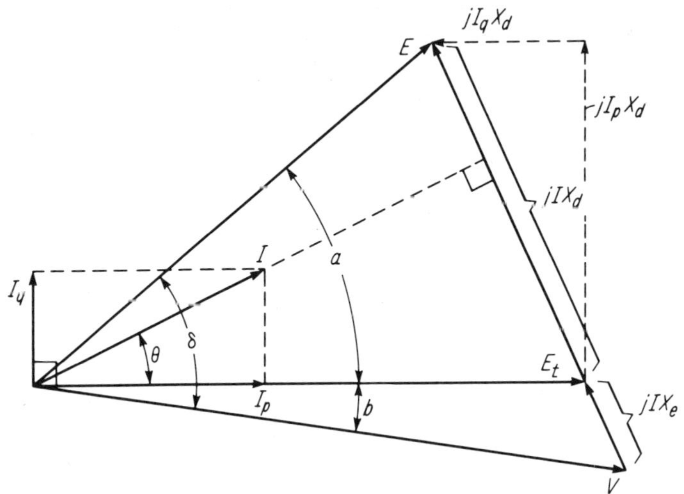

Fig. 14-12. Vector diagram for circuit of Fig. 14-11.

between P, Q, E_t, and δ, where V and E are allowed to swing. Eventually, for the P_{max} condition, δ will be set to 90°.

From Fig. 14-12,

$$\tan \delta = \frac{\tan a + \tan b}{1 - \tan a \tan b} \qquad (14\text{-}12)$$

$$\tan a = \frac{I_p X_d}{E_t - I_q X_d}; \quad \tan b = \frac{I_p X_e}{E_t + I_q X_e} \tag{14-13}$$

Substituting Eq. 14-13 into Eq. 14-12 and simplifying yields,

$$\tan \delta = \frac{E_t I_p (X_e + X_d)}{E_t^2 - E_t I_q (X_d - X_e) - I_q^2 X_d X_e - I_p^2 X_d X_e} \tag{14-14}$$

but

$$P = E_t I_p, \quad \text{and} \quad {}^{\circ}Q = -E_t I_q \tag{14-15}$$

Substituting Eq. 14-15 into Eq. 14-14,

$$\tan \delta = \frac{P(X_e + X_d)}{E_t^2 + Q(X_d - X_e) - \dfrac{Q^2}{E_t^2} X_d X_e - \dfrac{P^2}{E_t^2} X_d X_e} \tag{14-16}$$

Equation 14-16 can be rearranged in the form of the equation for a circle with P and Q as variables. The desired form is

$$(P - k_1)^2 + (Q - k_2)^2 = (\text{radius})^2 \tag{14-17}$$

This is accomplished by multiplying both sides of Eq. 14-16 by the denominator of the right-hand side, then completing the square for the P and Q terms. In order to save space, only the result of this maneuver is given here as

$$\left(P + \frac{E_t^2 (X_e + X_d)}{2 X_d X_e \tan \delta}\right)^2 + \left(Q - \frac{E_t^2 (X_d - X_e)}{2 X_d X_e}\right)^2 = \left(\frac{E_t^2 (X_d + X_e)}{2 X_d X_e \sin \delta}\right)^2 \tag{14-18}$$

Recall that the center of the circle represented by Eq. 14-17 is at k_1, k_2. Likewise, the center of the circle represented by Eq. 14-18 is found at

$$P = -\frac{E_t^2 (X_c + X_d)}{2 X_d X_e \tan \delta} \tag{14-19}$$

and

$$Q = \frac{E_t^2 (X_d - X_e)}{2 X_d X_e} = \frac{E_t^2}{2}\left(\frac{1}{X_e} - \frac{1}{X_d}\right) \tag{14-20}$$

The radius of the circle equation is

$$R = E_t^2 \frac{X_d + X_e}{2 X_d X_e \sin \delta} = \frac{E_t^2}{2 \sin \delta}\left(\frac{1}{X_d} + \frac{1}{X_e}\right) \tag{14-21}$$

To find the extreme pullout conditions for P and Q, δ will be set to 90° in Eqs. 14-19 through 14-21. This determines the pullout curve of Q vs. P which is sometimes referred to as the *close-hand regulated curve or static stability limit curve* for the machine. Actually, for automatic regulation, the maximum power point may go well beyond 90°. This is true because of the sudden increases in excitation which would accompany a terminal-voltage decrease as the machine responds to change. In order to account for the automatic regulator, it would necessitate that one determine the point at which $dP/d\delta = 0$. Such a curve has been referred to as the *dynamic stability limit curve*, and would be less restrictive than the *close-hand regulated pullout curve*. Then, setting $\delta = 90°$, the center of the circle for the pullout curve is found at

$$P_c = 0, \qquad Q_c = \frac{E_t^2}{2}\left(\frac{1}{X_e} - \frac{1}{X_d}\right) \tag{14-22}$$

The radius for the circle is at

$$R = \frac{E_t^2}{2}\left(\frac{1}{X_d} + \frac{1}{X_e}\right) \tag{14-23}$$

Again, refer to the vector diagram of Fig. 14-12. For a given P, the vertical component of E is constant and equals I_pX_d. When $\delta = 90°$, E will be at a minimum, and the machine cannot be excited below this level. Operation must be within the circle shown in Fig. 14-13. The Q intercept is

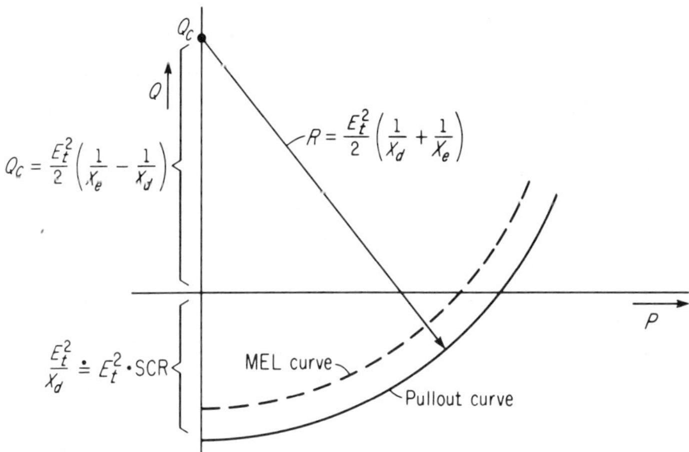

Fig. 14-13. Pullout curve of Q vs. P for a synchronous generator.

readily found by subtracting Q_c from R, which yields E_t^2/X_d. If E_t were at 1.0 per unit, then the intercept appears at $\dfrac{1}{X_d} \doteq \text{SCR}$ (short-circuit ratio).

Notice a second curve is shown above the pullout curve. This is referred to as the *minimum excitation limit* curve (MEL). An automatic excitation limit control will prevent the machine from going below this dotted curve. With any attempt on the machine's part to go below the MEL curve, the limit control would override the terminal voltage control, calling for an increased excitation to prevent instability. Note the MEL curve rides somewhat higher than the calculated pullout curve, thus providing an extra margin for stability. The amount of this margin will vary with the philosophy of the particular utility, but a typical application might back off the magnitude of the radius by 10–20 percent of the E_t^2/X_d intercept. Sandwiched in between the MEL and pullout curves may be placed a loss of field relay curve. If, for any reason, the reactive power should fall below this limit the relay will take appropriate action to sound an alarm, trip the machine, or whatever the particular application should call for.

For manual control of the field where the terminal voltage is not closely watched, a curve—which is even more restrictive than the pullout curve—is imposed. This additional margin may be necessary as any attemp of the generator to pick up load—without a corresponding and automatic increase in field—may force the power angle into the unstable region. This manual control curve (along with the MEL and pullout curves of Fig. 14-13) may be added to the other generator capability curves of Sec. 13-2 in providing guidelines for the plant operator.

14-7. Some Factors in Determining the Voltage Control Point of Generators

Voltage regulation may be to any of several points, just as was the case with automatic load-tap-changing (LTC) transformers of Chapter 11. Line-drop compensators again may be used to represent the transformer IZ drop if one is to regulate to the high side of the generator transformer. Or the voltage may be regulated at the generator terminals with zero compensation. A third possibility is to insert negative-reactance compensation, which would permit terminal voltage to drop slightly with an increased reactive output. This last method was treated in Chapter 11 and has the advantage of permitting units to parallel automatically by alleviating the problem of circulating current. If one compensates for generator transformer impedance drop and thus regulates to the transformer

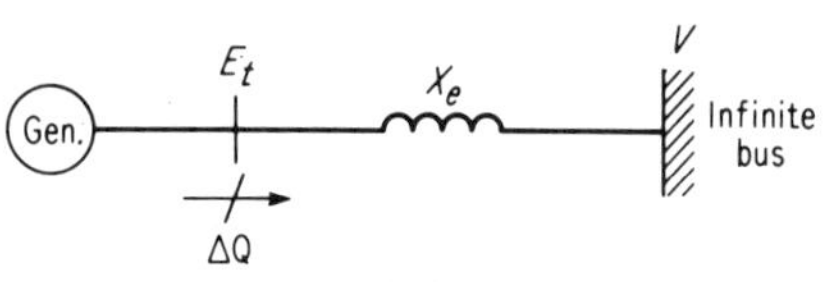

Fig. 14-14

secondary, it will take more vars per volt to regulate this point than would be required when regulating to the generator terminals. Refer to Fig. 14-14. Again, as explained in Chapter 11, the amount of voltage change ΔE experienced at some point in the system by injecting vars (Q) or lagging amperes at that point is approximated as

$$\%\Delta E_t \doteq pu\Delta I_{\text{lag}} \cdot \%X_e$$

or

$$\%\Delta E_t \doteq \text{pu}\Delta Q \cdot \%X_e \tag{14-24}$$

As seen from the approximate Eq. 14-24, the farther into the system to which the controls regulate, the smaller X_e and the smaller the effect of reactive vars upon voltage. A generator will take a greater swing in vars in attempting to control voltage to the more remote point.

14-8. Transient Stability and the Swing Equation

As indicated in the introduction, the subject of transient stability is concerned with sudden disturbances that might occur in the system (faults, line switching, sudden application or removal of loads). The remainder of this chapter is devoted primarily to this subject. Since the coverage must necessarily be limited, certain assumptions and simplifications will be made. System resistances have been considered negligible with respect to reactances. The cylindrical rotor machine is assumed and the direct and quadrature axis reactances are assumed equal. Much simplification is possible (in studying the stability of one particular machine) by reducing the system to that of an equivalent two-machine system, again by representing the machine as feeding to the infinite bus. Direct-axis transient reactance (X_d') as defined in Chapter 7 will be used for the machine representation. One of the assumptions often made for the first seconds following a disturbance is that the shaft input power to the machine remains constant. This is often done, considering that the mechanical system of governors, steam valves, and the like is relatively sluggish with respect to the rapidly changing electrical quantities. With the development of fast valving, such an assumption of constant input will not always be valid.

One of the basic problems of stability revolves around the determination of whether or not the machine power angle δ (or torque angle) will stabilize after a sudden disturbance. This torque angle is the angle between the generator excitation voltage and the system voltage as explained in Sec. 14-2. If a study reveals that δ continues to increase (or decrease) after the disturbance, the machine will of course go out of step with the system. On the other hand, if provision can be made to alleviate the extreme condition (such as faster relaying for isolation of a faulted line, or perhaps a more rapid field excitation response) the unstable situation may be avoided without loss of the machine. The rotor and angle of excitation voltage accelerates (or decelerates) in a manner proportional to the accelerating torque. Neglecting losses, this accelerating or decelerating torque T_a is always the difference between the mechanical or shaft input torque T_s and the electromagnetic output torque T_e, or

$$T_a = T_s - T_e \tag{14-25}$$

For a generator, when the shaft torque $T_s > T_e$, then T_a, is positive, tending to accelerate the rotor. Refer to Fig. 14-15 for the signs that have been

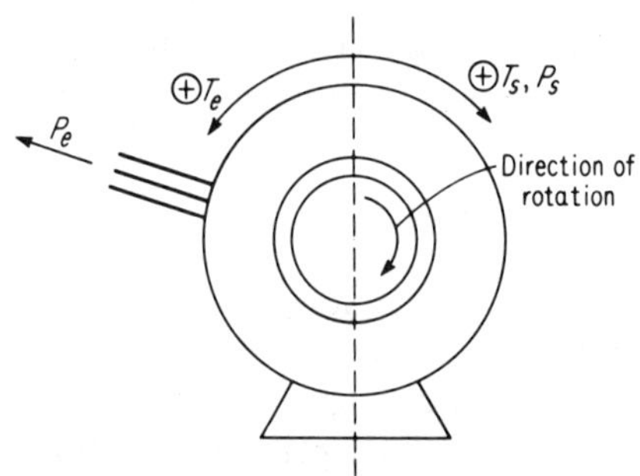

Fig. 14-15. Signs assumed for generator in Eq. 14-25.

assumed for generator action. For motor action, the signs of both T_s and T_e will reverse.

A number of symbols are useful at this point in reviewing the necessary equations of rotational mechanics, as well as in arriving at the so-called "swing" equation for the synchronous machine. Some of these symbols with MKS units are:

θ = angular displacement (radians)
ω = angular velocity (radians/second)
α = angular acceleration (radians/second^2)
T = torque in newton meters
I = moment of inertia (kg – meter^2)
M = angular momentum (joule-sec/radian)
M' = angular momentum (megajoule-sec/electrical degree)
(*Note:* M and M' are also termed *inertia constants.*)
G = machine rating in mva

$$\text{H} = \text{inertia constant } \frac{\text{Kinetic energy stored in megajoules (KE)}}{\text{Machine mva rating } (G)}$$

Unfortunately, both M and H are called *inertia constants*. Their differences will be pointed out later in more detail.

Accelerating power is related to torque by the familiar equation of rotational mechanics, where

$$P_a = T_a\omega \tag{14-26}$$

where ω is the angular velocity of the machine in mechanical radians per second. A power expression can be written which is similar to the torque Eq. 14-25:

$$P_a = T_a\omega = P_s - P_e \tag{14-27}$$

Two additional equations of rotational mechanics are given below as

$$T_a = I\alpha \tag{14-28}$$

and

$$M = I\omega \tag{14-29}$$

Substituting these equations into Eq. 14-27 yields

$$P_a = (I\alpha)\omega = M\alpha = P_s - P_e \tag{14-30}$$

Up to this point the angular quantities ω and α in Eqs. 14-26 through 14-30 have been in terms of mechanical radians of the rotor. However, at this point we will change from mechanical radians to electrical radians, remembering that the number of electrical radians = number of *mechanical* radians times the number of poles/2. For example, the unit on α in Eq. 14-30 changes from *mechanical* radians/sec^2 to *electrical* radians/sec^2. This is easily accomplished by being consistent with the unit on M, changing M from megajoule-sec per mechanical radian to megajoule-sec per electrical radian.

The angular acceleration (α) is written as

$$\alpha = \frac{d^2\theta}{dt^2} \text{ electrical radians/sec}^2 \tag{14-31}$$

However, θ can be expressed as the sum of (1) the time varying angle (ωt) on the rotating reference axis as described in Sec. 14-2, plus (2) the torque angle δ of the rotor with respect to the rotating reference axis. In other words,

$$\theta = \omega t + \delta \quad \text{electrical radians} \tag{14-32}$$

Substituting into Eq. 14-31,

$$\alpha = \frac{d^2(\omega t + \delta)}{dt^2} \tag{14-33}$$

or

$$\alpha = \frac{d^2\delta}{dt^2} \quad \text{electrical radians/sec}^2 \tag{14-34}$$

Now substituting this value of α back into Eq. 14-30 results in the "swing" equation for the machine:

$$P_a = M \frac{d^2 \delta}{dt^2} = P_s - P_e \tag{14-35}$$

Solving this differential equation and plotting δ as a function of t would result in the "swing curve" for a machine. This will be treated later in more detail, but in general, the swing curves will reveal any tendency of the torque angle to oscillate and/or increase beyond the point of return. Again, refer back to Sec. 14-2 for a more detailed coverage of the electrical and physical significance of δ.

Suppose in the swing Eq. 14-35 that M is in megajoule-seconds per electrical radian. Then P would be in megawatts and δ in electrical radians. From our previous definition, the kinetic energy (KE in megajoules) of the rotor can be expressed as the product of the mva rating (G) and the inertia constant H. In other words,

$$\text{KE} = GH = 1/2\, I\omega^2 \times 10^{-6} \tag{14-36}$$

or

$$GH = 1/2\, M\omega \tag{14-37}$$

Expressing the inertia constant or angular momentum M in terms of the inertia constant H and the mva rating (G):

$$M = \frac{2GH}{\omega} \tag{14-38}$$

since

$$\omega = 2\pi f$$

then

$$M = \frac{GH}{\pi f} \quad \text{megajoule-sec/elect. radian} \tag{14-39}$$

A more common unit of M would be that of megajoule seconds per electrical degree in which case the δ of the swing equation would be in electrical degrees. Converting M to the new quantity M' is accomplished as shown.

$$M' = M \times \frac{2\pi \text{ elect. radians}}{360 \text{ electrical degrees}}$$

$$= \frac{GH}{\pi f} \times \frac{2\pi}{360}$$

$$M' = \frac{GH}{180f} \quad \text{megajoule-sec/elect. degree} \tag{14-40}$$

At one location, several machines might be lumped together as one machine where the equivalent M constant (M'_e) becomes

$$M'_e = M'_1 + M'_2 + \cdots \tag{14-41}$$

Since both M' and H are often termed *inertia constants*, their basic differences should be explained. H is somewhat similar to a per unit quantity even though quantities in the ratio which it expresses do not have precisely the same units. H expresses a ratio (megajoules/mva) of kinetic energy stored at synchronous speed to the power rating of the machine. Machines of a given speed and prime mover type may have H constants which fall into a fairly narrow range, not varying widely with mva. However, the M constant will vary directly with both the H constant and the mva as Eq. 14-40 demonstrates.

The trend in large machine designs today is toward smaller H constants. This factor, if taken by itself, makes the transient stability problem more critical as the rotor will accelerate or decelerate faster. As pointed out in Sec. 14-5, which deals with the trend toward lower short-circuit ratios—what is lost (from the transient stability viewpoint) in the lower H and SCR values, must be made up with such features as faster excitation systems, faster breakers, lower reactance lines or faster steam valving. As previously indicated, the H constant will vary considerably with the machine type. However, for purpose of example only, a typical H constant for a steam unit today might be on the order of 5, while a hydro unit might be on the order of 3.

When only one particular machine (and system) is under study, for transient stability, a simpler model applies and a conventional method of dealing with this problem utilizes the *equal-area criterion*, as covered in the next section. However, for multi-machine studies, the conventional solution has used a step-by-step, incremental solution to determine the movement of the individual machine angles. This latter method uses the swing equation and a linear approximation and the method has been applied to both the a-c analyzer and the digital computer. There are other approaches to the multimachine problem which will be briefly pointed out as space permits.

14-9. The Single-Machine Problem and the Equal-Area Criterion

It is possible to determine as to whether a single machine will remain stable, should a sudden disturbance occur, without resorting to the calculation of the swing equation. For a single machine to infinite bus problem, the simpler approach is termed the *equal-area criterion* and, while the approach is more or less intuitive, there are various ways of proving its validity. Refer to Fig. 14-16a for a simple transient model of

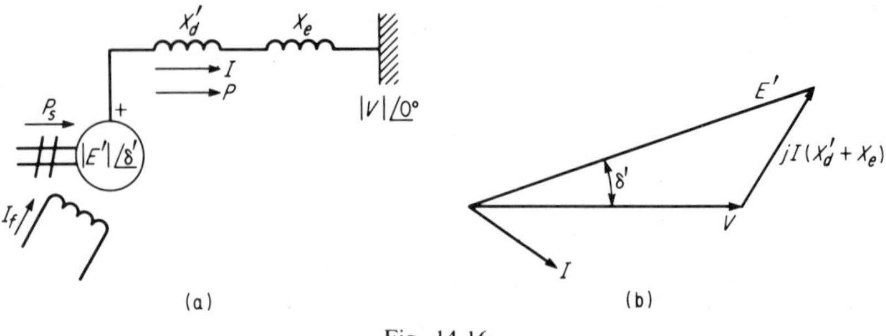

Fig. 14-16

a single machine tied to an infinite bus. The electrical power P_e transferred from this machine can be expressed in terms of the power angle δ' in much the same manner as in Eq. 14-4. However, the direct axis synchronous reactance X_d of the steady-state problem is now replaced by the direct-axis transient reactance X_d. P_e is then written as

$$P_e = \frac{|E'|\,|V|}{X_d' + X_e} \sin \delta' \tag{14-42}$$

where

$$P_e(\max) = \frac{|E'|\,|V|}{X_d' + X_e} \tag{14-43}$$

The E' of Eq. 14-42 is the excitation voltage behind the transient reactance. Notice E' and δ' differ from the steady-state E and δ values as was demonstrated in Fig. 7-8b. However, as a matter of convenience, the prime notation will be dropped on E and δ from this point on in the transient coverage. Again, several types of disturbances that might occur include:

1. Short circuits
2. Line switching
3. Sudden load changes

Most disturbances of a critical nature entail the sudden change of electrical output (during a disturbance) while the mechanical input remains relatively constant. However, in order best to demonstrate the intuitive reasoning behind the equal area criterion, the shaft input of the generator of Fig. 14-16a will be suddenly increased as the first disturbance noted. The infinite bus is again considered large enough to absorb any change in P_e which results—while V and f remain constant. The plot of P_e vs. δ is given in Fig. 14-17. The initial value of shaft power P_{so} and P_e is represented by point a. With the sudden increase of shaft power to P_{s1}

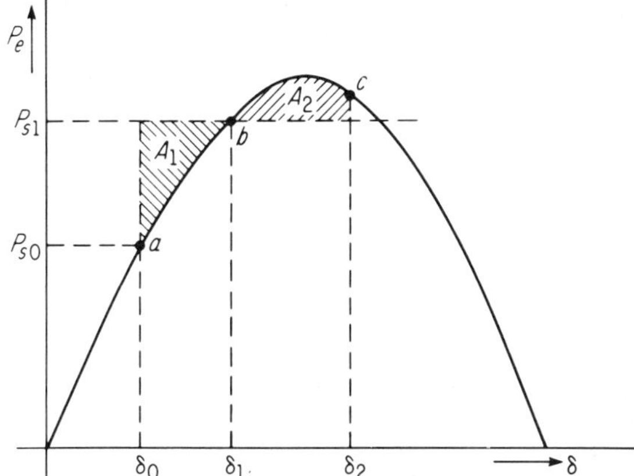

Fig. 14-17. P_e vs. δ for a generator where P_{shaft} is suddenly increased from P_{s0} to P_{s1}.

there is momentarily more shaft input than electrical output. Accelerating power ($P_a = P_{s1} - P_e$) is now available to accelerate the rotor to some new angle δ_1 where $P_e = P_{s1}$. At this point the rotor stops accelerating but is still at a speed slightly above synchronous. δ continues to advance, overshooting point b and on into the region where output P_e exceeds input P_{s1}. The rotor decelerates until it reaches some maximum point c, whereupon its speed is again at synchronous. Having returned all of its extra kinetic energy back to the electric circuit, the rotor continues to decelerate ($P_e > P_{s1}$) falling through point b and back toward a. Oscillations will continue for a time about point b, finally coming to rest at b.

While the rotor was moving from point a to b, work was being done on the rotor in increasing the kinetic energy of the revolving mass, since $P_{s1} > P_e$. This amount of extra work done on the rotor will be shown to be represented by the shaded area A_1, whereas the area A_2 represents an equal amount of energy which the rotor returns to the circuit. It will also be shown that sufficient area must be available above the P_{s1} line to permit $A_2 = A_1$. While this fact may appear intuitive, a closer look will be taken at this equal area criterion.

Again referring to Fig. 14-17, the extra work ΔW_1 done to accelerate the rotating mass by an accelerating torque T_a acting from δ_0 to δ_1 is given as

$$\Delta W_1 = \int_{\delta_0}^{\delta_1} T_a \, d\delta \tag{14-44}$$

but

$$T_a = P_a \div 2\pi n \tag{14-45}$$

Since the speed n is nearly constant, Eq. 14-45 is substituted into Eq. 14-44, where $K = \dfrac{1}{2\pi n}$:

$$\Delta W_1 = K \int_{\delta_0}^{\delta_1} P_a \, d\delta = K \int_{\delta_0}^{\delta_1} (P_s - P_e) \, d\delta \qquad (14\text{-}46)$$

or

$$\Delta W_1 = KA_1 \qquad (14\text{-}47)$$

When the rotor reaches point c, the rotor is again at synchronous speed and has given back all of the extra kinetic energy to the circuit. This extra energy ΔW_2 is represented by A_2 as shown:

$$\Delta W_2 = \int_{\delta 1}^{\delta_2} T_a \, d\delta \qquad (14\text{-}48)$$

$$\Delta W_2 = K \int_{\delta 1}^{\delta_2} P_a \, d\delta = KA_2 \qquad (14\text{-}49)$$

Accelerating power is now negative, since $P_{s1} < P_e$. Since energy is conserved,

$$\Delta W_1 = \Delta W_2 = KA_1 = KA_2 \qquad (14\text{-}50)$$

or

$$A_1 = A_2 \qquad (14\text{-}51)$$

Not only does this indicate the maximum point to which δ will swing, but it also establishes the criterion that sufficient area must be available above the P_{s1} line and below the P_e curve to satisfy the A_2 requirement. If the area available for $A_2 < A_1$, then the excess kinetic energy will cause δ to continue increasing in an effort to return power back to the electric circuit. This however would be impossible, since P_e is now decreasing with an increase in δ (slope is negative). Instability would result as the machine angle moves farther apart from the system angle.

Another observation should be made from Fig. 14-17. Notice that it is permissible for the rotor to oscillate past the point where $\delta = 90°$, as long as the equal-area criterion is met.

In studying the effects of faults upon generator stability, one is often looking for the critical clearing time (in cycles or seconds) during which time the breakers must isolate the faulted line from the system. Should the breakers fail to open within this period, the criterion for stability would not be met. The equal-area criterion will not of itself determine this critical clearing time, but it does offer a means for determining the critical clearing angle corresponding to this time. The most severe fault to be considered for transient stability studies is the symmetrical three-phase fault. This is also the fault which is more easily calculated. However, it may be permissible to extend the clearing time to accommodate a less severe fault, thus improving the economics of protection. The criterion for such studies is often the double line-to-ground fault, though we will be considering three-phase faults in this chapter. Several problems will be analyzed briefly followed by an Example. In every case, this section will deal with one machine and the infinite bus. The multimachine problem will be discussed in Sec. 14-10.

Next, consider the circuit of Fig. 14-18. The far end of one circuit

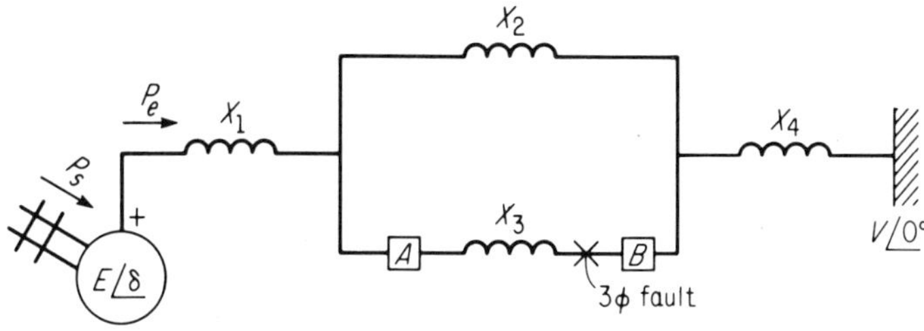

Fig. 14-18

is faulted as shown, whereupon P_e drops suddenly to zero. Breakers at A and B then open the line thus isolating the fault. P_e then picks up to some new value as the generator transfers power to the system over the one remaining line. Three separate conditions are represented in this sequence

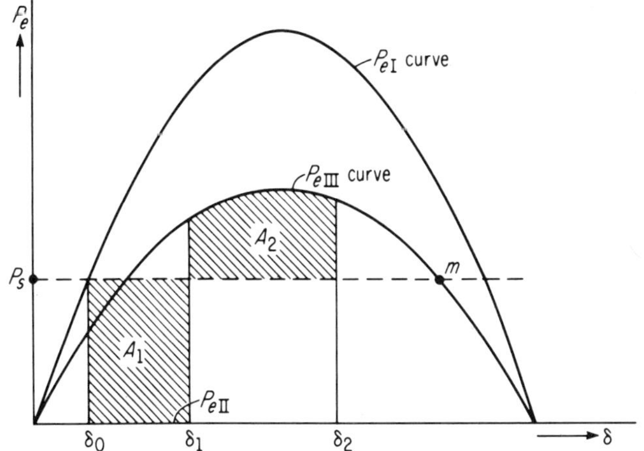

Fig. 14-19. Demonstrating the sequence corresponding to Fig. 14-18. I—System normal. II—Fault applied. III—Fault isolated.

and are demonstrated in Fig. 14-19. The conditions are:

I. System normal—both circuits are in. The P_e relation is given as

$$P_{eI} = \frac{|E|\,|V|\,\sin\delta}{X_1 + X_4 + \dfrac{X_2 X_3}{X_2 + X_3}}$$

P_e and P_s are, of course, considered equal before the fault.

II. System faulted as shown in Fig. 14-18. This effectively isolates the machine from the rest of the system as illustrated in Fig. 14-20. P_e

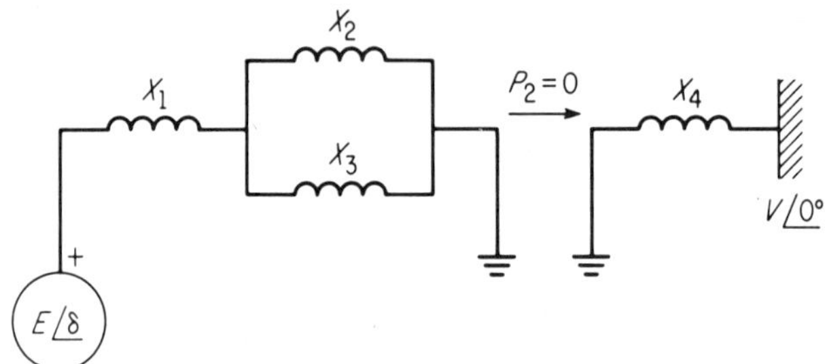

Fig. 14-20. Illustrating condition II for Fig. 14-19.

then falls to zero. The Area A_1 of Fig. 14-19 is again proportional to the kinetic energy stored in the rotor as it accelerates, taking on the new angle δ_1.

III. Faulted line open. Assume here that breakers A and B open simultaneously. The new P_{eIII} relation is

$$P_{eIII} = \frac{|E|\,|V|}{X_1 + X_2 + X_4}\sin\delta$$

After opening the faulted line, the rotor overshoots δ_1, decelerating but increasing in angle until δ_2 is reached, whereupon the rotor is again at synchronous speed. At this point the extra kinetic energy (represented by A_1) has been returned to the circuit in the form of electrical energy (represented by A_2). Again $A_1 = A_2$ where

$$A_1 = (\delta_1 - \delta_0)\,P_s \tag{14-52}$$

and

$$A_2 = \int_{\delta_1}^{\delta_2} (P_{eIII} - P_s)\,d\delta \tag{14-53}$$

If one were to determine the critical clearing angle (δ_1) for this problem, δ_2 of Eq. 14-53 would be set to correspond with the intersection (m) of the P_{eIII} curve and the P_s line. Equation 14-52 for A_1 would then be set equal to A_2 and the value of δ_1 would be determined.

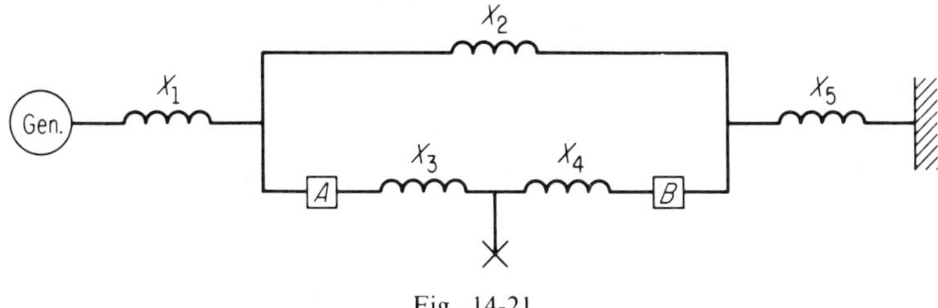

Fig. 14-21

Refer to Fig. 14-21 for the situation of a three-phase fault placed somewhere between the ends of one line as shown. Three different equivalent reactance values exist between the generator and infinite bus. They correspond to the following conditions.

I. System normal, where

$$X_{eI} = X_1 + X_5 + \frac{X_2(X_3 + X_4)}{X_2 + X_3 + X_4}$$

II. Three-phase fault as shown. The equivalent reactance is found as follows:

(a) Convert the delta of Fig. 14-22a to its wye (Y) equivalent elements of X_a, X_b and X_c (see dashed lines).

(b) Combine the series impedances of Fig. 14-22a; $X_1 + X_a$, $X_b + X_5$.

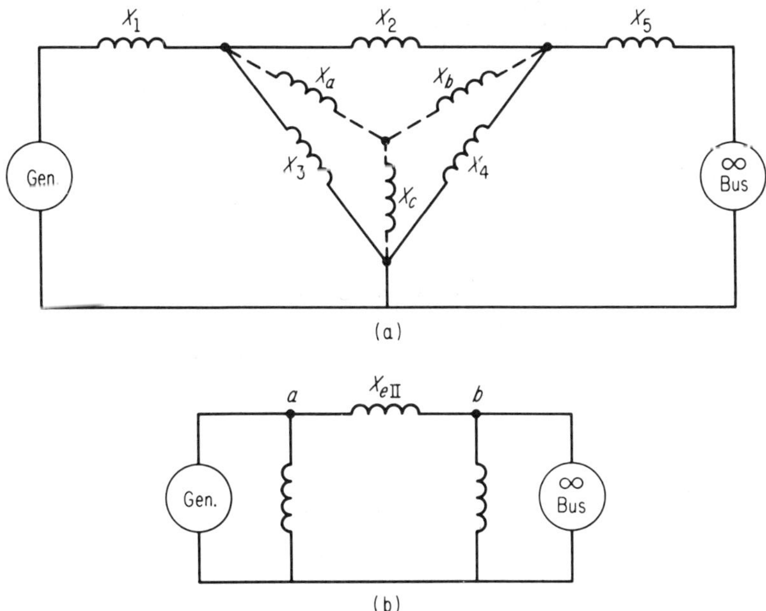

Fig. 14-22

Convert the resulting wye to the equivalent delta of Fig. 14-22b. The reactance between points *a* and *b* obviously becomes the reactance to be used in the power transfer equation.

III. Fault isolated by opening breakers *A* and *B*. The reactance used in the power transfer equation is now equal to the series combination of X_1, X_2 and X_5, or

$$X_{e\text{III}} = X_1 + X_2 + X_5$$

The equal-area criteria will, in general, be applied as shown in Fig. 14-23. Again, it entails the determination of the critical clearing angle δ_c that will guarantee that $A_1 = A_2$.

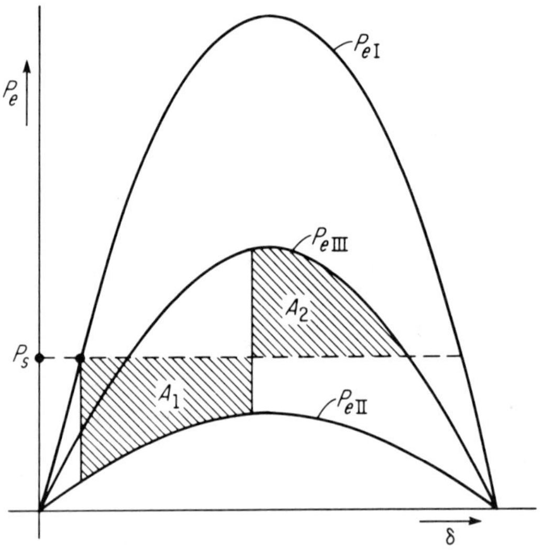

Fig. 14-23

Example 14-1. Given the circuit of Fig. 14-24 where a three-phase fault is applied on one end of a line near breaker *B*, as shown. Find the critical fault-clearing angle for clearing the fault with the simultaneous

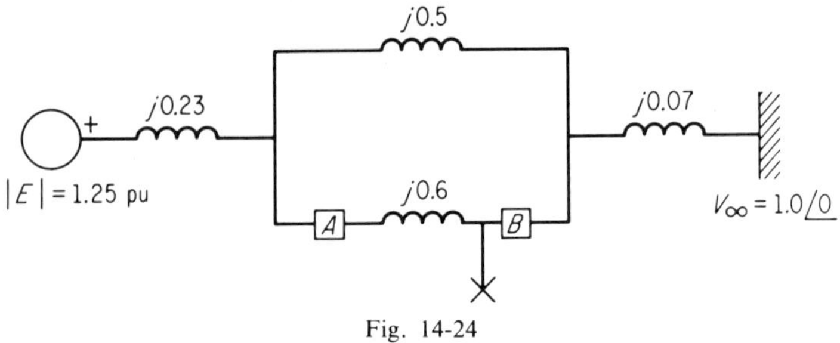

Fig. 14-24

opening of breakers A and B. The generator is delivering 1.0 per unit megawatts at the instant preceding the fault. The three separate power-transfer curves involved are,

I. Normal case, where,

$$X_{eI} = 0.23 + \frac{0.5 \times 0.6}{0.5 + 0.6} + 0.07$$

$$= \underline{\underline{0.573}} \text{ pu ohms}$$

$$P_{eI} = \frac{|E|\,|V|}{X_{eI}} \sin \delta$$

$$= \frac{1.25 \times 1.0}{0.573} \sin \delta = \underline{\underline{2.18 \sin \delta}}$$

II. Faulted case, where $P_{eII} = 0$.

III. One line open, where

$$X_{eIII} = 0.23 + 0.5 + 0.05 = \underline{\underline{0.78}}$$

$$P_{eIII} = \frac{1.25 \times 1.0}{0.78} \sin \delta$$

$$= \underline{\underline{1.603 \sin \delta}}$$

The initial operating angle δ_0 can now be determined:

$$P_{\text{shaft}} = P_e = P_{\text{maxI}} \sin \delta_0$$

or

$$1.0 = 2.18 \sin \delta_0$$

$$\sin \delta_0 = 0.458$$

$$\delta_0 = 27.3^\circ \equiv \underline{\underline{0.476}} \text{ radian}$$

The maximum allowable angle on the P_{eIII} curve of Fig. 14-25 is found as follows:

$$P_{eIII} = P_{e\,\max} \sin \delta_{\max}$$

$$1.0 = 1.603 \sin \delta_{\max}$$

$$\sin \delta_{\max} = 0.624$$

$$\delta_{\max} = 180 - 38.6^\circ = \underline{\underline{141.4^\circ}}$$

$$= \underline{\underline{2.47}} \text{ radians}$$

The equal-area criterion for this problem is graphically demon-

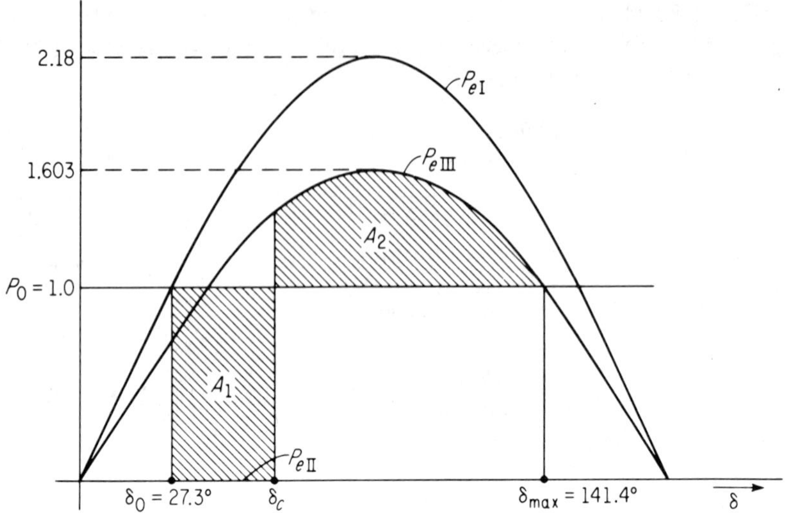

Fig. 14-25. Application of the equal-area criterion to Prob. 14-1.

strated in Fig. 14-25, where

$$\begin{aligned}
A_1 &= P_0(\delta_c - \delta_0) = 1.0(\delta_c - 0.476) \\
&= \delta_c - 0.476 \\
A_2 &= \int_{\delta_c}^{\delta_m} (P_{eIII} - P_0)\, d\delta \\
&= \int_{\delta_c}^{2.47} (1.603 \sin \delta - 1.0)\, d\delta \\
&= -1.603 \cos \delta - 1.0\delta \Big]_{\delta_c}^{2.47} \\
&= -1.603(\cos 141.4° - \cos \delta_c) - 1.0(2.47 - \delta_c) \\
&= -1.603(-0.782 - \cos \delta_c) - 2.47 + \delta_c \\
&= 1.253 + 1.603 \cos \delta_c - 2.47 + \delta_c \\
&= -1.22 + 1.603 \cos \delta_c + \delta_c
\end{aligned}$$

Setting $A_1 = A_2$ and solving for δ_c,

$$\begin{aligned}
\delta_c - 0.476 &= \delta_c - 1.22 + 1.603 \cos \delta_c \\
1.603 \cos \delta_c &= 0.744 \\
\cos \delta_c &= 0.464 \\
\delta_c &= \underline{\underline{62.3°}}
\end{aligned}$$

14-10. Transient Stability and the Point-by-Point Method

The equal-area criterion is useful in the determination of the critical fault clearing angle. The point-by-point (or step-by-step) method of solution is used for the solution of the critical fault clearing time associated with this angle. This method may also be adapted to the multimachine problem. A swing curve of δ vs. time may then be plotted for each machine under study, thus revealing any tendency for a machine to swing out of step with the rest of the system. Relay time plus breaker time is often expressed in cycles, and this time must of course be within the critical clearing time calculated. Standard interrupting times of 2, 3, 5, and 8 cycles are common and include both the relay plus breaker time. Step-by-step calculations may be made from (1) the digital computer, (2) the a-c analyzer board, or (3) hand calculation. In any case, the calculations are based upon the solution of the swing Eq. 14-35. This is not to say that the step-by-step method is the only numerical method for solving this differential equation (or equations, in a multimachine study). It is a conventional, approximate method, but it is a proven one, involving separate consecutive calculations of the rotor angles as time is incremented. Accuracy is naturally improved as the time increment is made smaller.

We will first consider the general procedure for calculating points on the swing curve for one machine tied to an infinite bus. A sudden disturbance has occured, lowering the electrical output as the transfer reactance increases. The assumption is made that both the accelerating power P_a and the angular acceleration (α) are constant from the middle of a preceding interval to the middle of an interval under study. Both of these values are calculated at the beginning of this interval. In addition, it is assumed that the angular velocity ω, computed at the middle of an interval, remains constant from the beginning to the end of the interval under study. Sufficient accuracy is generally maintained in the choice of a 0.05-sec (three cycle) interval, although 0.1 sec may be acceptable for some problems. Refer to Fig. 14-26 to aid in the understanding of the procedure. The procedure itself is outlined for the first interval as follows:

1. Calculate the initial accelerating power, where

$$\begin{aligned} P_{a(0+)} &= P_{\text{shaft}} - P_{e(0+)} \\ &= P_{\text{shaft}} - \frac{EV}{X} \sin \delta_0 \end{aligned}$$

Machine losses have been neglected.

2. Recall from the swing Eq. 14-35 that

$$P_a = M \frac{d^2 \delta}{dt^2}$$

or

$$P_a = M\alpha$$

Then the initial angular acceleration α_{0+} immediately following the disturbance is

$$\alpha_{(0+)} = \frac{P_{a(0+)}}{M} \tag{14-54}$$

If α is to be calculated in electrical degrees/sec^2, then P_a will be in mega-

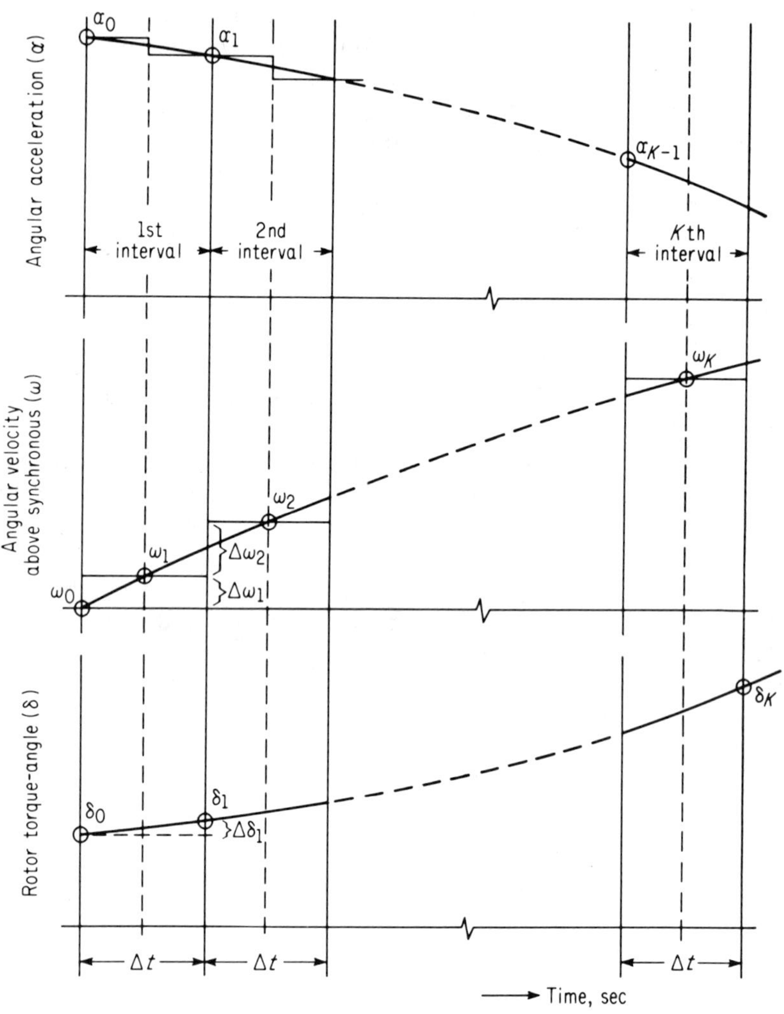

Fig. 14-26. Demonstrating the approximations used in the incremental calculations of α, ω, and δ.

joules and M takes on the value of the M' of Eq. 14-40 where $M' = \dfrac{GH}{180f}$ megajoule-sec/elect. degree.

3. The change in angular velocity for the first half of the first interval, as depicted in Fig. 14-26, can be expressed as equaling the product of the acceleration times 1/2 the regular time interval, or

$$\Delta\omega_1 = \alpha_{(0+)} \frac{\Delta t}{2} \tag{14-55}$$

For the first increment, the angular velocity (ω_1) above synchronous speed is

$$\omega_1 = \omega_0 + \Delta\omega_1$$

or

$$\omega_1 = \omega_0 + \alpha_{(0+)} \frac{\Delta t}{2} \tag{14-56}$$

4. The change in rotor angle for the first interval equals the product of this angular velocity times the time or

$$\Delta\delta_1 = \omega_1(\Delta t) \tag{14-57}$$

and the new δ at the end of the first interval is

$$\delta_1 = \delta_0 + \Delta\delta_1$$

or

$$\delta_1 = \delta_0 + \omega_1\Delta t \tag{14-58}$$

This value of δ_1 becomes the first calculated point on the swing curve of δ vs. time. For the second interval, the new δ_1 value will be used to find the new accelerating power, $P_{a(1)}$. The entire procedure is repeated for each interval in sequence.

The general equations used in interval K are:

$$P_{a(k-1)} = P_s - P_{e(k-1)} \tag{14-59}$$

where

$$P_{e(k-1)} = \frac{|E|\,|V|}{X} \sin \delta_{(k-1)} \tag{14-60}$$

$$\alpha_{k-1} = \frac{P_{a(k-1)}}{M} \tag{14-61}$$

$$^{*}\Delta\omega_k = \alpha_{k-1}\Delta t \tag{14-62}$$

$$\omega_k = \omega_{k-1} + \Delta\omega_k \tag{14-63}$$

$$\Delta\delta_k = \omega_k\Delta t \tag{14-64}$$

$$\delta_k = \delta_{k-1} + \Delta\delta_k \tag{14-65}$$

Recall from Eq. 14-55 that the time interval used for the first calculation of $\Delta\omega_1$ is only $\Delta t/2$.

If ω is not required, it is possible to eliminate Eq. 14-62 and 14-63, going directly from the calculation of acceleration to the determination of $\Delta\delta_k$. This is accomplished by substitution as follows:

$$\begin{aligned}\Delta\delta_k &= \omega_k \Delta t \\ &= (\omega_{k-1)} + \Delta\omega_k)\ \Delta t \\ &= \omega_{k-1}\Delta t + (\alpha_{k-1}\Delta t)\Delta t\end{aligned}$$

$$\Delta\delta_k = \Delta\delta_{k-1} + \frac{P_{a(k-1)}}{M}(\Delta t)^2 \tag{14-66}$$

A more detailed study of this point-by-point method is offered in Example 14-2. Values for ω will be retained in this example.

Example 14-2. Given the same circuit as was used in Example 14-1 (Fig. 14-24), again with a sudden three-phase fault at the end of one line. Shaft power input (P_s) is again 1.0 per unit. The initial machine angle δ_0 at the instant of fault was found to be 27.3°. The critical clearing angle found from Example 14-1 was 62.3°. The inertia constant $H = 4.0$. Find the critical clearing time and plot the swing curves for both (a) the sustained fault, and (b) clearing the fault at some time before the critical clearing time.

A time increment of 0.05 sec will be used. Calculations for the first interval are given in detail.

1. The accelerating power available immediately following the fault ($t = 0^+$) is

$$P_{a(0^+)} = P_s - P_e = 1.0 - 0 = \underline{\underline{1.0}}\ \text{pu}$$

2. The initial angular acceleration is

$$\alpha_{(0^+)} = P_{a(0^+)}/M'$$

where

$$M' = \frac{GH}{180f} = \frac{1.0 \times 4.0}{180 \times 60} = 3.7 \times 10^{-4}\ \text{pu}$$

then

$$\alpha_{(0^+)} = \frac{1.0}{3.7 \times 10^{-4}} = \underline{\underline{2.7 \times 10^3}}\ \text{deg/sec}^2$$

3. The change in angular velocity for the first half of the interval is:

$$\Delta\omega_1 = \alpha_{(0^+)}\,(\Delta t/2) = 2.7 \times 10^3 \times \frac{.05}{2} = 67.5\ \text{deg/sec}$$

then

$$\omega_1 = \omega_0 + \Delta\omega_1 = 0 + \underline{\underline{67.5}} \text{ deg/sec}$$

4. The change in the rotor torque angle δ for the first interval is

$$\Delta\delta_1 = \omega_1 \Delta t = 67.5 \times .05 = 3.375^\circ$$

$$\delta_1 = \delta_0 + \Delta\delta_1 = 27.3^\circ + 3.375^\circ = \underline{\underline{30.7^\circ}}$$

The calculations for the 2nd interval are,

$$P_{a(1)} = 1.0 - 0.0 = 1.0$$

$$\alpha_1 = \frac{1.0}{3.7 \times 10^{-4}} = 2.7 \times 10^3$$

$$\Delta\omega_2 = \alpha_1 \Delta t = 2.7 \times 10^3 \times .05 = 135 \text{ deg/sec}$$

$$\omega_2 = \omega_1 + \Delta\omega_2 = 67.5 + 135 = 202.5 \text{ deg/sec}$$

$$\Delta\delta_2 = 202.5 \times .05 = 10.13^\circ$$

$$\delta_2 = \delta_1 + \Delta\delta_2 = 30.7^\circ + 10.13 = \underline{\underline{40.83^\circ}}$$

Results of each interval are now placed in tabular form in Table 14-1.

TABLE 14-1
RESULTS OF SWING CURVE CALCULATIONS FOR A SUSTAINED FAULT

Interval Under Study	Time at Beginning of Interval	$\delta_0 = 27.3^\circ$ $P_e = 0$						
		P_e	$P_{a(k-1)}$	$\alpha_{(k-1)}$	$\Delta\omega_k$	ω_k	$\Delta\delta_k$	δ_k
#1	0.00	0	1.0	2.7×10^3	67.5	67.5	3.375	30.7
2	0.05	↓	↓	↓	135.0	202.5	10.13	40.8
3	0.10				↓	337.5	16.9	57.7
4	0.15					472.5	23.6	81.3
5	0.20					607.5	30.4	111.7
6	0.25					742.5	37.1	148.8
7	0.30					877.5	43.9	192.7

The swing curve for the sustained fault is plotted in Fig. 14-27. Notice that the critical angle of 62.3° (obtained in Example 14-1) is reached at $t = \underline{0.17}$ sec, corresponding to 10.4 cycles. An 8-cycle breaker should isolate the fault in time.

Calculations are next made for a new swing curve for isolating the fault in 0.15 sec or 9 cycles. Refer to Table 14-2 for results of these calculations. Notice that the accelerating power just before the fault is isolated (at 0.15 sec) is 1.0 pu, while P_a immediately following the clearing of the fault is -0.355 pu. As indicated on the table $P_e = 0$ for $t < 0.15$, and $P_e = 1.603 \sin \delta$ for $t > 0.15$ sec. The latter expression was determined in Example 14-1 for the system where the fault is cleared by opening the

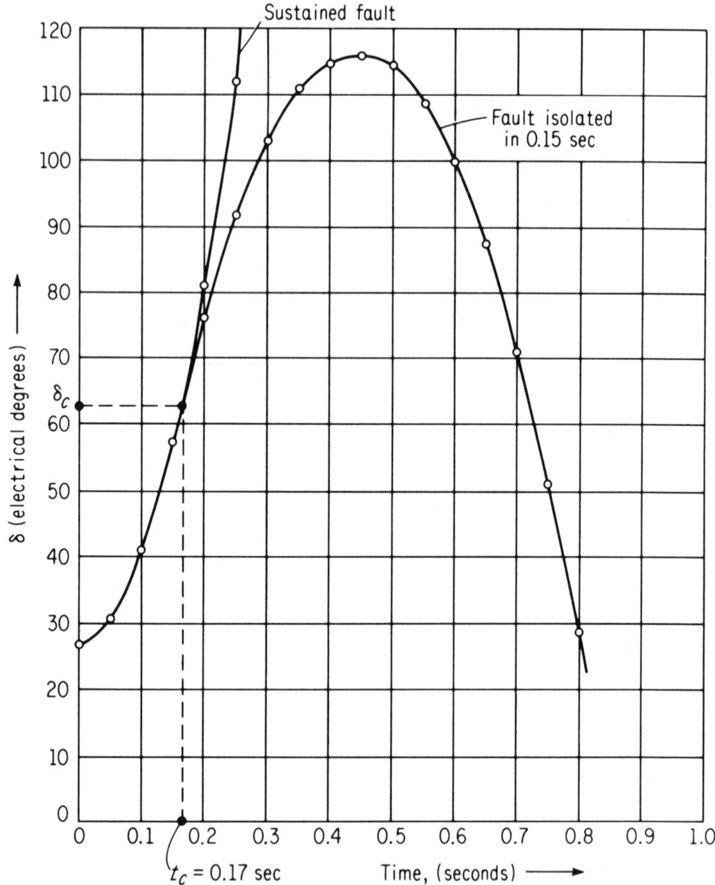

Fig. 14-27. Swing curves for Example 14-2.

faulted line at both ends. In calculating the angular velocity (ω_4) for the fourth interval, an average accelerating power is used, where

$$P_{a(\text{ave})} = \frac{1.0 - 0.355}{2} = 0.323$$

Results for the swing curve are plotted in Fig. 14-27 for comparison with the sustained fault curve.

Notice in Table 14-2 that intervals above the sixth have omitted columns of α, $\Delta\omega_k$ and ω_k. This was done in order to demonstrate that $\Delta\delta_k$ can be calculated directly from a knowledge of P_a. This was pointed out in Eq. 14-66. For example, in interval 7,

$$\Delta\delta_7 = \Delta\delta_6 + P_{a(6)}\frac{(\Delta t)^2}{M}$$

Table 14-2
RESULTS OF SWING CURVE CALCULATIONS FOR ISOLATING THE FAULT IN 0.15 SEC

Interval Under Study	Time at Beginning of Interval	$\delta_0 = 27.3, P_e = 0$ for $t < 0.15$; $P_e = 1.603 \sin \delta$ for $t > 0.15$						
		P_e	$P_{a(k-1)}$	$\alpha_{(k-1)}$	$\Delta\omega_k$	ω_k	$\Delta\delta_k$	δ_k
1	0.00	0	1.0	2.7×10^3	67.5	67.5	3.375	30.7
2	0.05	0	1.0	2.7×10^3	135	202.5	10.13	40.8
3	0.10	0	1.0	2.7×10^3	135	337.5	16.9	57.7
4	0.15−	0	1.0					
4	0.15	1.355	−0.355					
4	0.15 ave	—	0.323	8.7×10^2	43.6	381.1	19.05	76.8
5	0.20	1.56	−0.56	-1.514×10^3	−75.7	305.4	15.27	92.1
6	0.25	1.60	−0.60	-1.62×10^3	−81.0	224.4	11.22	103.3
7	0.30	1.56	−0.56	$6.76\, P_{a(k-1)} = -3.79$			7.43	110.7
8	0.35	1.50	−0.50	−3.38			4.05	114.8
9	0.40	1.455	−0.455	−3.07			0.98	115.8
10	0.45	1.444	−0.444	−3.00			−2.02	113.8
11	0.50	1.468	−0.468	−3.16			−5.18	108.6
12	0.55	1.52	−0.52	−3.51			−8.69	99.9
13	0.60	1.58	−0.58	−3.82			−12.51	87.4
14	0.65	1.60	−0.60	−4.05			−16.56	70.8
15	0.70	1.515	−0.515	−3.47			−20.03	50.8
16	0.75	1.245	−0.245	−1.65			21.68	29.1

where

$$\frac{(\Delta t)^2}{M} = \frac{(0.05)^2}{3.7 \times 10^{-4}} = 6.75$$

or

$$\begin{aligned}\Delta\delta_7 &= \Delta\delta_6 + 6.76 \times P_{a(6)} \\ &= 11.22 + 6.76 \times (-0.56) \\ &= 11.22 - 3.79 = \underline{\underline{7.43}}\end{aligned}$$

Problems

14-1. Referring to Fig. 14-2a, assume the steady-state synchronous reactance X_s is 130 percent on the machine's base of 300 mva and X_e = 6 percent on a 100-mva base. X_e includes a transformer of 10 percent reactance on a 300-mva base plus the system impedance to the infinite bus. The terminal voltage (E_t) is 1.05 pu and the current is 1.0 pu at 0.85 p.f., lagging E_t. Consider E_t as reference here.

(a) Find the infinite bus voltage V, the transformer secondary (high side) voltage E_h, the voltage E behind synchronous reactance and the power angle δ. Sketch the phasor diagram.

(b) Repeat part (a), changing I to 0.90 p.f. leading and E_t to 0.95 pu volts.

14-2. Repeat Prob. 14-1 for the transient case, where X_s is replaced with an X'_d of 20 percent on a 300-mva base.

14-3. The generator of Fig. 14-6 has a synchronous reactance of 1.3 pu on a 100-mva base. The equivalent system reactance $X_e = 0.2$ pu, also on a 100-mva base. The infinite bus voltage is 1.0 pu.

(a) The generator is on the verge of instability, delivering a maximum permissible output of 120 mw. What is the voltage behind synchronous reactance?

(b) Show the resulting phasor diagram for part (a) and find the resulting current, using the infinite bus as reference.

(c) The power output of the machine in part (a) is gradually reduced to 70 mw without touching the field excitation. Find the new current and power angle δ. Also show the new phasor diagram.

14-4. Refer to Fig. 14-11. Given a generator and a system with reactances of $X_d = 1.3$ and $X_e = 0.2$, both on a 100-mva base. Assume a generator terminal voltage of 0.95 pu. Infinite bus voltage is unknown. Find the center and radius for the pullout curve of Fig. 14-13. Draw the curve to scale and determine, from it, the minimum permissible output vars for the following pu output watts:

0.0; 0.25; 0.50; 0.75; 1.0

14-5. Given the machine and system reactances of Prob. 14-4. Find the minimum Q permissible (from a stability viewpoint) corresponding to an output of 1.0 pu mw. In place of the pullout-curve technique, use the simple vector diagram approach, only assume $|V| = 1.0$ and E_t is unknown. Compare the Q of this problem with that found in Prob. 14-4 for 1.0 pu mw output.

14-6. A 100-mva generator has a synchronous reactance of 1.2 pu ohms and feeds a system of $j15$ percent ohms on a 100-mva base. The generator has been operating at 90 mw and -10-mvar output. It is desired to increase the reactance output from -10 to $+40$ mvar by an increase in field current. Using the approximate methods of Sec. 14-7, what is the resulting change in machine terminal voltage?

14-7. Assume the machine of Prob. 14-6 is operating at 95 percent of the nameplate voltage rating with a reactive output of -10 mvar. What is the maximum reactive output in mvar permitted before the machine exceeds 105 percent of its nameplate rating (assume infinite bus voltage stays at some constant value). While the machine voltage varies 10 percent with the mvar change, how much does the generator transformer secondary voltage change? Its impedance is 10 percent on a 100-mva base.

14-8. What kinetic energy is stored in a 50-mva, 60-Hz, two-pole generator with an inertia constant (H) of 4.5 kw-sec per kva? The machine is running at synchronous speed.

14-9. Assume the machine of Prob. 14-8 has been running steadily with a shaft input (minus rotational losses) of 65,000 hp when the electrical power developed suddenly changes from its normal value to a value of 40 mw. Find the rate of rotor acceleration or deceleration.

14-10. Change the value of $|E|$ in Example 14-1 to 1.0 pu, leaving all other given information the same. Find the critical fault clearing angle.

14-11. Given the circuit and data of Fig. 14-24 only with the three-phase fault point halfway between the breakers A and B. What is the critical fault clearing angle for this circuit?

14-12. What is the critical clearing time for Prob. 14-11? Plot the swing curve for the sustained-fault case.

appendix **A**

LINE PARAMETERS

A-1. Resistance Considerations

Many references are available which give a detailed coverage of transmission-line resistance, skin effect, etc. Only a very brief summary of some basic principles and formulas is given here.

The effective a-c resistance R_e of smaller diameter conductors (at frequencies of 60 Hz and under) approaches the value of the d-c resistance. For d-c currents, uniform current distribution is assumed throughout the cross section of conductor. However, as the frequency of current increases, nonuniform distribution of currents becomes more pronounced. The phenomenon responsible for this nonuniform distribution is called *skin effect.* Skin effect is present due to the fact that outer portions of the conductor's area are not linked by flux of the internal currents, whereas internal areas see greater flux linkages. Since the inductance of any element is proportional to the flux linkages per ampere, then inner areas of the conductor offer greater reactance to current flow, especially as frequency increases. The tendency is for current to follow the outer paths of lower reactance. This reduces the effective path area, thus increasing the effective resistance of the conductor.

Some of the basic equations effecting conductor resistance will be briefly reviewed.

The d-c resistance of a conductor is directly proportional to its length (L) and inversely proportional to its area A, according to the expression

$$R = \frac{\rho L}{A} \tag{A-1}$$

The constant ρ is termed the resistivity constant at some particular temperature (often 20°C). Terms of Eq. A-1 must be in consistent units. If L is in feet and A in circular mils, then ρ is in ohm-circular mils per foot (ohm-CM/ft). These are generally the preferred units for power work. Several values of ρ at 20°C are:

$$\begin{aligned}\rho &= 10.37 \text{ ohm-CM/ft for standard annealed copper}\\ &= 10.66 \text{ ohm-CM/ft for hard-drawn copper}\\ &= 17.00 \text{ ohm-CM/ft for hard-drawn aluminum}\end{aligned}$$

Recall that a circular mil for a cylindrical conductor is found by squaring the conductor's diameter, where the diameter has been expressed in mils (1 mil = 1/1,000 inch). The conversion from square mils to circular mils is

$$\text{Number of circular mils} = (\text{number of sq mils}) \times \frac{4}{\pi} \tag{A-2}$$

Other common units for resistivity are the ohm-meter^2 per meter (or ohm-meter) and the ohm-centimeter^2 per centimeter (or ohm-centimeter).

Resistivity of any given conductor does vary linearly with temperature, according to the equation

$$\rho = \rho_0(1 + \alpha t) \tag{A-3}$$

Here ρ_0 is the resistivity taken at 0°C and t is the temperature in °C where ρ is to be found. α is known as the *temperature coefficient of resistance.* Since $R \propto \rho$ from Eq. A-1, then it follows that

$$R = R_0(1 + \alpha t) \tag{A-4}$$

Another method of expressing relative resistance in terms of temperature is given in the expression

$$\frac{R_2}{R_1} = \frac{T_0 + t_2}{T_0 + t_1} \tag{A-5}$$

Here R_1 and R_2 are resistances corresponding to the temperatures t_1 and t_2 (°C) respectively, while T_0 is a constant varying with the material. For annealed copper, T_0 is taken as 234.5°.

Skin effect, as mentioned previously will increase the effective resistance R_e. It can also have the effect of decreasing reactance as internal flux linkages become smaller. At any rate, with conductors in parallel, the current will divide inversely as the impedance of the separate lines. The overall equivalent can be written in terms of the total watts loss and the total current through the parallel combination, the basic equation being

$$R_e = \frac{P_{\text{loss}}}{I^2} \tag{A-6}$$

A-2. Dielectric Field Relationships for Parallel Plates, Point Charges, and Cylindrical Conductors

As a prelude to the investigation of transmission-line capacitance, first refer to Fig. A-1a. Recall that for parallel plates, the accumulated

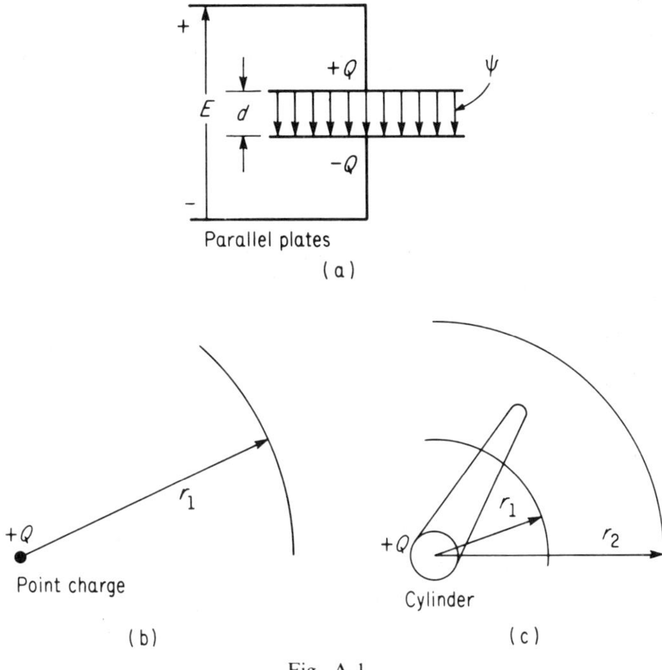

Fig. A-1

charge (Q) on the plates is proportional to the product of plate capacitance and applied voltage, or

$$Q = CE \quad \text{coulombs} \tag{A-7}$$

Also the dielectric fiux ψ which exists between the plates is equal in value to the charge Q on the plates. Dielectric flux density (D) at any point between the plates (where fringing is neglected) is

$$D = Q/A \quad \text{coulombs/meter}^2 \tag{A-8}$$

The field intensity ϵ in volts per meter is related to this density by

$$\mathcal{E} = \frac{D}{\epsilon} = \frac{Q}{\epsilon A} \tag{A-9}$$

where ϵ is the permittivity constant of the dielectric medium. Here, recall that

$$\epsilon = \epsilon_r \times \epsilon_0 \tag{A-10}$$

where ϵ_r is the relative permittivity of the medium (near unity in air) and ϵ_0 is the permittivity of free space. The value of $\epsilon_0 = 8.85 \times 10^{-12}$. The capacitance of *parallel plates* can be written in terms of the physical dimensions (area A and distance d) as

$$C = \frac{\epsilon A}{d} \tag{A-11}$$

In terms of charge and voltage, from Eq. A-7, capacitance is expressed as

$$C = Q/V \qquad \text{farads} \tag{A-12}$$

Next refer to Fig. A-1b where a *point charge* is demonstrated. Again, if the dielectric flux ψ is identical with the charge Q, then the charge density at some radial distance r is

$$D = \frac{Q}{A} = \frac{Q}{4\pi r^2} \qquad \text{coulombs/meter}^2 \tag{A-13}$$

and

$$\mathcal{E} = \frac{D}{\epsilon} = \frac{Q}{4\pi r^2 \epsilon} \qquad \text{volts/meter} \tag{A-14}$$

Finally, consider the *cylinder* of Fig. A-1c, which is more nearly akin to the transmission-line problem. The flux density and field intensity at a distance r_1 (for one unit of line length) is

$$D = \frac{Q}{A} = \frac{Q}{2\pi r_1 \times 1} \tag{A-15}$$

and

$$\mathcal{E} = \frac{D}{\epsilon} = \frac{Q}{2\pi r \epsilon} \tag{A-16}$$

In determining the voltage between points from r_1 to r_2 outside of the cylinder of Fig. A-1c, we sum (integrate) the incremental voltages between r_1 and r_2 as

$$E_{12} = -\int_{r_1}^{r_2} \mathcal{E}\, dr$$

$$= -\int_{r_1}^{r_2} \frac{Q}{2\pi \epsilon r}\, dr$$

$$E_{12} = \frac{-Q}{2\pi\epsilon} \cdot \ln \frac{r_2}{r_1} \tag{A-17}$$

A-3. Single-Phase Line Capacitance

Consider the electric field between the two parallel lines in Fig. A-2. In order to determine the voltage between conductors a and b of Fig. A-2,

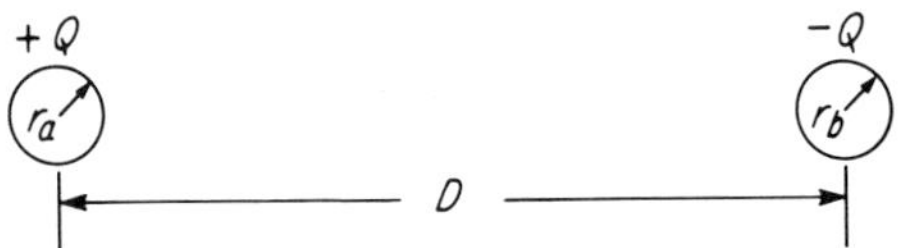

Fig. A-2. Single-phase line

one can add the separate effects of the a to b voltages obtained (1) assuming only conductor a charged, then (2) assuming only conductor b charged. First, with a positive charge on a, the voltage out to conductor b using Eq. A-17 is

$$E'_{ab} = -\frac{Q}{2\pi\epsilon} \ln \frac{D}{r_a} \tag{A-18}$$

Next with a negative charge on conductor b,

$$E''_{ba} = \frac{+Q}{2\pi\epsilon} \ln \frac{D}{r_b} \tag{A-19}$$

Adding the two voltages,

$$\begin{aligned} E_{ba} &= E'_{ba} + E''_{ba} \\ &= -E'_{ba} + E''_{ba} \\ &= -\left[\frac{-Q}{2\pi\epsilon} \ln \frac{D}{r_a}\right] + \frac{Q}{2\pi\epsilon} \ln \frac{D}{r_b} \end{aligned}$$

$$E_{ba} = \frac{Q}{2\pi\epsilon} \ln \frac{D^2}{r_a r_b} \tag{A-20}$$

Since $C = Q/E$, then the capacitance per meter of this single phase line is

$$C = \frac{2\pi\epsilon}{\ln D^2/r_a r_b} \quad \text{farad/meter} \tag{A-21}$$

but

$$\epsilon_{\text{air}} = \epsilon_r \cdot \epsilon_0 = 1 \times 8.85 \times 10^{-12}$$

and

$$\ln N = 2.303 \log N$$

and

$$\text{Number of meters} = 1609 \times \text{Number of miles}$$

Substituting into Eq. A-21, the capacitance may be expressed in farads per mile as

$$C = \frac{2\pi \times 8.85 \times 10^{-12} \times 1609}{2.3 \times \log D^2/r_a r_b}$$

$$= \frac{3.88 \times 10^{-8}}{\log D^2/r_a r_b} \quad \text{farad/mile} \qquad \text{(A-22)}$$

If $r_a = r_b$ Eq. A-22 becomes:

$$\text{Single-phase line capacitance } C_{ab} = \frac{1.94 \times 10^{-8}}{\log D/r} \quad \text{farad/mile} \qquad \text{(A-23)}$$

Capacitance to a neutral point of a two-wire line is twice that of C_{ab}, or

$$C_{an} = 2C_{ab} = \frac{3.88 \times 10^{-8}}{\log D/r} \quad \text{farad/mile, to neutral} \qquad \text{(A-24)}$$

This is easily determined by considering two capacitors in series, each equal to twice the total equivalent capacitance. Equation A-23 has assumed uniform charge distribution over the surface of the conductor, which is not quite the case. Another slight error is introduced in the case of the stranded conductor, but for most purposes it is sufficient to use the outside radius of the stranded conductor in Eqs. A-22 or A-23. Since

$$X_c = \frac{1}{2\pi f C}$$

From Eq. A-23, capacitive reactance for the 60-Hz single-phase line becomes

$$X_c = 1.37 \times 10^5 \times \log \frac{D}{r} \quad \text{ohms/mile} \qquad \text{(A-25)}$$

A-4. Capacitance of the Symmetrical Three-Phase Line

Figure A-3a depicts a symmetrical three-phase line. The voltage between any two conductors can be expressed as the sum of voltages due to each charged conductor. The basic Eq. A-17 is used in each case to determine the voltage corresponding to a particular charge. This section will determine the capacitance to neutral for the symmetrical line of Fig. A-3a. Notice from the phasors of Fig. A-3b that the voltage to neutral E_{an} can be written in terms of line to line values as

$$3E_{an} = E_{ab} + E_{ac} \qquad \text{(A-26)}$$

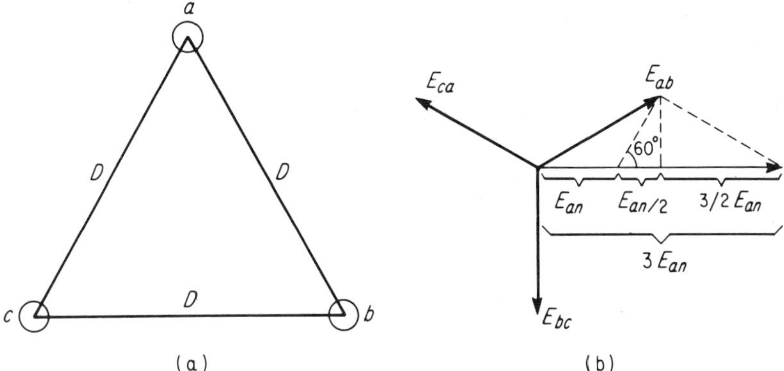

Fig. A-3. (a) Symmetrical three-phase line. (b) Vector diagram demonstrating that $E_{ab} + E_{ac} = 3\ E_{an}$.

The equations for E_{ab} and E_{ac} are

$$E_{ab} = -\frac{1}{2\pi\epsilon}\left[Q_a \ln \frac{D}{r} + Q_b \ln \frac{r}{D} + Q_c \ln \frac{D}{D}\right] \tag{A-27}$$

$$E_{ca} = -\frac{1}{2\pi\epsilon}\left[Q_a \ln \frac{r}{D} + Q_b \ln \frac{D}{D} + Q_c \ln \frac{D}{r}\right] \tag{A-28}$$

In obtaining E_{an} subtract E_{ca} of Eq. A-28 from E_{ab} of Eq. A-27. Then

$$E_{ab} + E_{ac} = -\frac{1}{2\pi\epsilon}\left[2Q_a \cdot \ln \frac{D}{r} + (Q_b + Q_c) \ln \frac{r}{D}\right] \tag{A-29}$$

If we assume no other charges in the vicinity, then $(Q_b + Q_c) = -Q_a$. Substituting into Eq. A-29 yields

$$E_{ab} + E_{ac} = -\frac{3Q_a}{2\pi\epsilon} \ln \frac{D}{r} \tag{A-30}$$

From Eq. A-26,

$$E_{an} = \frac{1}{3}(E_{ab} + E_{ac}) = \frac{1}{3}\left[-\frac{3Q_a \ln D/r}{2\pi\epsilon}\right] \tag{A-31}$$

or

$$E_{an} = -\frac{Q_a}{2\pi\epsilon} \ln \frac{D}{r} \tag{A-32}$$

but

$$C_{an} = \frac{Q_a}{E_{an}} = \frac{2\pi\epsilon}{\ln D/r} \quad \text{farad/meter} \tag{A-33}$$

Again, if more convenient units are used as was done for Eq. A-22, the resulting capacitance to neutral for the symmetrical three-phase line is

$$C_{an} = \frac{3.88 \times 10^{-8}}{\log D/r} \quad \text{farad/mile} \tag{A-34}$$

Equation A-34 is a convenient form when dealing with a system on a per phase basis. Notice Eq. A-34 is identical to the Eq. A-24 for a single-phase line.

Line-charging current I_{ch} per phase of a transmission line is expressed in amperes per mile, per unit amperes per mile, total amperes, megavars

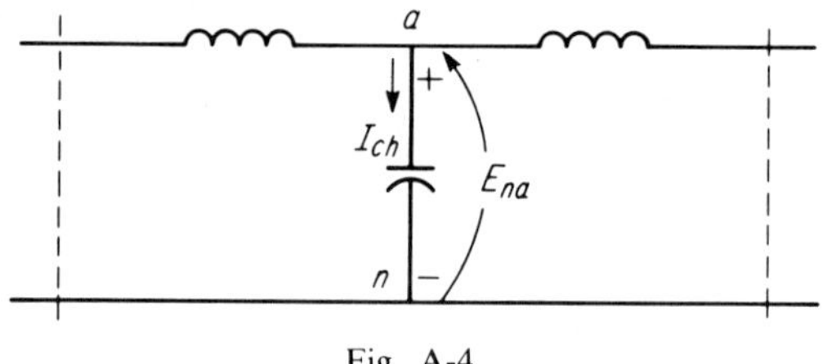

Fig. A-4

per mile, total megavars, etc. Expressed as amperes per mile, the current of Fig. A-4 is

$$I_{ch} = \frac{E_{na}}{-Jx_{an}}$$

or

$$I_{ch} = E_{na}(j\omega C_{an}) \quad \text{amp/mile} \tag{A-35}$$

Equations for unsymmetrically spaced lines are not treated here in detail. If conductors are unsymmetrically spaced, lines may be transposed to equalize line to neutral susceptance and impedance values. The equations for obtaining the average capacitance per phase are somewhat tedious. It will merely be pointed out that Eq. A-34 would necessarily be modified by replacing the distance D with an equivalent value (D_{eq}) where

$$D_{eq} = \sqrt[3]{D_{ab} D_{bc} D_{ca}} \tag{A-36}$$

D_{ab}, D_{bc}, and D_{ca} are the individual distances between conductors a, b, and c.

The effect of the earth upon the capacitance Eq. A-34 may be found for a three-phase line by assuming a mirror image of the line below the earth. Charges on the image line are of opposite sign to that of the actual line. An analysis of this effect will not be given. A general observation is made, however. When conductors are spaced high above the ground (with respect to conductor spacing) the correction applied to

Eq. A-34 is slight. The general effect of the ground is to increase the line-to-neutral capacitance.

A-5. Basic Inductance Relationships

Reviewing ohm's law for the magnetic circuit,

$$\phi = \frac{NI}{\mathcal{R}}$$

Stating this equation in words, the magnetic flux flowing within a magnetic circuit of reluctance $\mathcal{R}$ is proportional (and in phase) with the mmf and inversely proportional to $\mathcal{R}$. The familiar right-hand rule for the coil of Fig. A-5a applies. With this rule, if one grasps the coil with finger of

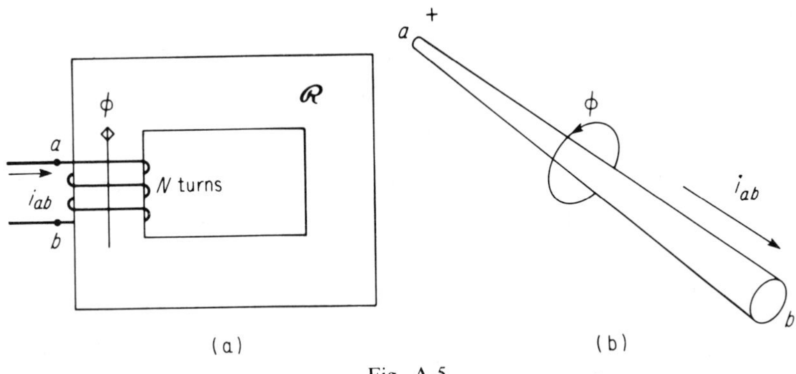

Fig. A-5

the right hand pointing in the direction of current flow, the thumb points in the direction of the flux. The right-hand rule for the single conductor of Fig. A-5b states that if the thumb of the right hand points in the direction of current flow, the fingers encompass the conductor in the direction of flux flow.

Basic voltage equations for the single line of Fig. A-5b are

$$e_{ab} = \frac{-L di_{ab}}{dt} = -\frac{d(N\phi)}{dt} \tag{A-37}$$

The term $N\phi$ is also referred to as τ with the unit of weber turns. Solving Eq. A-37 for the inductance L,

$$L = \frac{d(N\phi)}{di} = \frac{d\tau}{di} \quad \text{henrys} \tag{A-38}$$

When the magnetic medium has a straight-line saturation curve, Eq. A-38 becomes

$$L = \frac{\tau}{i} \quad \text{henrys} \tag{A-39}$$

where τ is the total flux linkage in weber turns ($N\phi$) or, in terms of L,

$$\tau = Li \tag{A-40}$$

MKS units will be used here, where ϕ is in webers, i is in amperes, and L is in henrys. For the a-c line with sinusoidal current, Eq. A-39 becomes

$$L = \psi_t / I \tag{A-41}$$

and

$$X = 2\pi f L = \frac{2\pi f \psi_t}{I_{\text{rms}}} \quad \text{ohms} \tag{A-42}$$

where ψ_T represents the equivalent rms flux linkages of the line. The rms phaser voltage from b to a of Fig. A-5b is

$$E_{ba} = I_{ab} Z = I_{ab}(jx)$$

$$= I_{ab} \left(\frac{j2\pi f \psi_t}{I_{ab}} \right)$$

$$E_{ba} = j2\pi f \psi_t \quad \text{volts} \tag{A-43}$$

ψ_t is an rms phasor which is in phase with I_{ab}.

Next refer to Fig. A-6. By the right-hand rule, current flowing into the paper would cause clockwise flux in each element dx as shown.

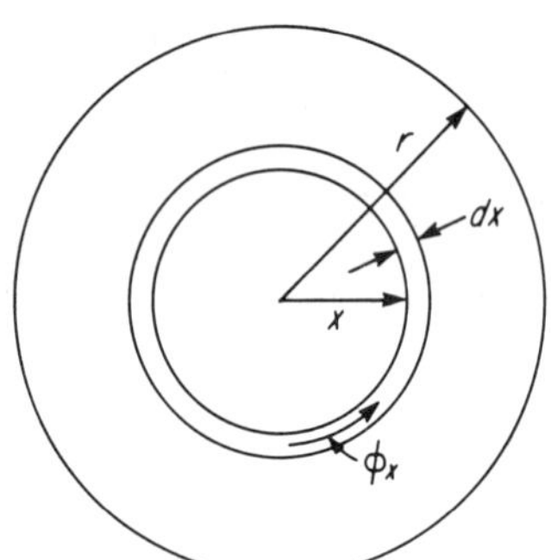

Fig. A-6. Cylindrical conductor cross section.

Assume uniform current density. The total flux linkages are made up of the external linkages (ψ_{ex}) which link all the current and the partial or internal flux linkages (ψ_{int}), which link only part of the current. In order to apply Eq. A-41, ψ_t is broken up so that

$$\psi_t = \psi_{\text{ex}} + \psi_{\text{int}} \tag{A-44}$$

A-6. Partial Flux Linkages

First, consider the internal flux linkages of Fig. A-6. Taking a tubular cylinder of infinitesimal thickness dx, the total flux linkages will be summed to determine ψ_{int}. The flux per meter of length in the tube is

$$d\phi_x = B_x A_x = B_x \cdot dx \cdot 1 \tag{A-45}$$

$$d\phi_x = \mu H_x \cdot dx \tag{A-46}$$

$$d\phi_x = \frac{\mu I_x}{2\pi x}\, dx \tag{A-47}$$

The current I_x is all the current inside of the tube. Again, assuming equal current density, I_x is a fraction of the total current, or

$$I_x = \left(\frac{\pi x^2}{\pi r^2}\right) \cdot I = \left(\frac{x^2}{r^2}\right) I \tag{A-48}$$

Substituting results of Eq. A-48 into Eq. A-47,

$$d\phi_x = \mu \cdot \frac{I \cdot x}{2\pi r^2}\, dx \tag{A-49}$$

But the flux $(d\phi_x)$ links only a fraction of the total current in the conductor:

$$d\psi = \frac{A_x}{A_{\text{tot}}}\, d\phi = \frac{x^2}{r^2} \cdot \frac{\mu I x}{2\pi r^2}\, dx \tag{A-50}$$

$$d\psi = \frac{\mu I x^3}{2\pi r^4}\, dx$$

Now summing all of these flux linkages from the center outward to where $x = r$,

$$\psi_{\text{int}} = \int_0^r \frac{\mu I x^3\, dx}{2\pi r^4} \tag{A-51}$$

$$\psi_{\text{int}} = \frac{\mu I}{8\pi} \tag{A-52}$$

Given a medium with a relative permeability of unity. In the MKS system, $\mu = 4\pi \times 10^{-7}$. Then

$$\psi_{\text{int}} = \frac{I}{2} \times 10^{-7} \quad \text{weber turns/meter} \tag{A-53}$$

from Eq. A-41

$$L_{\text{int}} = \frac{1}{2} \times 10^{-7} \quad \text{henry/meter} \tag{A-54}$$

A-7. External Flux Linkages

Refer to Fig. A-7. The tube of thickness dx outside the conductor

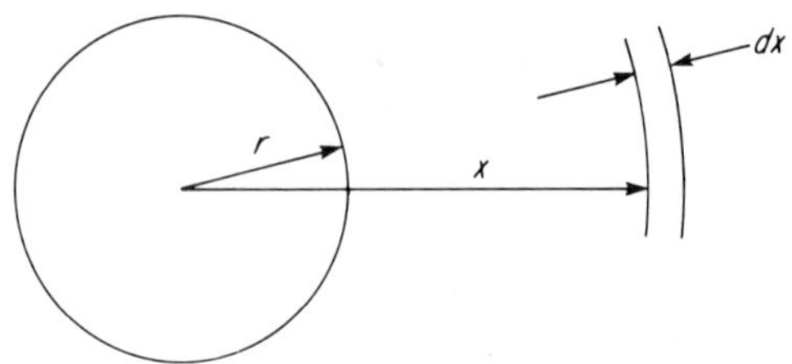

Fig. A-7

has a flux density of

$$B_x = \mu H_x = \frac{\mu I}{2\pi x}$$

The external linkages up to a distance D outside of the conductor are

$$\psi_{\text{ext}} = \int_r^D B_x \, dx = \int_r^D \left(\frac{\mu I}{2\pi x}\right) dx \qquad \text{(A-55)}$$

$$\psi_{\text{ext}} = \frac{\mu I}{2\pi} \ln \frac{D}{r} \qquad \text{(A-56)}$$

Again for a value of $\mu = 4\pi \times 10^{-7}$, Eq. A-56 becomes

$$\psi_{\text{ext}} = 2 \times 10^{-7} I \ln \frac{D}{r} \quad \text{weber-turns/meter} \qquad \text{(A-57)}$$

$$L_{\text{ext}} = 2 \times 10^{-7} \ln \frac{D}{r} \quad \text{henrys/meter} \qquad \text{(A-58)}$$

Equations A-57 and A-58 for external flux and linkage can be applied to the flux linkages and inductance taken between two points, 1 and 2, if D is replaced by D_2 and r is replaced by D_1, or

$$\left.\begin{aligned} \psi_{12} &= 2 \times 10^{-7} I \ln \frac{D_2}{D_1} \\ L_{12} &= 2 \times 10^{-7} \ln \frac{D_2}{D_1} \end{aligned}\right\} \qquad \text{(A-59)}$$

A-8. Total Inductance and Conductor GMR

The total flux linkages from Eqs. A-52 and A-56 are

$$\psi_{\text{tot}} = \psi_{\text{int}} + \psi_{\text{ext}} \qquad \text{(A-60)}$$

$$\psi_{tot} = \frac{\mu I}{8\pi} + \frac{\mu I}{2\pi} \ln \frac{D}{r} \tag{A-61}$$

and

$$L_{tot} = \frac{\mu}{8\pi} + \frac{\mu}{2\pi} \ln \frac{D}{r} \tag{A-62}$$

again where

$$\mu = 4\pi \times 10^{-7}$$

$$L_{tot} = \left(\frac{1}{2} + 2 \ln \frac{D}{r}\right) 10^{-7} \qquad \text{henrys/meter} \tag{A-63}$$

The two terms of Eq. A-63 may be combined as follows:

$$\begin{aligned} L_{tot} &= 2 \times 10^{-7} \left(\frac{1}{4} + \ln \frac{D}{r}\right) \\ &= 2 \times 10^{-7} \left(\ln \epsilon^{0.25} + \ln \frac{D}{r}\right) \\ &= 2 \times 10^{-7} \left(\ln \frac{D}{\epsilon^{-0.25} \times r}\right) \end{aligned}$$

$$L_{tot} = 2 \times 10^{-7} \ln \frac{D}{r'} \qquad \text{henrys/meter} \tag{A-64}$$

where

$$r' = \epsilon^{-0.25} \times r = \underline{\underline{0.779r}} \tag{A-65}$$

(Compare Eq. A-64 with Eq. A-58.)

r' is known as the geometric mean radius (GMR), representing the radius of a hollow tube which replaces the solid conductor. The hollow tube is of such small thickness that there are no internal flux linkages. This procedure is possible where permeability of the conductor and the air are the same.

A-9. Single-Phase Line Inductance

The single-line inductance of Eq. A-64 can readily be applied to the single-phase line of Fig. A-8, where one conductor is the return path for the other ($I_a = -I_b$).

Let ψ_1 of Fig. A-8 represent the flux linkages linking only the current I_a. The inductance corresponding to these linkages is

$$L_{\psi_1} = 2 \times 10^{-7} \ln \frac{D}{r'_a} \tag{A-66}$$

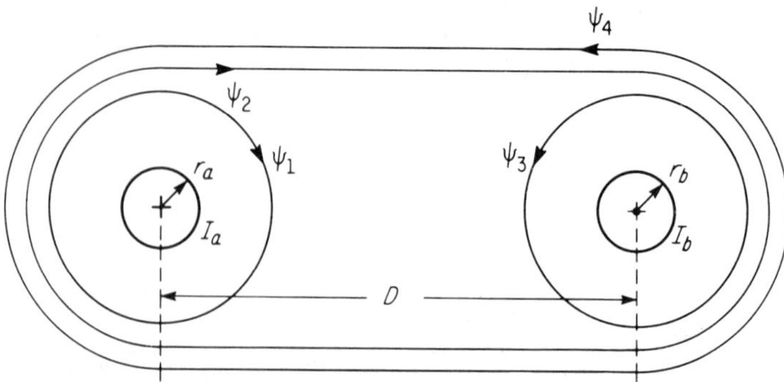

Fig. A-8. Cross section of a single-phase line.

ψ_3 represents the flux linkages which link only I_b. The inductance associated with these linkages is

$$L_{\psi_3} = 2 \times 10^{-7} \ln \frac{D}{r'_b} \tag{A-67}$$

ψ_2 and ψ_4 represent the flux linkages, due to I_a and I_b (totaling to zero current). The effective inductance to the single-phase series circuit due to each of these two linkages (ψ_2 and ψ_4) is zero.

The total inductance of the two-wire line is then the sum of L_{ψ_1} and L_{ψ_3}, or

$$L = 2 \times 10^{-7}\left(\ln \frac{D}{r'_a} + \ln \frac{D}{r'_b}\right)$$

$$= 2 \times 10^{-7} \ln \frac{D^2}{r'_a r'_b} \tag{A-68}$$

$$L = 4 \times 10^{-7} \ln \frac{D}{\sqrt{r'_a r'_b}} \quad \text{henrys/meter} \tag{A-69}$$

or

$$L = 1.482 \log \frac{D}{\sqrt{r'_a r'_b}} \quad \text{mh/mile} \tag{A-70}$$

A-10. Three-Phase Line Inductance

Given the four-wire, three-phase line of Fig. A-9. The flux linkages of phase a are first to be determined out to point p. Applying Eq. A-58, the linkage due to I_a is

$$\psi_{apa} = 2 \times 10^{-7} I_a \ln \frac{D_{ap}}{r'_a} \tag{A-71}$$

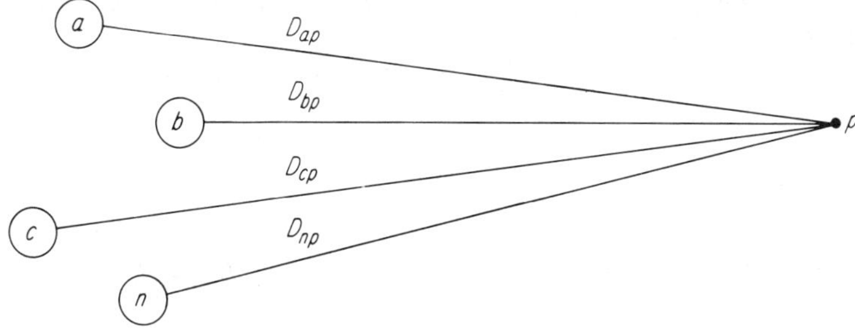

Fig. A-9

Phase a linkage due to I_b is

$$\psi_{apb} = 2 \times 10^{-7} I_b \ln \frac{D_{bp}}{D_{ab}} \tag{A-72}$$

Similar equations exist for ψ_{apc} and ψ_{apn}. Adding these linkages, we obtain the total linkage for phase a out to point p as

$$\psi_{ap} = 2 \times 10^{-7} \left(I_a \ln \frac{D_{ap}}{r'_a} + I_b \ln \frac{D_{bp}}{D_{ab}} + I_c \ln \frac{D_{cp}}{D_{ac}} + I_n \ln \frac{D_{np}}{D_{an}} \right) \tag{A-73}$$

Rearranging Eq. A-73 and substituting $I_n = -(I_a + I_b + I_c)$,

$$\psi_{ap} - 2 \times 10^{-7} \left[I_a \ln \frac{1}{r'_a} + I_b \ln \frac{1}{D_{ab}} + I_c \ln \frac{1}{D_{ac}} + I_n \ln \frac{1}{D_{an}} + I_a \ln \frac{D_{ap}}{D_{np}} + I_b \ln \frac{D_{bp}}{D_{np}} + I_c \ln \frac{D_{cp}}{D_{np}} \right] \tag{A-74}$$

As the point p is moved far away from the group of conductors, the last three terms of Eq. A-73 drop out. Equations for the flux linkages of phases a, b, and c are

$$\left.\begin{aligned}
\psi_a &= 2 \times 10^{-7} \left[I_a \ln \frac{1}{r'_a} + I_b \ln \frac{1}{D_{ab}} + I_c \ln \frac{1}{D_{ac}} + I_n \ln \frac{1}{D_{an}} \right] \\
\psi_b &= 2 \times 10^{-7} \left[I_b \ln \frac{1}{r'_b} + I_a \ln \frac{1}{D_{ab}} + I_c \ln \frac{1}{D_{bc}} + I_n \ln \frac{1}{D_{bn}} \right] \\
\psi_c &= 2 \times 10^{-7} \left[I_c \ln \frac{1}{r'_c} + I_a \ln \frac{1}{D_{ac}} + I_b \ln \frac{1}{D_{bc}} + I_n \ln \frac{1}{D_{cn}} \right]
\end{aligned}\right\} \tag{A-75}$$

If complex values of I_a, I_b, and I_c are known, inductances L_a, L_b, and L_c can be found, even for the untransposed line. For example, ψ_a could be determined and divided by the a phase current to obtain the inductance for phase a.

For the *transposed three-wire line*, the average flux linkage for phase a could be found by adding the equations for ψ_a which are obtained by interchanging D_{ab}, D_{ac}, and D_{an}. The result would be divided by three, yielding

$$\psi_{a(\text{ave})} = \frac{2 \times 10^{-7}}{3}\left[3I_a \ln \frac{1}{r'_a} + (I_b + I_c) \ln \frac{1}{D_{ab} D_{ac} D_{cb}}\right] \qquad \text{(A-76)}$$

Substituting $I_b + I_c = -I_a$,

$$\psi_{a(\text{ave})} = 2 \times 10^{-7}\left[I_a \ln \frac{\sqrt[3]{D_{ab} D_{ac} D_{cb}}}{r'_a}\right] \qquad \text{(A-77)}$$

and

$$L_a = 2 \times 10^{-7} \ln \frac{\sqrt[3]{D_{ab} D_{ac} D_{cb}}}{r'_a} \quad \text{henry/meter} \qquad \text{(A-78)}$$

The equations for L_b and L_c would be identical. The term $\sqrt[3]{D_{ab} D_{ac} D_{cb}}$ is sometimes referred to as the effective spacing (D_e). Ohmic reactance for each line is given as

$$X = 2\pi f L$$

or

$$X = 4\pi \times 10^{-7} f \ln \frac{D_e}{r'} \quad \text{ohms/meter/phase} \qquad \text{(A-79)}$$

Much could be said regarding inductance of parallel-circuit lines, inductance of stranded conductors, inductance by geometric mean distances, etc. Space does not permit such coverage here.

appendix **B**

CIRCLE DIAGRAMS

B-1. Introduction

A number of electrical problems lend themselves to graphical analysis, involving circle diagrams, where trends and solutions can be found over a wide range of the variable. As an example, refer to Fig. B-1 which

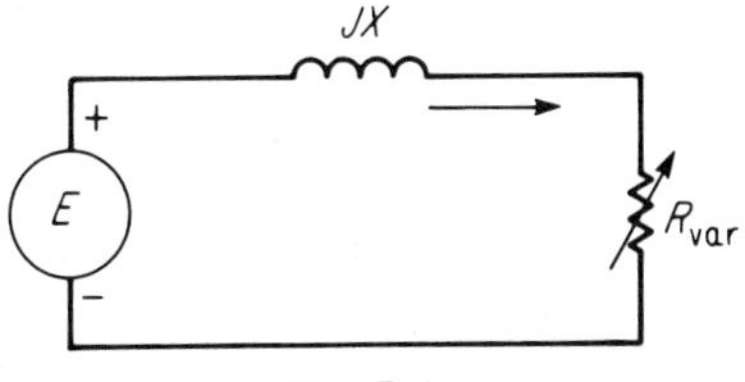

Fig. B-1

demonstrates a simple series circuit with variable resistance. The equation for this simple circuit is

$$I = \frac{E}{Z} = \frac{E \sin \theta}{X} \tag{B-1}$$

The last form represents the polar form of an equation for a circle. The rectangular form is found by multiplying Eq. B-1 by I, yielding

$$I^2 = \frac{EI}{X} \sin \theta \tag{B-2}$$

Rewriting in terms of horizontal and vertical current components, where $I_x = I \sin \theta$,

$$I_y^2 + I_x^2 = \frac{EI_x}{X} \tag{B-3}$$

The standard rectangular form of the equation of a circle is

$$(x - h)^2 + (y - k)^2 = a^2 \tag{B-4}$$

The center of this circle represented by Eq. B-4 is at h,k with a radius equal to a. Now rewriting Eq. B-3 and placing in standard form,

$$\left(I_y^2 - \frac{EI_x}{X}\right) + I_y^2 = 0$$

$$\left(I_x^2 - \frac{EI_x}{X} + \left(\frac{E}{2X}\right)^2\right) + I_y^2 = \left(\frac{E}{2X}\right)^2$$

$$\left(I_x - \frac{E}{2X}\right)^2 + (I_y)^2 = \left(\frac{E}{2X}\right)^2 \tag{B-5}$$

The center of this circle is at $(E/2X, 0)$ and the radius is at $E/2X$. The diagram is given in Fig. B-2. I_y can be considered the power component

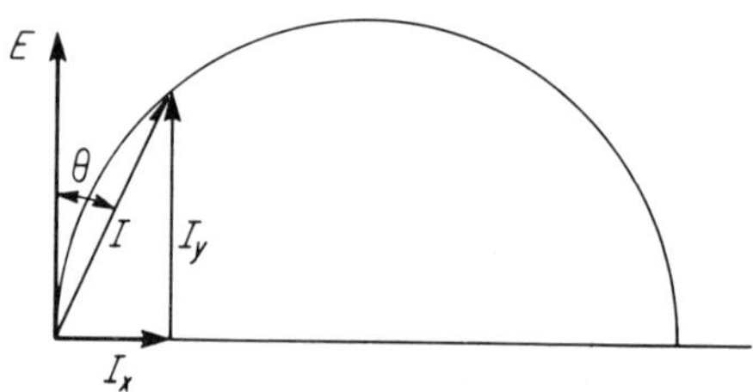

Fig. B-2. Circle diagram for circuit of Fig. B-1.

of current (in phase with E) while I_x is the reactive component. An outstanding example of the application of such a diagram is found in the three-phase induction motor application. Its approximate equivalent circuit also contains a fixed reactance in series with a resistance which varies with the motor slip. Space does not permit detailed coverage of this application. Rather, it will merely be pointed out that, without the circle diagram, separate calculations are necessary for each value of motor slip in order to determine motor performance. However, *with* the circle diagram approach, it is a simple matter (by measuring line segments) to obtain all pertinent performance information such as power factor, horsepower, efficiency, and torque.

Another application of the circle diagram more akin to the network problem is that of the sending and receiving circle diagrams for a four-terminal network.

B-2. Sending and Receiving Circles

Refer to the four-terminal network of Fig. B-3, where

$$E_s = AE_R + BI_R \tag{B-6}$$

$$I_s = CE_R + DI_R \tag{B-7}$$

$$A = |A| \underline{/\alpha}, \quad B = |B| \underline{/\beta}, \quad D = |D| \underline{/\Delta}$$

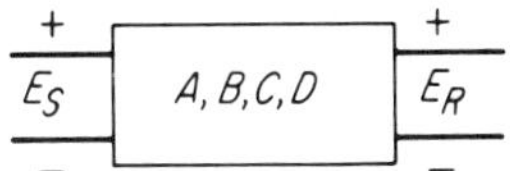

Fig. B-3. Four-terminal network.

The phasor diagram for Eq. B-6 is given in Fig. B-4. We have taken the phasor E_R as reference. Now if $|E_R|$ and $|E_s|$ are held constant, E_s will swing about the origin as indicated by the arc. In order to convert the

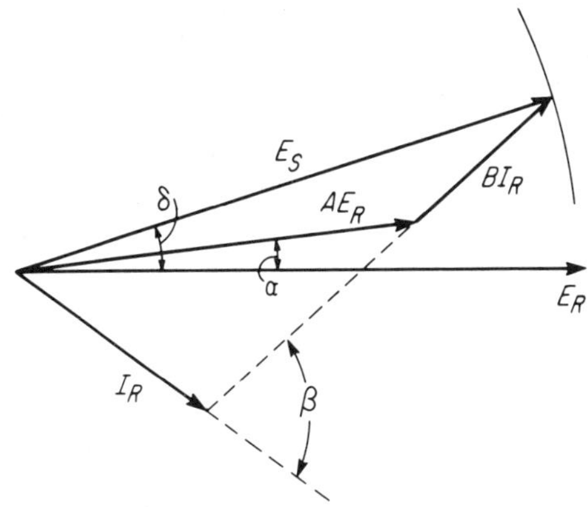

Fig. B-4. Phasor diagram for Eq. B-6.

voltage phasor diagram of Fig. B-4 to a useful *receiving-end* power diagram, we multiply all phasors of Eq. B-6 by E_R/B, which yields

$$\frac{E_s E_R}{B} = \frac{A E_R^2}{B} + E_R \cdot I_R \tag{B-8}$$

The factor $(|E_R| \underline{/0^\circ})/|B| \underline{/\beta}$ has shifted all phasors of Fig. B-4 by the angle of $-\beta°$. Figure B-5 is the new receiving-end power phasor diagram where the desired value of $|E_R|\,|I_R| \underline{/\theta_R}$ is now apparent with its components of real power (horizontal component) and reactive power (vertical component).

The phasor *mn* has a constant length of $(|E_s||E_R|)/|B|$ and a variable angle of $\beta - \delta$, since the torque angle δ (between E_s and E_R) will vary with the load. This phasor *mn* will swing about point *m*, while the phasor AE_R^2/B will remain fixed in both magnitude and phase angle. Inductive vars are normally taken as positive (with current lagging the voltage). In Fig. B-5 we have allowed the phasor $E_R I_R$ to take on the angle of the current I_R, lagging the voltage in this case. Actually,

$$E_R I_R^* = P + jQ \tag{B-9}$$

where the asterisk denotes the conjugate of I_R. One may either (1) realize that lagging reactive power on the diagram corresponds to a positive Q or (2) redraw the phasor diagram by rotating Fig. B-5 about its horizontal axis. We have chosen here to leave the diagram as is.

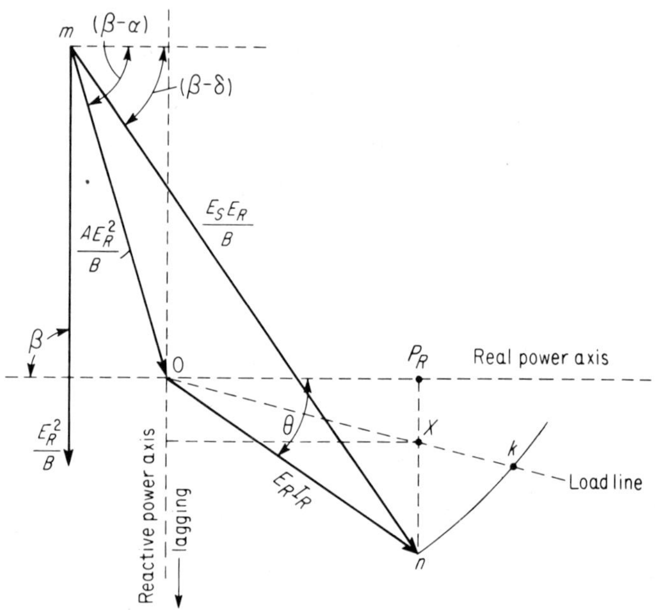

Fig. B-5. Receiving-end circle diagram with $|E_s|$ and $|E_R|$ constant. E_R is reference.

A load line is included on Fig. B-5 for a given load power factor. If a synchronous condenser is placed at the receiving end (automatically holding E_R constant) then load magnitude is free to vary. The synchronous condenser will adjust to provide the Q necessary to satisfy the diagram. Otherwise (without the presence of synchronous condenser, shunt capacitors or automatic tap-changing transformer), only one load point (k) will satisfy the diagram. If for a given load and no synchronous condensers, the diagram is not satisfied for a given $|E_R|$, then the value of $|E_s|$ would have to be adjusted at the sending end.

For the given load point x, the receiver power is found on the real axis as P_R and the load reactive power as Q_R. The synchronous condensers must make up the difference by absorbing vars equal to the line xn.

If the sending voltage $|E_s|$ is altered, with $|E_R|$ remaining constant, a family of curves is possible as mn varies in length. The center of the circles remains at point m.

If $|E_s|$ is held constant and $|E_R|$ is varied, the center point (m) is moved along the line mo, since the angles of β and α are unchanged.

SENDING CIRCLES

In Chapter 2 it was shown that when Eq. 2-10 is rearranged by inverting the a matrix, we obtain

$$\begin{bmatrix} E_R \\ I_R \end{bmatrix} = \begin{bmatrix} D & -B \\ -C & A \end{bmatrix} \begin{bmatrix} E_s \\ I_s \end{bmatrix} \qquad \text{(B-10)}$$

Let

$$E_s = |E_s| \angle 0^\circ, \quad I_s = |I_s| \angle \theta_s$$
$$B = |B| \angle \beta, \quad D = |D| \angle \Delta, \quad E_R = |E_R| \angle \delta \qquad \text{(B-11)}$$

From matrix Eq. B-10 we obtain

$$E_R = DE_s - BI_s$$

or

$$DE_s = E_R + BI_s \qquad \text{(B-12)}$$

The phasor diagram for Eq. B-12 is seen in Fig. B-6. Multiplying Eq. B-12 by the factor of E_s/B [or $(|E_s| \angle 0^\circ)/|B| \angle \beta$] will rotate all phasors

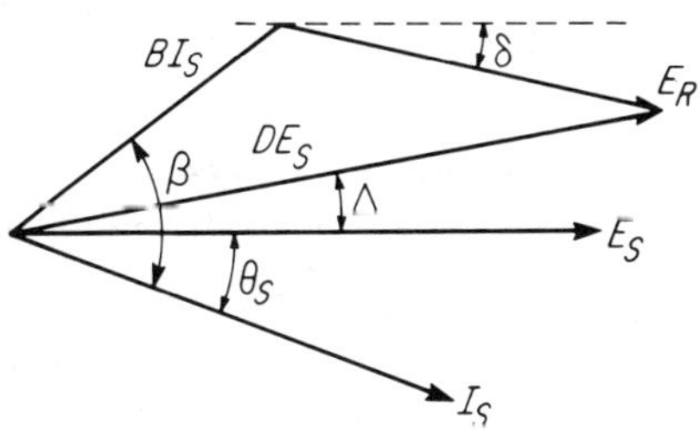

Fig. B-6. Phasor diagram for Eq. B-12.

of Fig. B-6 by $-\beta^\circ$. The new sending-end power equation becomes

$$\frac{DE_s^2}{B} = \frac{E_s E_R}{B} + E_s I_s \qquad \text{(B-13)}$$

The sending-end power diagram is produced in Fig. B-7 with the desired value of $|E_s||I_s| \angle \theta_s$ apparent. Here the phasor om' remains fixed while $m'n'$ swings about point m' with a constant magnitude of $(|E_s||E_R|)/|B|$. Again the horizontal and vertical components of the $E_s I_s$ phasor represent the real and reactive components of sending power, respectively. In this diagram, a lagging reactive component corresponds to a positive Q delivered by the sending end.

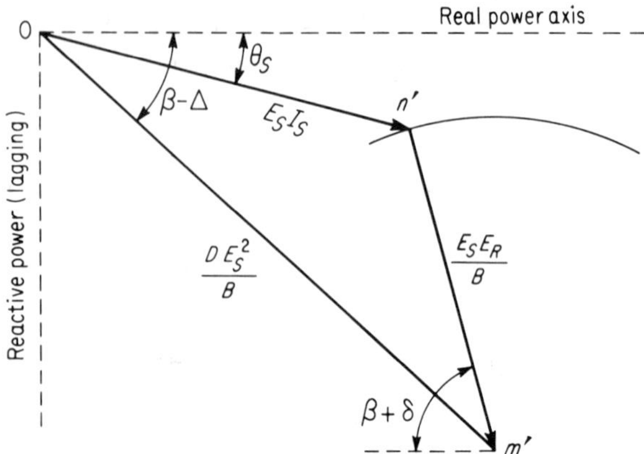

Fig. B-7. Sending-end circle diagram with $|E_s|$ and $|E_R|$ constant. E_s is reference.

The sending and receiving circles may be joined together according to Fig. B-8. The arc of the receiving circle revolves about point m, while that of the sending circle moves about m'. The load point is shown beyond the receiving-end arc. A synchronous generator on the receiving end would then be required to draw leading vars in the amount of xn in order to satisfy the conditions of the diagram. Once the angle of $\beta - \delta$ (and δ) is established, the angle of $\beta + \delta$ can be constructed on the sending circle, which fixes the line $m'n'$. The difference between P_s and P_R represents transmission line losses.

The *maximum power* which can be transmitted over a four-terminal network of $ABCD$ constants is also the steady-state stability limit for the network. For the receiving end, the power received may be written by observing the phasors of Fig. B-8. It is

$$P_R = \frac{|E_s||E_R|}{|B|} \cos(\beta - \delta) - \frac{|A||E_R|^2}{|B|} \cos(\beta - \alpha) \quad \text{(B-14)}$$

The torque angle δ is variable and the maximum value of P_R occurs when $\cos(\beta - \delta) = 1$, or when $\beta = \delta$. Then

$$P_{R\,\max} = \frac{|E_s||E_R|}{|B|} - \frac{|A||E_R|^2}{|B|} \cos(\beta - \alpha) \quad \text{(B-15)}$$

A SPECIAL CASE

Refer to the special case of Fig. B-9 in which the transmission line is a simple series reactance. From Chapter 3, Eq. 3-2, $A = 1 \angle 0^\circ$, $B =$

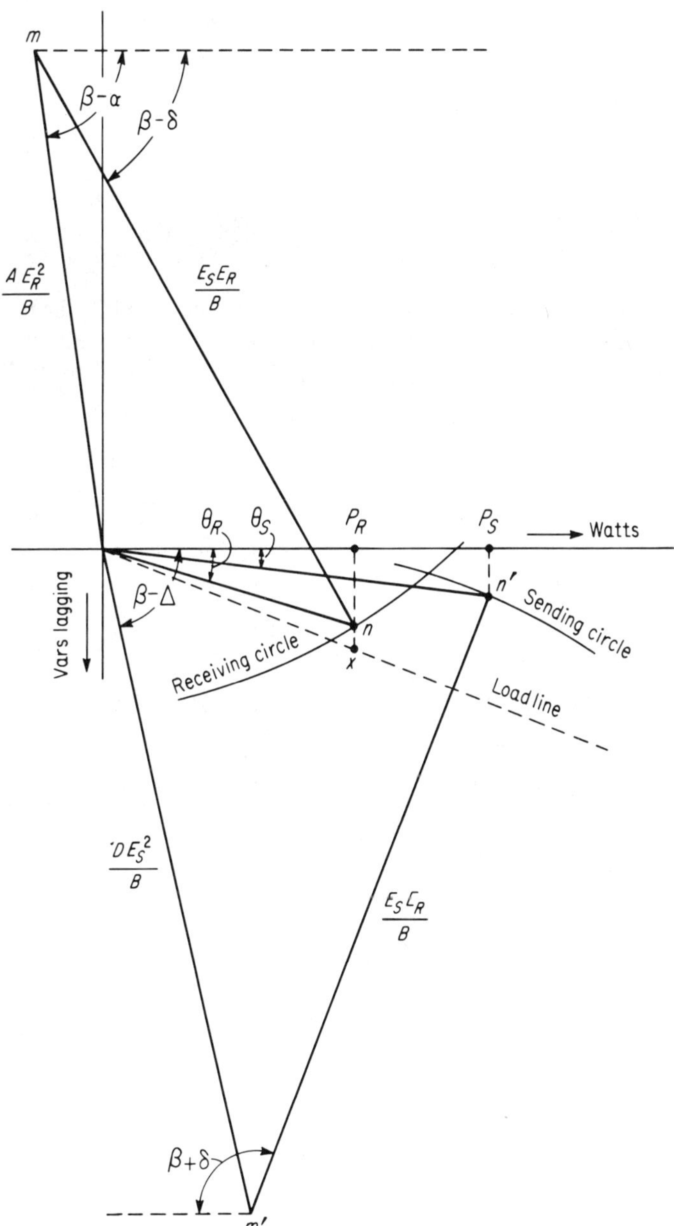

Fig. B-8. Sending and receiving circles diagrams combined.

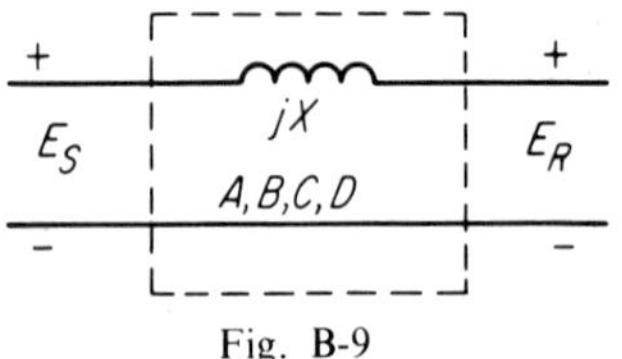

Fig. B-9

$X \angle 90°$, $C = 0$, $D = 1 \angle 0°$. Substituting these values into Eq. B-15 gives

$$P_{\max} = \frac{|E_s||E_R|}{X} \tag{B-16}$$

This result agrees with Eqs. 4-1 and 14-15 where $\delta = 90°$. The receiving phasor diagram for this special case may be found by substituting the appropriate *ABCD* constants on Fig. B-5, which results in Fig. B-10a.

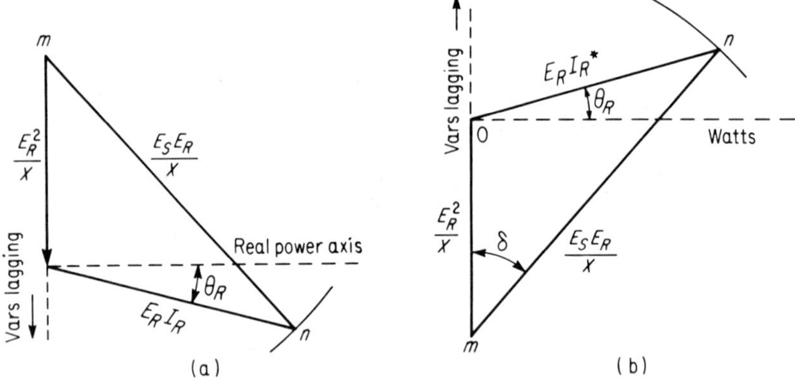

Fig. B-10. (a) Receiving-end circle for special line of reactance only. (b) Lagging reactive power appears as positive vars.

Figure B-10b has rotated the diagram about the real power axis in order for the lagging reactive power to appear as positive vars.

appendix C

FUNDAMENTALS OF DETERMINANTS AND MATRICES

C-1. Determinants

One major application of determinants for the engineer is found in the solution of problems involving linear, simultaneous equations. The determinant is a square array (n rows and n columns) of numbers or quantities.

Given a system of three linear equations in three unknowns I_1, I_2, and I_3.

$$\left.\begin{aligned} (1)\quad Z_{11}I_1 + Z_{12}I_2 + Z_{13}I_3 &= E_1 \\ (2)\quad Z_{21}I_1 + Z_{22}I_2 + Z_{23}I_3 &= E_2 \\ (3)\quad Z_{31}I_1 + Z_{32}I_2 + Z_{33}I_3 &= E_3 \end{aligned}\right\} \tag{C-1}$$

This system may be solved by means of Cramer's rule, where

$$I_1 = \frac{D_1}{D}; \quad I_2 = \frac{D_2}{D}; \quad I_3 = \frac{D_3}{D} \tag{C-2}$$

Equation C-2 is true when the determinant $D \neq 0$. Four determinants are involved in the solution: D, D_1, D_2, and D_3. A numerical value is associated with each determinant as will be shown. The denominator determinant D common to all unknowns is formed by the coefficients of the unknowns, or

$$D = \begin{vmatrix} Z_{11} & Z_{12} & Z_{13} \\ Z_{21} & Z_{22} & Z_{23} \\ Z_{31} & Z_{32} & Z_{33} \end{vmatrix} \tag{C-3}$$

The numerator determinant (D_k) in the solution of I_k is formed by replacing the kth row of coefficients by the corresponding constant elements on the right-hand side of the equal sign in Eq. C-1. These determinants are

$$D_1 = \begin{vmatrix} E_1 & Z_{12} & Z_{13} \\ E_2 & Z_{22} & Z_{23} \\ E_3 & Z_{32} & Z_{33} \end{vmatrix}; \qquad D_2 = \begin{vmatrix} Z_{11} & E_1 & Z_{13} \\ Z_{21} & E_2 & Z_{23} \\ Z_{31} & E_3 & Z_{33} \end{vmatrix}$$

$$D_3 = \begin{vmatrix} Z_{11} & Z_{12} & E_1 \\ Z_{21} & Z_{22} & E_2 \\ Z_{31} & Z_{32} & E_3 \end{vmatrix}$$

Two definitions should be offered here. A *minor* of an element of a determinant is defined as another determinant obtained by deleting the row and column where the element is found. A *cofactor* of an element in the ith row and kth column is defined as $(-1)^{i+k}$ times the minor of the element.

EVALUATING A DETERMINANT—GENERAL

In general, a determinant is equal to the sum of the products of the elements of any row (or column) and their respective cofactors.

EVALUATION OF A SECOND-ORDER DETERMINANT

Given a two order determinant below

$$D = \begin{vmatrix} a_{11} & a_{12} \\ a_{21} & a_{22} \end{vmatrix} \tag{C-4}$$

The value of D is found as

$$D = a_{11}a_{22} - a_{12}a_{21} \tag{C-5}$$

EVALUATION OF A THIRD-ORDER DETERMINANT

The general rule for evaluation of a determinant will be applied to the third-order determinant of Eq. C-3.

$$D = \begin{vmatrix} Z_{11} & Z_{12} & Z_{13} \\ Z_{21} & Z_{22} & Z_{23} \\ Z_{31} & Z_{32} & Z_{33} \end{vmatrix} \qquad [\text{C-3}]$$

The minor of element Z_{11}, has been circled as an example where the minor determinant has been obtained by striking row 1 and column 1. Elements of the first row will be chosen with their corresponding cofactors, although any row or column would serve the purpose. The value of D in Eq. C-3 is then

$$D = \oplus Z_{11} \begin{vmatrix} Z_{22} & Z_{23} \\ Z_{32} & Z_{33} \end{vmatrix} \ominus Z_{12} \begin{vmatrix} Z_{21} & Z_{23} \\ Z_{31} & Z_{33} \end{vmatrix} \oplus Z_{13} \begin{vmatrix} Z_{21} & Z_{22} \\ Z_{31} & Z_{32} \end{vmatrix}$$

$$= Z_{11}(Z_{22}Z_{33} - Z_{23}Z_{32}) - Z_{12}(Z_{21}Z_{33} - Z_{23}Z_{31}) + Z_{13}(Z_{21}Z_{32} - Z_{22}Z_{31})$$

The same procedure of expansion by minors may be applied to higher-order determinants.

Several additional properties of determinants will be briefly given:

1. If any two rows (or columns) of a determinant are interchanged, the value of the determinant is multiplied by -1.
2. The value of a determinant is not changed if rows and columns are interchanged in the same order.
3. If corresponding elements of two rows (or columns) of a determinant are proportional, the value of the determinant is zero.
4. The value of a determinant is unchanged if the elements of one row (or column) are altered by adding to them any constant multiple of the corresponding elements of any other row (or column).

Some of the foregoing properties are useful in manipulating determinants to simplify their expansion and evaluation. For example, if all elements but one of a row (or column) can be made zero, then only one element and its cofactor need be considered in evaluating the determinant.

As an example of the use of determinants in the solution of an electrical network, refer to Fig. C-1. Loop equations are arranged so

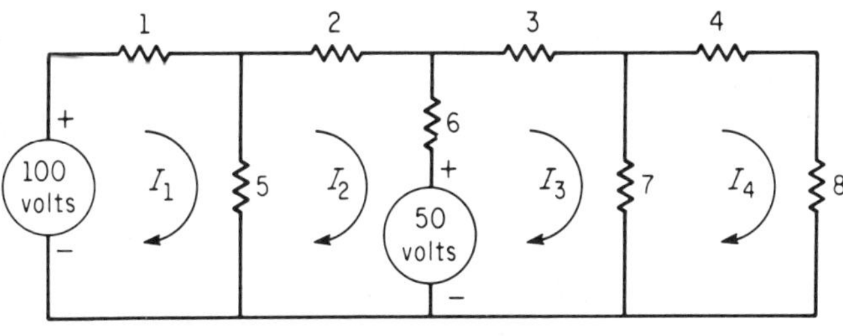

Fig. C-1

that coefficients of like currents are vertically aligned and voltage constants appear on the right-hand side of the equal sign. Loop equations are:

$$\begin{array}{llcl} (1) & 6I_1 - 5I_2 & = & 100 \\ (2) & -5I + 13I_2 - 6I_3 & = & -50 \\ (3) & -6I_2 + 16I_3 - 7I_4 & = & 50 \\ (4) & -7I_3 + 19I_4 & = & 0 \end{array}$$

Again, using Cramer's rule,

$$I_1 = \frac{D_1}{D}; \qquad I_2 = \frac{D_2}{D}; \quad \text{etc.}$$

As indicated in Eq. C-3, the denominator determinant is

$$D = \begin{vmatrix} 6 & -5 & 0 & 0 \\ -5 & 13 & -6 & 0 \\ 0 & -6 & 16 & -7 \\ 0 & 0 & -7 & 19 \end{vmatrix}$$

Expanding D by use of the first-row elements and their cofactors,

$$D = 6 \begin{vmatrix} 13 & -6 & 0 \\ -6 & 16 & -7 \\ 0 & -7 & 19 \end{vmatrix} + 5 \begin{vmatrix} -5 & -6 & 0 \\ 0 & 16 & -7 \\ 0 & -7 & 19 \end{vmatrix}$$

The two third-order determinants can be expanded further as follows:

$$\begin{aligned} D &= 6\left\{13 \begin{vmatrix} 16 & -7 \\ -7 & 19 \end{vmatrix} + 6 \begin{vmatrix} -6 & -7 \\ 0 & 19 \end{vmatrix}\right\} + 5\left\{(-5) \begin{vmatrix} 16 & -7 \\ -7 & 19 \end{vmatrix}\right\} \\ &= 6\{13(16 \times 19 - 49) + 6(-6) \times 19)\} - 25(16 \times 19 - 49) \\ &= 9411 \end{aligned}$$

The unknown I_2 will be singled out in this solution where $I_2 = D_2/D = D_2/9411$. The determinant D_2 is

$$D_2 = \begin{vmatrix} 6 & 100 & 0 & 0 \\ -5 & -50 & -6 & 0 \\ 0 & 50 & 16 & -7 \\ 0 & 0 & -7 & 19 \end{vmatrix}$$

Note the second row of I_2 coefficients has been replaced by the voltage constants on the right-hand side of the equal sign. D_2 will not be expanded, but the procedure is identical to that of the D expansion. Should the student care to evaluate D_2, its value is 85,200. Therefore

$$I_2 = \frac{85{,}200}{9411} = \underline{\underline{9.05}} \quad \text{amp}$$

C-2. Basic Properties of Matrices

An $m \times n$ (or m,n) matrix is a rectangular array of elements arranged in m rows and n columns. Rectangular brackets will be used here to contain these quantities. The bar over an upper-case letter will serve as the symbol for a matrix ($\bar{B}$, for example). The lowercase symbol a_{ij} will represent a matrix element from the ith row and jth column of matrix $\bar{A}$.

ADDITION (OR SUBTRACTION) OF MATRICES

To add (or subtract) two matrices of the same order (same number of rows, and same number of columns), merely add (or subtract) corresponding elements of the two matrices. In other words, to add $\bar{A}$ and $\bar{B}$, where

$$\bar{A} + \bar{B} = \bar{C} \tag{C-6}$$

then

$$c_{ij} = a_{ij} + b_{ij} \tag{C-7}$$

Example. Let

$$A = \begin{bmatrix} 4 & 0 \\ 3 & -2 \end{bmatrix}; \qquad B = \begin{bmatrix} 1 & -2 \\ -1 & 3 \end{bmatrix}$$

then

$$C = \begin{bmatrix} (4+1) & (0-2) \\ (3-1) & (-2+3) \end{bmatrix} = \begin{bmatrix} 5 & -2 \\ 2 & 1 \end{bmatrix}$$

Addition and subtraction are defined only for matrices of the same order. The *commutative* law holds for addition, or

$$\bar{A} + \bar{B} = \bar{B} + \bar{A}$$

The *associative* law for addition holds, or

$$\bar{A} + (\bar{B} + \bar{C}) = (\bar{A} + \bar{B}) + \bar{C}$$

Multiplication of a matrix $\bar{A}$ by a constant k requires that each element of the matrix be multiplied by k as shown:

$$k\bar{A} = \bar{A}k = \begin{bmatrix} ka_{11} & \cdots ka_{1n} \\ \vdots & \vdots \\ ka_{m1} & \cdots ka_{mn} \end{bmatrix} \tag{C-8}$$

MATRIX MULTIPLICATION

The product of $\bar{A} \times \bar{B}$ is defined if $\bar{A}$ has the same number of columns as the number of rows in $\bar{B}$. The matrices are then said to be conformable. If $\bar{A}$ is an $m \times n$ matrix and $\bar{B}$ is an $n \times p$ matrix, the product $\bar{C} = \bar{A}\bar{B}$ will be an $m \times p$ matrix. An element in the ith row and jth column of the $\bar{C}$ matrix is

$$C_{ij} = \sum_{r=1}^{r=n} a_{ir} b_{rj} \tag{C-9}$$

If the product $\bar{A}\bar{B}$ is defined, the product $\bar{B}\bar{A}$ may or may not be defined. Even if $\bar{B}\bar{A}$ *is* defined, the resulting products of $\bar{A}\bar{B}$ and $\bar{B}\bar{A}$ are not, in general, equal. We say that matrix multiplication is *not commutative*, or

$$\bar{A}\bar{B} \neq \bar{B}\bar{A} \tag{C-10}$$

Matrix multiplication is *associative* and *distributive*, e.g.,

$$\text{Associative:} \qquad (\bar{A}\bar{B})\bar{C} = \bar{A}(\bar{B}\bar{C}) = \bar{A}\bar{B}\bar{C} \tag{C-11}$$

$$\text{Distributive:} \qquad \left.\begin{aligned} (\bar{A} + \bar{B})\bar{C} &= \bar{A}\bar{C} + \bar{B}\bar{C} \\ \bar{A}(\bar{B} + \bar{C}) &= \bar{A}\bar{B} + \bar{A}\bar{C} \end{aligned}\right\} \tag{C-12}$$

Equations C-11 and C-12 assume matrices are conformable. To better illustrate the matrix multiplication Eq. C-9, the following example is given. Given $\bar{A}$ and $\bar{B}$, find $\bar{A}\bar{B}$.

$$\bar{A} = \underset{(3 \times 2)}{\begin{bmatrix} 4 & 0 \\ 2 & 1 \\ 7 & 3 \end{bmatrix}}; \qquad \bar{B} = \underset{(2 \times 2)}{\begin{bmatrix} 5 & 6 \\ 1 & 9 \end{bmatrix}}$$

The matrices are conformable ($\bar{A}$ has two columns and $\bar{B}$ has two rows). $\bar{B}\bar{A}$ is not defined however. The product $\bar{C}$ is

$$\bar{C} = \bar{A}\bar{B} = \begin{bmatrix} \rightarrow & \\ 4 & 0 \\ 2 & 1 \\ 7 & 3 \end{bmatrix} \begin{bmatrix} 5 & 6 \\ 1 & 9 \end{bmatrix} \downarrow$$

The arrows indicate the order of assuming the products according to Eq. C-9:

$$\bar{C} = \begin{bmatrix} (4 \times 5 + 0 \times 1) & (4 \times 6 + 0 \times 9) \\ (2 \times 5 + 1 \times 1) & (2 \times 6 + 1 \times 9) \\ (7 \times 5 + 3 \times 1) & (7 \times 6 + 3 \times 9) \end{bmatrix} = \underset{(3 \times 2)}{\begin{bmatrix} 20 & 24 \\ 11 & 21 \\ 38 & 69 \end{bmatrix}}$$

SOME ADDITIONAL MATRIX PROPERTIES

1. The *transpose* of a matrix $\bar{A}$ is another matrix $\bar{A}^T$, where rows and columns of $\bar{A}$ have been interchanged. The transpose of an $m \times n$ matrix is an n row, m column $(n \times m)$ matrix.

Example

$$\begin{bmatrix} 2 & 3 & 0 \\ 1 & 7 & 8 \end{bmatrix}^T = \begin{bmatrix} 2 & 1 \\ 3 & 7 \\ 0 & 8 \end{bmatrix}$$

2. An $n \times n$ matrix (equal number of rows and columns) is termed a *square* matrix. A determinant value (D) is associated with a square matrix. If $D = 0$, then the square matrix is said to be *singular*. If $D \neq 0$ the matrix is said to be *nonsingular*.

3. A *diagonal* matrix is a square matrix whose off-diagonal elements are zero.

Example

$$\begin{bmatrix} 3 & 0 & 0 \\ 0 & 1 & 0 \\ 0 & 0 & 7 \end{bmatrix}$$

Should any of the diagonal elements of a diagonal matrix be zero, the determinant of the matrix is zero and the matrix is therefore singular. If all elements of the square matrix are zero, the matrix is a *null*, or zero, matrix. A *unit matrix* ($\bar{U}$) is a diagonal matrix with all diagonal elements equal to unity. If a unit matrix is multiplied by a constant (K), the resulting matrix is a diagonal matrix with all diagonal elements equal to K. This is termed a *scalar* matrix.

Examples follow of a 3×3 unit matrix and a 2×2 scalar matrix.

$$\bar{U} = \underbrace{\begin{bmatrix} 1 & 0 & 0 \\ 0 & 1 & 0 \\ 0 & 0 & 1 \end{bmatrix}}_{3 \times 3 \text{ unit matrix}}; \qquad 3 \begin{bmatrix} 1 & 0 \\ 0 & 1 \end{bmatrix} = \underbrace{\begin{bmatrix} 3 & 0 \\ 0 & 3 \end{bmatrix}}_{2 \times 2 \text{ scalar matrix}}$$

Multiplying a matrix $\overline{A}$ by the unit matrix $\overline{U}$ (assuming $\overline{U}$ is made conformable to $\overline{A}$) will always leave the matrix $\overline{A}$ unchanged, e.g.,

$$\overline{U}A = \overline{A}\,\overline{U} = \overline{A} \tag{C-13}$$

4. *The reversal rule.* If a matrix $\overline{C}$ is a product of $\overline{A}$ and $\bar{B}$, then the transpose of $\overline{C}$ equals the product of $\overline{A}^T$ and $\overline{B}^T$ in reverse order, i.e., when

$$\overline{C} = \overline{A}\,\overline{B}, \qquad \overline{C}^T = \overline{B}^T\overline{A}^T \tag{C-14}$$

5. A real, square matrix is said to be *symmetric* if it is equal to its transpose, or if $a_{ij} = a_{ji}$. A real, square matrix is said to be *skew-symmetric* when $a_{ij} = -a_{ji}$. For elements where $j = i$, this implies that the diagonal elements are zero in the skew-symmetric matrix.

6. *Inverse of a matrix.* Division does not exist as such in matrix algebra. However, if A is a square nonsingular matrix, $\overline{A}$'s inverse ($\overline{A}^{-1}$) can be found such that

$$\overline{A}^{-1}\overline{A} = \overline{A}\,\overline{A}^{-1} = \overline{U} \tag{C-15}$$

Only a square and nonsingular matrix has an inverse. The conventional method for obtaining an inverse is given in the three-step sequence that follows.

(a) Form the transpose of the original matrix by interchanging rows and columns.

(b) Replace each element of the transposed matrix with its cofactor, where an element's cofactor $= (-1)^{i+j} \times$ minor.

(c) Divide each element of the matrix resulting from part (b) by the determinant D of the original matrix.

Example. Given $\overline{A}$, find $\overline{A}^{-1}$.

$$\overline{A} = \begin{bmatrix} 1 & 1 & 0 \\ 2 & 1 & 1 \\ 3 & -1 & 1 \end{bmatrix}$$

Step (a) Form the transpose:

$$\overline{A}^T = \begin{bmatrix} 1 & 2 & 3 \\ 1 & 1 & -1 \\ 0 & 1 & 1 \end{bmatrix}$$

Step (b) Replacing elements of A^T by their cofactors: cofactor K_{11} of element a_{11} is

$$K_{11} = (-1)^{1+1}\begin{vmatrix} 1 & -1 \\ 1 & 1 \end{vmatrix} = 2$$

$$K_{12} = (-1)^{1+2}\begin{vmatrix} 1 & -1 \\ 0 & 1 \end{vmatrix} = -1$$

Continuing in this manner, the resulting matrix from step (b) becomes

$$\begin{bmatrix} 2 & -1 & 1 \\ 1 & 1 & -1 \\ -5 & 4 & -1 \end{bmatrix}$$

Step (c) Divide each element of step (b) by the determinant of A, where

$$D = \begin{vmatrix} 1 & 1 & 0 \\ 2 & 1 & 1 \\ 3 & -1 & 1 \end{vmatrix} = 1 \times \begin{vmatrix} 1 & 1 \\ -1 & 1 \end{vmatrix} - 1\times \begin{vmatrix} 2 & 1 \\ 3 & 1 \end{vmatrix}$$

$$= 3$$

Then finally the inverse is

$$\bar{A}^{-1} = \frac{1}{3}\begin{bmatrix} 2 & -1 & 1 \\ 1 & 1 & -1 \\ -5 & 4 & -1 \end{bmatrix} = \begin{bmatrix} 2/3 & -1/3 & 1/3 \\ 1/3 & 1/3 & -1/3 \\ -5/3 & 4/3 & -1/3 \end{bmatrix}$$

As a check, the product of $\bar{A}\bar{A}^{-1}$ will yield the unit matrix according to Eq. C-15, or

$$\bar{A}^{-1}\bar{A} = \begin{bmatrix} 1 & 0 & 0 \\ 0 & 1 & 0 \\ 0 & 0 & 1 \end{bmatrix}$$

A typical application of matrix manipulation is found in the solution of the unknown loop currents in Fig. C-2. Equations for the system are

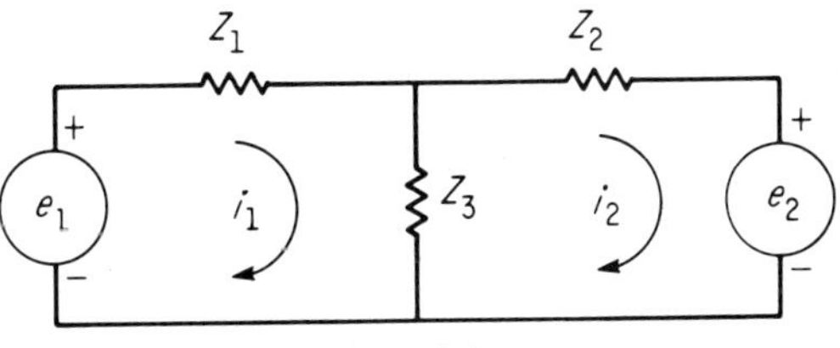

Fig. C-2

$$(1) \quad e_1 = (Z_1 + Z_3)i_1 - Z_3 i_2$$
$$(2) \quad -e_2 = -Z_3 i_1 + (Z_2 + Z_3) i_2$$

Equations (1) and (2) can be written quickly in matrix form as

$$\begin{bmatrix} e_1 \\ -e_2 \end{bmatrix} = \begin{bmatrix} (Z_1 + Z_3) & -Z_3 \\ -Z_3 & (Z_2 + Z_3) \end{bmatrix} \begin{bmatrix} i_1 \\ i_2 \end{bmatrix} \begin{matrix} \longrightarrow \\ \downarrow \end{matrix}$$

or

$$\overline{E} = \overline{Z}\overline{I} \tag{C-16}$$

Arrows are again shown to indicate the order of reading the products. Notice for this problem that the Z_{kk} diagonal element includes all of the series-loop impedance of loop K. The off-diagonal element Z_{ij} here includes impedance common to both loops i and j. This fact is helpful in the writing of the $\overline{Z}$ matrix by inspection. However, more detail is given in Chapters 5 and 6 for the setting up of network matrices by various methods. Note the typical square $\overline{Z}$ matrix and the typical $\overline{E}$ and $\overline{I}$ column matrices or vectors. In this problem, $\overline{E}$ and $\overline{Z}$ are known with elements in the $\overline{I}$ vector to be found. In order to find $\overline{I}$, one can invert the $\overline{Z}$ matrix and multiply both sides of Eq. C-16 by $\overline{Z}^{-1}$, yielding

$$\overline{Z}^{-1}\overline{E} = \overline{Z}^{-1}\overline{Z}\overline{I} = \overline{U}\overline{I}$$

or

$$\overline{Z}^{-1}\overline{E} = \overline{I} \tag{C-17}$$

where

$$\overline{I} = \begin{bmatrix} i_1 \\ i_2 \end{bmatrix}$$

Numerous other methods for "smoking out" the unknown vector are covered in Chapter 12 in the sections regarding numerical methods.

INDEX